INTRODUCTION TO PHYSICAL GEOLOGY

INTRODUCTION TO PHYSICAL GEOLOGY

W. KENNETH HAMBLIN
Brigham Young University

MACMILLAN PUBLISHING COMPANY
New York

COLLIER MACMILLAN CANADA, INC.
Toronto

MAXWELL MACMILLAN
INTERNATIONAL PUBLISHING GROUP
New York   Oxford   Singapore   Sydney

To SCOTT and NANCY,
whose lives have been an inspiration.

Editor: Robert A. McConnin
Developmental Editor: Madalyn Stone
Production Supervisor: Dora Rizzuto
Production Manager: Pamela Kennedy Oborski
Text and Cover Designer: Laura C. Ierardi
Cover background photograph: H. D. Thoreau/West Light

This book was set in Else Medium by York Graphic Services, Inc.,
printed and bound by Von Hoffmann Press.
The cover was printed by Lehigh Press.

Macmillan Publishing Company
866 Third Avenue, New York, New York   10022

Collier Macmillan Canada, Inc.
1200 Eglinton Avenue East, Suite 200
Don Mills, Ontario, M3C 3N1

Library of Congress Cataloging-in-Publication Data

Hamblin, W. Kenneth (William Kenneth)
   Introduction to physical geology / W. Kenneth Hamblin.
      p.   cm.
   Includes index.
   ISBN 0-02-349341-0
   1. Physical geology.   I. Title.
QE28.2.H37   1991
550—dc20   90-37379
                                                              CIP

Printing: 1 2 3 4 5 6 7 8   Year: 1 2 3 4 5 6 7 8 9 0

PREFACE

Geology is an incredibly fascinating subject. It concerns Earth and all that is in it: its origin, history, and dynamics of how it changes. New technology has allowed us to map the surface of the ocean floor, measure the rates at which continents split and move apart, and "X-ray" Earth's interior. We can also study Earth from space, compare it to other planetary bodies, and recognize some of the reasons for climatic change. With this enlightenment comes a new awareness of how fragile our planet is, and how impossible it is for exponential growth to continue indefinitely. Our society is experiencing profound and rapid change. Successful adjustments to this change will require a clear understanding of the nature of our Earth and how it works.

This book was written for an introductory course in physical geology for nonmajors to give them a greater understanding of the planet on which we live. The text is especially designed for a one-term course appropriate for junior and community colleges on a level where it is impractical to cover all that is normally included in the textbook, *Earth's Dynamic Systems*. However, the same approach is used as in that text and many of its illustrations are included here. I have eliminated some chapters such as "Planetary Geology," combined the material in others, and attempted to streamline the discussions without compromising the content.

W. Kenneth Hamblin

ACKNOWLEDGMENTS

I want to express my sincere thanks to the following individuals who reviewed the manuscript for this book and offered many important helpful comments and constructive criticisms:

Philip C. Hewitt, SUNY at Brockport
David T. King, Jr., Auburn University
Marshall D. Malcolm, Brevard Community College
Ellen P. Metzger, San Jose State University
John J. Renton, West Virginia University

Because large sections of the text and many illustrations are taken from my larger work, *Earth's Dynamic Systems,* 5th edition, acknowledgments are due to many of the same reviewers who offered many helpful suggestions and comments. The principal reviewers were:

M. James Aldrich, North Carolina State University at Raleigh
Burton A. Amundson, Sacramento City College
James L. Baer, Brigham Young University
David M. Best, Northern Arizona University
Myron G. Best, Brigham Young University
Willis G. Brimhall, Brigham Young University
Douglas G. Brookins, University of New Mexico
H. C. Clark, Rice University
Peter S. Dahl, Kent State University
E. Julius Dasch, Oregon State University
George H. Davis, University of Arizona
H. M. Davis, Brigham Young University
C. Patrick Ervin, Northern Illinois University
William H. Forbes, University of Maine
Donald C. Haney, Eastern Kentucky University
C. Woodbridge Hickcox, University of Georgia
Roger D. Hoggen, Ricks College
Roger Hooke, University of Minnesota
Calvin James, University of Notre Dame
Ted V. Jennings, Purdue University
Chester O. Johnson, Gustavus Adolphus College
Cornelis Klein, Indiana University at Bloomington
H. Wayne Leimer, Tennessee Technical University
Paul Lowman, Goddard Space Flight Center
Erwin J. Mantei, Southwest Missouri State University
David A. Mustart, San Francisco State University
David Nash, University of Cincinnati
Hallan C. Noltimier, Ohio State University
Morris S. Peterson, Brigham Young University
Arthur L. Reesman, Vanderbilt University
John R. Reid, University of North Dakota
James A. Rhodes, Stoffer Chemical Company
Steven M. Richardson, Iowa State University
Kenneth Van Dellen, Macomb Community College

I also want to especially thank the following individuals who reviewed the manuscript for the fifth edition of *The Earth's Dynamic Systems:*

David M. Best, Northern Arizona University
E. Julius Dasch, Oregon State University
James R. Firby, University of Nevada at Reno
Thomas M. C. Hobbs, Northern Harris County College
Roger D. Hoggen, Ricks College
Warren D. Huff, University of Cincinnati
Cornelis Klein, Indiana University at Bloomington
Karl J. Koenig, Texas A&M University
George W. Moore, Oregon State University
John R. Reid, University of North Dakota
Steven M. Richardson, Iowa State University
Sam B. Upchurch, University of South Florida
Kenneth Van Dellen, Macomb Community College
John R. Wagner, Clemson University

Two artists who contributed to previous editions of *The Earth's Dynamic Systems* deserve special recognition. William L. Chesser executed the drawings for the first edition. Many of his fine illustrations reappear here. Robert Pack prepared new figures for the third edition, many of which reappear here as well. In addition, Dale Claflin has revised some of the artwork and prepared new illustrations for this book. The combined talents and imaginative work of these three artists form a major contribution to this book.

My secretary, Sherrie Heywood, has my sincere thanks and appreciation for not only typing the manuscript but also for her helpful discussions of this project.

I must also express my appreciation to the development and production staff of Macmillan Publishing Company. Bob McConnin, Senior Editor, supervised the project; without his guidance and urging, this book would never have been completed. Madalyn Stone worked closely with this project as Developmental Editor and helped immeasurably with her constructive criticism and keen insight. Laura Ierardi developed the design and supervised the layout so that the beauty of geology could be presented in an attractive and pleasing way. I also appreciate the work of Dora Rizzuto, who was in charge of production and transformed the manuscript into a finished book.

W. K. H.

TO THE INSTRUCTOR

Supplementary Material

1. *Instructor's Guide.* This guide was prepared to help the instructor use the text and related supplemental material efficiently. It contains suggestions for lecture preparation, discussion material, and a test bank of over 700 questions keyed to the text. This thoroughly class-tested test bank is also available on personal computer.
2. *Slide Set.* A set of slides to complement both lecture and laboratory presentations has been carefully selected as an aid to instructors.
3. *Laboratory Manual.* The seventh edition of the laboratory manual, *Exercises in Physical Geology,* written by Hamblin and Howard is available for laboratory work associated with a typical course in physical geology. It includes exercises on rocks, minerals, topographic maps, stereo-aerial photographs, landsat images, and geologic maps.

TO THE STUDENT

Learning Aids

One of the most difficult problems you face in beginning a course in a new subject is to identify basic facts and concepts and to separate them from supportive material. This problem is often expressed by the question, "What do I need to know for the test?" I have attempted to overcome this problem by presenting the material in each chapter in a manner that will help you to recognize immediately the essential concepts and to guide you in your studies. The following briefly explains how you can best use this book as an effective tool for learning geology.

1. *Outline of Major Concepts.* Since this is your first introduction to geology, you cannot be expected to identify at once all the key concepts in page after page of text. To help you focus on the basic points of each chapter, I have identified them at the beginning of each chapter under the title of "Major Concepts."

2. *Statement–Discussion Method.* In each chapter the major subdivisions are introduced under the subheading *Statement.* The *Statement* presents major concepts, principles, or theories. Although not intended to be a summary or abstract of the section, it expresses the facts and concepts of the subject in one all-embracing view. These may be difficult for you to comprehend fully at the first reading, but you will gain further insight about the *Statement* from the paragraphs under the subheading *Discussion.* Here, a more complete discussion of the major concepts and principles, designed to help you grasp the ideas in the initial *Statement,* is presented and illustrated, and evidence and examples supporting the statement of principles are given. A brief history of how a concept or principle was developed is included if the history is pertinent to an understanding of a concept. The great value of this system is that you can focus on the main facts and concepts in the *Statement* and

clearly understand it by the elaboration presented in the *Discussion*. In this way you can make your study time effective. In a review for a test, for example, you need only make sure that you understand the *Statement*. Thus the critical material for review is reduced by about 70 percent.

3. *Guiding Questions*. Experience has shown that the most successful students are those who read with a specific purpose: to answer a question. Therefore, a major teaching tool in this book is represented by the guiding questions at the beginning of each major topic. These questions are intended to guide you in your study and to point out a few of the many fascinating issues with which geology is concerned.

4. *Summary*. The major concepts are further reinforced by end-of-chapter summaries that are presented in outline form. You can also use the summary to easily locate pertinent text discussion for more information.

5. *Key Terms*. Key terms appearing in the text are listed at the end of the chapter after the summary in a section called "Key Terms." Page references are given to locate the definition of these terms in the chapter; they are also defined in the glossary at the end of the book.

6. *Review Questions and Sample Tests*. Instead of preparing a separate student study guide, I have included at the end of each chapter a series of review questions to further help you focus on important concepts. There is also a "Sample Test" that contains questions in a variety of formats such as multiple choice, true and false, and so on. The answers are in the back of the book so that you can evaluate your mastery of the material in each chapter.

7. *Glossary*. At the end of the book you will find a glossary that defines many important geologic terms.

CONTENTS

1

THE PLANET EARTH 1

Earth from a Perspective in Space 2
The Earth Compared with Other Terrestrial
Planets 7

2

THE STRUCTURE OF EARTH 16

The Major Structural Units of Earth 18
Continents and Ocean Basins 20
Major Features of the Continents 21
Major Features of the Ocean Floor 22

3

EARTH'S DYNAMIC SYSTEMS 28

The Hydrologic System 30
The Tectonic System 36
Gravity and Isostasy 45

4

MINERALS 48

Matter 50
The Nature of Minerals 53
Silicate Minerals 58
Rock-forming Minerals 59

5

IGNEOUS ROCKS 66

Nature of Magma 68
Rock Textures 69
Kinds of Igneous Rocks 71
Extrusive Rock Bodies 73
Intrusive Rock Bodies 79
The Origin of Magma 82

6

SEDIMENTARY ROCKS 86

Nature of Sedimentary Rocks 88
Types of Sedimentary Rocks 91
Sedimentary Structures 93
Origin of Sedimentary Rocks 96
Significance of Sedimentary Rocks 99

7

METAMORPHIC ROCKS 102

The Nature and Distribution of Metamorphic
Rocks 104
Metamorphic Processes 106
Kinds of Metamorphic Rocks 107
Metamorphic Zones 109
Metamorphic Rocks and Plate Tectonics 110

8

GEOLOGIC TIME 114

The Discovery of Time 116
Unconformities 117
Relative Dating 118
Radiometric Measurements of Absolute Time 123
Magnitude of Geologic Time 125

9

WEATHERING 128

Mechanical Weathering 130
Chemical Weathering 132
Geometry of Rock Disintegration 133
Products of Weathering 136
Climate and Weathering 139
Rate of Weathering 140

10

RIVER SYSTEMS 144

The Major Characteristic of a River System 146
The Dynamics of Stream Flow 146
Equilibrium in River Systems 148
Processes of Stream Erosion 150
Differential Erosion 158
Processes of Stream Deposition 160

11

GROUNDWATER SYSTEMS 168

Characteristics of the Groundwater System 170
Artesian Water 176
Thermal Springs and Geysers 176
Erosion by Groundwater 178
Deposition by Groundwater 181
Alteration of Groundwater Systems 184

12

GLACIAL SYSTEMS   190

Glacial System   192
Valley Glaciers   194
Continental Glaciers   198
Pleistocene Glaciation   203
Causes of Glaciation   211

13

SHORELINE SYSTEMS   214

Waves   216
Erosion Along Coasts   221
Deposition Along Coasts   224
Evolution of Shorelines   227
Reefs   228
Types of Coasts   231
Tides   233
Tsunamis   235

14

EOLIAN SYSTEMS   238

Wind as a Geologic Agent   240
Wind Erosion   240
Transportation of Sediment by Wind   244
Migration of Sand Dunes   245
Types of Sand Dunes   247
Desertification   250
Loess   250

15

PLATE TECTONICS   254

Continental Drift   256
Development of the Theory of Plate Tectonics   260
Plate Geography   266
Plate Boundaries   268
Rates of Plate Motion   272
Mantle Convection   273
Mantle Plumes   274

16

EARTH'S SEISMICITY   278

Characteristics of Earthquakes   280
Earthquake Prediction   283
Earthquakes and Plate Tectonics   284
Seismic Waves as Probes of Earth's Interior   288

17

VOLCANISM   296

Global Patterns of Volcanism   298
Volcanism at Divergent Plate Margins   299
Volcanism at Convergent Plate Margins   302
Intraplate Volcanic Activity   308

18

EVOLUTIONS OF THE OCEAN BASINS   312

The Landscape of the Seafloor   314
The Oceanic Ridge   314
The Abyssal Floor   316
Trenches   318
Islands and Seamounts   318
Continental Margins   318
Submarine Canyons   320
Gravity Mapping of the Oean Floor   320
Composition and Structure of the Oceanic Crust   324
The Ocean Basins   324
History of the Oceans   327

19

EVOLUTION OF THE CONTINENTS   332

The Continental Crust   334
Mountain Building   337
Types of Mountain Building   346
Origin and Evolution of Continents   350

GLOSSARY   355

ANSWERS TO QUESTIONS   369

ILLUSTRATION CREDITS   371

INDEX   373

1

THE PLANET EARTH

One of the most significant advances made in the twentieth century concerns our view of Earth and the dynamics of how it changes. Until recently, geologists were earthbound, obtaining most of their knowledge of Earth by looking horizontally from viewpoints at or near its surface. This perspective was nearsighted and limited, and in an attempt to overcome it, local observations were plotted on maps and charts, which served as models of the real world. Rivers, mountains, shorelines, and weather patterns were surveyed and studied from hundreds of observation points, but observers were never able to see Earth in a regional, panoramic view or to observe how it functions as a planet in space. Now, for the first time in all of time, humans have seen Earth from the depths of space—seen it whole and round and beautiful and small. The horizontal view no longer dominates; the new vertical dimension of space photography has extended the limits of vision so that we can look down on Earth in a way that seems almost godlike.

Space photography permits us to view the large-scale structures of the atmosphere, the oceans, and the continents, and to observe the regional relationships of river systems, mountain belts, and coastal features. We are now able to see the actual physical features of the planet and can view the Earth functioning as a dynamic system on a scale that may at first be difficult to grasp.

To help you understand and appreciate Earth, this chapter includes a section of space photographs that illustrate many of its landforms and structural features. These photographs are, in essence, a pictorial essay on Earth's dynamic systems. They are intended to be a visual reference for material discussed in this and all of the following chapters.

MAJOR CONCEPTS

1. A study of space photographs of Earth provides an important perspective concerning its distinguishing characteristics and planetary dynamics.
2. Earth's atmosphere is a thin shell surrounding the planet and is in constant motion. It is unique in that it is composed of 78 percent nitrogen and 21 percent oxygen.
3. The hydrosphere is one of the things that makes Earth unique. It moves from the ocean to the atmosphere and over the surface in a great cycle.
4. The biosphere exists because of water. Although it is relatively small compared to other major layers of Earth, it has been a major geologic force operating at the surface.
5. Earth is a dynamic planet that is continually changing as a result of its internal heat and the hydrologic cycle.
6. Other terrestrial planets are less dynamic because they lack sufficient internal energy and surface water.

EARTH FROM A PERSPECTIVE IN SPACE

How has space photography changed our view of Earth and its geologic systems?

Statement

Views of Earth from space, like those shown in Figures 1.1–1.4, reveal many features that make Earth unique and provides an insight into our planet's history and how it changes. The atmosphere is the thin, gaseous envelope that surrounds Earth, and its bright, swirling clouds are the most conspicuous features seen from space. The lithosphere, the outer, solid part of Earth, is visible in continents and islands. The hydrosphere, the planet's discontinuous water layer, is seen in the vast surface of the oceans. Even parts of the biosphere, the organic realm, which includes all of Earth's living things, can be seen from space, in the dark green tropical forest of equatorial Africa.

One of the unique features of Earth is that each of these major realms of the planet's surface is in constant motion and changes continually. The atmosphere and the hydrosphere move in ways that are dramatic and obvious. Movement, growth, and change in the biosphere are readily appreciated and easily understood, but Earth's crust is also in motion and has been throughout most of the planet's history.

Earth, therefore, is quite different from other planetary bodies in the solar system. The most obvious differences are that Earth has an extremely mobile and dynamic atmosphere and a hydrosphere that involves the movement of enormous volumes of water over its surface. But that is not all. Earth is unique in that continents and ocean basins have evolved on its crust, and they are also in motion.

Discussion

The Atmosphere

Perhaps the most conspicuous features of Earth as seen from space are the brilliant white swirling clouds of the atmosphere (Figures 1.1–1.4). Although this envelope of gas constitutes an insignificantly small fraction of the planet (less than 0.01 percent of the mass), it is particularly significant because it moves easily and is constantly interacting with the ocean and land. It plays a part in the evolution of most features of the landscape and is essential for life. On the scale of the illustration in Figure 1.2, most of the atmosphere would be concentrated in a layer as thin as the ink with which the photo is printed.

The atmosphere is in constant motion; its circulation patterns are clearly seen in Figures 1.1–1.4 by the shape and orientation of the clouds. At first glance, the patterns may appear confusing, but upon close examination of Figure 1.3 we find that they are well organized. If we smooth out the details of local weather systems, the global atmospheric circulation becomes apparent. Solar heat, the driving force of atmospheric circulation, is greatest in the equatorial regions; it causes water in the oceans to evaporate and the moist air to rise. The warm, humid air forms an equatorial cloud belt above this low pressure system, bordered on the north and south by relatively high-pressure zones that are cloud-free in the middle latitudes, where air descends. At higher latitudes, low-pressure systems develop where the warm air from the low latitudes meets the polar fronts. The pattern of circulation is around the resulting low pressure and produces counterclockwise winds in the Northern Hemisphere and clockwise winds in the Southern Hemisphere. (Figure 1.3).

Our atmosphere is unique in the solar system. It is composed of 78 percent nitrogen, 21 percent oxygen, and minor amounts of other gases. The earliest atmosphere was much different and consisted largely of hydrogen, carbon dioxide, and water vapor. The present atmosphere began to form as soon as organisms evolved and through photosynthesis developed the ability to extract carbon dioxide from the air and expel oxygen. Thus, the oxygen in today's atmosphere is and was produced by life.

The atmosphere is divided into several layers. The part closest to Earth is called the troposphere (Greek *tropos,* meaning "turn" or "change") and extends from the surface to a height of about 13 km. This thin layer close to the surface contains almost all the water vapor, and therefore nearly all the clouds, storms, and precipitation. The overlying layer is the stratosphere, which extends to about 55 km above the surface. Today, jet air flight is mainly in the stratosphere, in order to avoid the many weather hazards in the troposphere. The stratosphere is also important because it contains the ozone layer, which absorbs much of the sun's stronger ultraviolet rays that would otherwise destroy exposed bacteria and severely burn animal tissue. No sharp boundary can be placed on the outer limits of the atmosphere. The density of gas molecules decreases almost imperceptibly into interplanetary space. Most scientists consider the outer boundary of the atmosphere to be about 9,500 km above Earth, a distance nearly as great as the diameter of Earth itself. The atmosphere, together with the hydrosphere, is a vital agent in maintaining a suitable temperature for the majority of life in the biosphere. Movement of water in the atmosphere and its precipitation on land are responsible for sculpturing many of the Earth's landforms.

Figure 1.1
Throughout history it seems that the wisest philosophers and the most brilliant minds had trouble estimating the real size of our planet. It once appeared that the size of Earth was immeasurably great, almost infinite. It was only after we ventured into space that we looked back and saw with surprise and disbelief just how small our planet really is. Some astronauts saw Earth as an island in a limitless ocean of space. Others compared it to a spaceship with a crew of more than six billion.

This is a view of a very slim, crescent Earth seen by the Apollo II astronauts on July 23, 1969, the day before splashdown.

Figure 1.2
One of the things every astronaut has emphasized is the fragility of Earth and the tiny little shell of air in which we live. Seen from space, the atmosphere appears as a thin line. It is obviously not the huge, infinite ocean of air we have been told about so many times. Above this tiny sphere is the black cold emptiness of space.

4

The Hydrosphere

The hydrosphere is the total mass of water on the surface of our planet. About 98 percent of the water is in the oceans; 2 percent is in streams, lakes, groundwater, and glaciers. Approximately 71 percent of Earth is covered with water. Thus, it is for good reason that Earth has been called the water planet. From any distance in space, Earth's outer surface is seen as a predominantly blue sphere, mottled with swirling white clouds (Figure 1.3). It has been estimated that if all the irregularities of the Earth's surface were smoothed out to form a perfect sphere, a global ocean would cover Earth to a depth of 2.25 km.

It is this great mass of water that makes Earth unique. Water permitted life to evolve and flourish. Every inhabitant on Earth is directly or indirectly controlled by it. All of Earth's weather patterns, climate, rainfall, and the extremely important carbon dioxide content of the atmosphere are influenced by the seas and oceans. The hydrosphere is in constant motion—evaporating from the oceans and moving through the atmosphere, precipitating as rain and snow, and returning to the sea in rivers, glaciers, and groundwater. As water moves over Earth's surface it erodes, transports, and deposits weathered rock material, constantly modifying Earth's landscape. Many of the distinctive surface features of Earth are due to the hydrosphere.

The Biosphere

The biosphere may be defined as the part of the Earth where life exists. It includes the forests, grasslands, and the familiar animals of the land, together with the numerous creatures that inhabit the sea and atmosphere. As a terrestrial covering, the biosphere is discontinuous and has an irregular shape; it is a single, interwoven web of life existing within and reacting with the atmosphere, hydrosphere, and lithosphere. It consists of more than 1.5 million described species, and perhaps as many as 3 million not yet described. Each species lives within its own limited environmental setting. By far, most of the biosphere exists in a narrow zone extending from the depth to which sunlight penetrates the oceans (about 200 m) to the snowline in the tropical and subtropical mountain ranges (about 6,000 m above sea level). At the scale of the photograph in Figure 1.4, the biosphere, all of the known life in the solar system, would occur in a thin layer no thicker than the paper on which the image is printed.

The main factors controlling the distribution of life on our planet are temperature, pressure, and chemistry. However, the range of environmental conditions in which life is possible is truly amazing, especially the range of environments in which microorganisms can exist. Also remarkable is the range of species that can exist in these varied environments.

Although the biosphere is relatively small compared with the other major layers of Earth (atmosphere, hydrosphere, and lithosphere), it has been a major geologic force operating at the surface. Essentially all of the present atmosphere has been produced by the chemical activity of the biosphere. The composition of the oceans is similarly affected by the activity of organisms; most marine organisms extract calcium carbonate from seawater to make their shells and hard parts. When the organisms die, their shells settle to the seafloor and accumulate as beds of limestone. The continued extraction of calcium carbonate from the ocean has a major effect upon the composition of the atmosphere, as well. In addition, all of the coal, oil, and gas in the Earth was formed by the biosphere, so large parts of the rocks near the surface of the Earth's crust originated in some way from organic activity.

The record of the biosphere (life on Earth) is preserved, sometimes in remarkable detail, by fossils that occur in most sedimentary rock and even in some igneous and metamorphic rocks. Indeed, the number of living species today represents only about 10 percent of the number of species that have existed since life first developed on Earth.

The Lithosphere

The lithosphere (rock sphere) is the outermost layer of the solid Earth. It is a layer of rock, about 70 km thick, that rests upon soft, weak material. Two types of crust occur within the lithosphere: (1) continental crust, which can be thought of as a raftlike slab of granitic rock, and (2) oceanic crust, which is composed of denser volcanic rock. In Figure 1.4 the lithosphere is exposed as dry land, but it forms a complete outer solid shell of Earth.

Concluding Statement

This then is Earth as it appears in space: small, delicate, and complex. It is a dynamic planet in which the various materials (air, water, and solid rock) are in constant motion, initiating constant change. The study of geology is the study of this planet Earth: its origin, history, and the dynamics of how it changes. Geologists study such diverse phenomena as the structure of continents and how they move, and the evolution of ocean basins and why they all die at a youthful age. They study volcanoes, glaciers, rivers, beaches, earthquakes, and the history of life.

From a perspective in space, our Earth is but a small, insignificant sphere in the cosmos; yet, from a human perspective, it is a vast and complex system that has evolved over billions of years. Learning about Earth and the forces that change it-the intellectual journey upon which you are about to embark—is a journey we hope you will never forget.

Figure 1.3
The hydrosphere, the thin film of water that makes the planet Earth unique, is essential for life. Earth is just the right distance from the sun so water can exist as a liquid, solid, and gas. If it were closer to the sun, our oceans would evaporate; if it were farther from the sun, the oceans would freeze solid. But water on Earth is in abundance and is extremely mobile. Drawing energy from the sun, it moves in great cycles from the oceans to the atmosphere, over the landscape in river systems, and ultimately back to the oceans again.

In this view of the Pacific Ocean, a large cyclonic storm can be seen just south of Alaska. From a human perspective, the Pacific is big. Astronauts remarked that you cannot comprehend it by looking at a globe, but when you're traveling at four miles a second and it takes 25 minutes to cross it, you know it's big.

THE EARTH COMPARED WITH THE OTHER TERRESTRIAL PLANETS

Statement

One of the best ways to understand why Earth is unique is to compare it with other terrestrial planets— the Moon, Mercury, Venus, and Mars. These planets constitute the inner solar system and are similar because they are composed of rocky material that gravitated toward the sun before the planets began to form. All are roughly the same size and have the same general mass and composition. In the beginning they were all probably quite similar, but later they evolved along different paths because of differences in size, internal energy, and distance from the sun.

Discussion

The Moon

We know from telescopic views, as well as from space photographs, that the Moon's surface is practically saturated with craters. We thus conclude that the major geologic process on the Moon has been the impact of meteorites. The Moon does not have water or an atmosphere, so there are no stream valleys or windblown sand. The surface of the Moon has been modified only by the impact of meteorites coming from space and by the internal energy that once deformed its crust and produced volcanic features. From rock samples brought back by the Apollo missions and from close-up satellite photographs, we know that the smooth, dark areas called maria (Latin, "seas") are covered by vast floods of lava. As shown in Figure 1.5, lava flows filled large, circular basins and adjacent lowlands and spread out over parts of the rugged, densely cratered terrain. Clearly, the densely cratered terrain was formed before the lava flows occurred—an obvious, but fundamental, temporal relationship between two major events in lunar history.

Two additional facts are apparent from Figure 1.5. First, the maria, or lava plains, have relatively few craters. The rate of meteorite impact must have decreased greatly after the lava was extruded. Second, the outlines of all craters, young and old, remain circular and undeformed. We therefore conclude that the crust of the Moon has not undergone enough compression to produce major buckling and folding.

From these observations of the Moon's surface features, we can recognize the following sequence of events in lunar history:

1. A period of intense meteorite bombardment, when the densely cratered terrain was formed
2. A major thermal event, resulting in widespread volcanic activity and great floods of lava
3. A period of light meteorite bombardment, with relatively few impacts

Except for the impact of a few meteorites, the Moon has not changed significantly since the lava plains were formed almost four billion years ago. Its crust has not been significantly deformed by internal forces. The Moon has no atmosphere or hydrosphere. No water or wind exists to modify its surface. Nothing moves on the Moon, except on rare occasions when a meteorite strikes the surface. No motion was observed by the Apollo astronauts, except for what they themselves produced. The footprints they left behind will remain fresh and unaltered for literally millions of years. Study Figure 1.5 for a moment and imagine the Moon with air and water. How would its surface be different?

Mercury

The surface of Mercury is strikingly similar to that of the Moon, as is readily apparent by comparing Figures 1.5 and 1.6. Each is pockmarked with craters that record the last stages in the birth of the solar system; each has younger areas of dark terrain formed by lava flows, but neither has an atmosphere, a hydrosphere, or a biosphere. Mercury and the Moon were too small to retain much internal energy. They did heat up initially and become differentiated with a core and a relatively thick, ridged outer crust (lithosphere). Some of their internal heat produced a short episode of volcanic activity, during which significant volumes of lava were extruded over parts of their surfaces, covering some of the early formed, densely cratered terrain. In a sense this early thermal event could be considered the initial stage of planetary dynamics that died in an embryonic stage. Both Mercury and the Moon thus remain "fossils" of the early stages in planetary development. They are important in that they provide a valuable record of the first chapters in the history of our corner of space.

Venus

Venus is about the same size as Earth. Both are much larger than Mercury and the Moon and as a result have more internal heat; they have remained dynamic with volcanic eruptions, crustal deformation, and atmospheric circulation. They are commonly considered the twins of the solar system, but not identical twins. They are composed of roughly the same mix of iron and rocky material and are comparably endowed with carbon and water. At first, they probably followed the same general path of planetary development, but later went their separate ways. The dense atmosphere that surrounds Venus has kept its surface hidden from view, but recently radar imagery

Figure 1.4

Viewed from space, Earth appears as a delicate blue ball wrapped in a swirl of filmy white clouds. This is a visual embrace of our physical world and the life upon it. In a photograph we see Earth motionless, frozen in a moment of time, but in this new view of our planet, there is much more action than one might imagine. The swirling patterns of clouds that dominate the scene underline the importance of water in Earth's system. Huge quantities of water are in constant motion, in the sea, the air, and on land. Several complete cyclonic storms, spiraling over hundreds of square kilometers, can be seen pumping vast amounts of water from the continents that erodes the land as it flows back to the sea.

Large parts of Africa and Antarctica are visible in this view, and the major climatic zones of our planet are clearly delineated. Much of the vast tropical forest of central Africa is seen beneath the discontinuous cloud cover. Also, large portions of the South Polar ice cap, a glacier more than 3,000 m thick that covers the continent of Antarctica, are clearly visible. Of particular interest in this view is the rift system of East Africa, which extends in a north-south direction across most of the continent. It is mostly obscured by clouds in the equatorial region but is well expressed in the sea. The Red Sea is a large fracture in the African continent, separating the Arabian Peninsula from the rest of Africa. One of the animals that evolved from the East African rift valleys was a creature that learned to live in all of the varied landscapes of the planet. His first home was here, but he recently walked on the Moon.

Figure 1.5
The surface of the Moon shows two contrasting types of landforms: densely cratered highlands, called terrae, and dark, smooth areas of lava plain, called maria. We know from rock samples brought back by the Apollo missions that the maria resulted from the great floods of lava that filled many large craters and spread out over the surrounding area. The volcanic activity thus occurred after the formation of the densely cratered terrain. These relationships between surface features imply that the Moon's history involves three major events: (1) a period of intense bombardment by meteorites, (2) a period of volcanic activity, and (3) a subsequent period of relatively light meteorite bombardment (resulting in young, bright-rayed craters). The lunar surface has a very low level of erosion and has not been modified by wind, water, or glaciers.

9

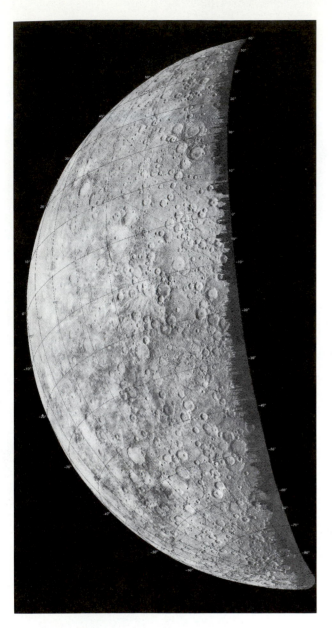

Figure 1.6
A photomosaic of Mercury, made from photographs taken at a distance of 234,000 km, shows that Mercury and the Moon are strikingly similar. Each has a densely cratered terrain, multiringed basins, a younger area of dark plains (maria), and young, rayed craters.

has given us a first clear glimpse of the planet's surface. On a regional scale, Venus looks somewhat like Earth drained of its seas (Figure 1.7). Two large highlands, resembling Earth's continents, are surrounded by low, smooth plains like the seafloor on Earth. High conical volcanic mountains occur on both "continents," and there are fascinating belts of folded mountains similar to those on Earth. The great difference between Venus and Earth is in the amount of

liquid water on the surface and the amount of carbon dioxide in the atmosphere. But what a difference!

It is likely that Venus started out much like Earth, with massive amounts of liquid water. Because Venus is closer to the sun, it receives more solar heat, and its water soon began to evaporate and rise as vapor to the upper atmosphere. There the sun's ultraviolet rays broke the water molecules apart and much of the fried hydrogen escaped into space. Most of the water on Venus was therefore lost forever. Without water, Venus became less and less like Earth in profound ways you might not first expect. Venus, like Earth, had a significant amount of internal heat, much more than the Moon and Mercury, and continued to be geologically active as volcanic eruptions continually spewed carbon dioxide into the atmosphere. On Earth, water directly or indirectly cycled much of the carbon back to Earth to form limestone rock. Carbon is therefore trapped in the crust. Because Venus lost its water early in its history, it could not return carbon to the crust as limestone; as a result, the level of carbon dioxide in the atmosphere of Venus rose unchecked. This created an intense greenhouse effect because, like a blanket, carbon dioxide holds the solar energy that falls on Venus and causes the temperature of Venus' surface to rise to a level high enough to melt lead.

Earth, in contrast, has been able to retain its water, and it is water more than anything else that makes Earth unique. There are many reasons why Earth has remained an aqueous planet. The Earth is just the right distance from the sun so that temperature ranges are such that water can exist as a liquid, solid, and a gas (Figure 1.4). If it were closer to the sun, our oceans would evaporate; if it were farther from the sun, the oceans would freeze solid. But there is plenty of water on Earth and it is extremely mobile. Heated by the sun, water moves in great cycles from the oceans to the atmosphere, over the landscape in river systems, and ultimately back to the oceans. As a result, Earth's surface has been continually changed and eroded into delicate systems of river valleys—a remarkable contrast to other planetary bodies, whose surfaces are dominated by craters.

The presence of water as a liquid on Earth's surface throughout its history has enabled marine organisms to evolve, which had a surprisingly profound effect on the atmosphere. Organic activity removes large quantities of carbon dioxide from the atmosphere as plants exchange oxygen for carbon dioxide. Many forms of oceanic life also remove carbon dioxide to make their shells, which later fall to the seafloor and solidify into limestone rock. Thus, while the atmosphere of Venus became locked in a runaway greenhouse effect, organic activity on Earth created a new atmosphere with abundant oxygen, and equilibrium became established by recycling carbon compounds through the atmosphere, oceans, and rocks in the crust. This resulted in a moderate climate throughout most of Earth's history.

Furthermore, Earth has its own internal source of heat which produces the crustal deformation, ocean basins, continents, mountain belts, and volcanic ac-

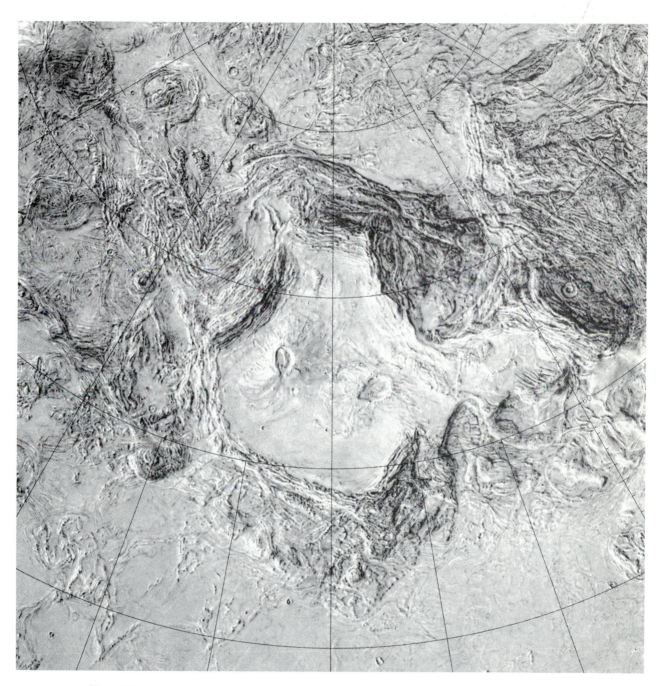

Figure 1.7
A shaded relief map constructed from radar images of part of the north polar region of Venus dramatically portrays the nature of that planet's surface. The most striking features are its high plateaus and fringing mountain belts, which rise 10 km above the plains. The plateau is thought to be an area of major compression, with resulting folded mountain belts and volcanic activity. The high elevation and steep cliffs are reminiscent of continental blocks on Earth. The surrounding lowlands might be compared to Earth's ocean floor. Compared with Earth, one of the most significant differences is that Venus lacks oceans and valleys produced by stream erosion. Surface temperatures are far too high for water to exist as a liquid; moreover, the present atmosphere of Venus contains very little water vapor. Thus, most of the surface features on Venus are produced by deformation and volcanism but are unmodified by stream erosion.

tivity that continually change and modify its surface. Earth's rigid outer layer (the lithosphere) is broken into huge fragments, or plates, like a cracked eggshell; there is enough internal heat to cause the plates to move, so the Earth's surface is constantly changing as a result of its own internal heat.

Does water play a role in the internal dynamics of a planet? This is a question we must understand in order not to stress the environment on Earth beyond its limits. Perhaps if nothing more, our studies of the diversity of compositions and conditions of the bodies in the solar system should remind us of the delicate balance that allows us to exist at all.

Mars

Beyond Earth and the Moon we come to one of the most fascinating worlds in the solar system: the red planet, Mars. For years it was considered to be a planet of mystery and intrigue—a planet possibly populated with life. As it turns out, this fanciful theory was all wrong. But the Mars we have just explored is even more exciting than anyone could imagine. Many of its Earthlike features are not only large, but gigantic. There are huge canyons, giant volcanoes, global dust storms, polar ice caps, and dry river beds (Figure 1.8). Mars has water, but at present temperatures and atmospheric pressures are so low that water can exist only as vapor or ice. Thus, virtually all of it is frozen as ice caps or as ground ice, like the permafrost regions on Earth. Now, wind alone is the major process altering the landscape of Mars. Some dust storms grow to such proportions that they blanket the entire planet. The huge river channels on Mars create an intriguing mystery. Under what ancient conditions did liquid water flow in great floods across the surface of Mars? Do dust storms play a role in atmospheric conditions and surface temperature? Did life evolve when water was abundant? Mars has experienced significant changes during its history; changes recorded on the surface of its landscape; changes still occurring.

It is believed that Mars started out much like Venus and Earth did, developing a dense atmosphere early in its history. At one time, Mars probably had a moderate climate and abundant liquid surface water. Rainfall and flooding produced significant erosion then, creating erosional features similar in some respects to those on Earth. But Mars is much smaller than Earth, has less internal heat, and was less tectonically active. Consequently, it could not recycle the carbon compounds. Carbon dioxide was removed from its atmosphere but not returned to it through continued volcanic activity. Thus, the greenhouse effect weakened, the surface cooled, and water froze. Mars was left cold and dry. The great dust storms that rage on Mars alter its surface and atmospheric temperatures. Earth's atmosphere acts not only as a window for sunlight, but as a blanket for heat. Unlike carbon dioxide, which thickens the blanket and creates a warming greenhouse, dust thrown into the atmosphere tends to close the window—it blocks solar energy from reaching the surface below. Mars, therefore, tends to remain cold, with all its water locked in its ice caps or frozen in the pore spaces in its rock and soil.

Conclusion

Studies of the terrestrial planets indicate that Earth, the Moon, Mercury, Venus, and Mars all started out in pretty much the same way. They were formed at the same time from the accumulation of iron and rocky material that condensed in the hot regions of the inner solar system. As they grew into planetary bodies orbiting the sun, they soon swept up most of the remaining debris in their orbital path. However, they are not all the same size. The diameters of Mercury and the Moon are about half that of Mars. Mars has a diameter of about half that of Earth and Venus (Figure 1.9). The surfaces of the terrestrial planets were saturated with billions of craters, and all of them heated up because of the energy from impacting meteorites and from the disintegration of radioactive material in their interiors. Their internal heat created volcanic activity. Mercury and the Moon, being relatively small, generated less heat and lost it faster than the larger planets. The lava flows that form the smooth plains on these planets are evidence of early geologic activity that soon died. The Moon and Mercury thus remain primitive bodies in that their surfaces haven't been changed by stream erosion or continued volcanism and crustal deformation.

Mars has about twice the diameter of the Moon and Mercury but only half that of Venus and Earth. It might be considered to represent an intermediate stage of planetary development. Compared to the Moon and Mercury it has had more internal energy, more volcanic activity, and more crustal deformation. Mars has a thin atmosphere and some surface water (now frozen). Compared to Earth, however, the geologic activity on Mars has been much less.

Earth, in contrast to the Moon, Mercury, and Mars, has been geologically active throughout all of its history. Its unique atmosphere, abundant water, and complex biosphere constantly modify and change its surface. It has significant internal heat, which continues to produce volcanic activity and crustal deformation. It is the dynamics of Earth that make it unique and inhabitable.

In the next chapter we will consider some of the characteristics of the major surface features of Earth and what we know about Earth's internal structure. With this information, we will be able to understand the global aspects of the dynamic systems of Earth introduced in Chapter 3.

Figure 1.8
The planet Mars, as photographed from the Viking spacecraft, shows a surface somewhat similar to that of the Moon; yet, significant differences exist. Mars has two major provinces: densely cratered highlands in the Southern Hemisphere and smooth plains (presumably vast floods of lava) covering the lowlands in the Northern Hemisphere and the floors of large basins. Unlike the Moon, however, Mars has many fascinating geologic features, indicating that its surface has been dynamic, having been modified by atmospheric processes, recent volcanic activity, and crustal deformation. Almost every geologic feature is gigantic. Three huge volcanoes, one more than 28 km high, can be seen in the left part of this image. An enormous canyon extends across the entire hemisphere, a distance roughly equal to that from New York to San Francisco. In addition, Mars has been eroded by enormous floods of water that once flowed across its surface but is now locked up as ground ice and ice caps in the polar regions.

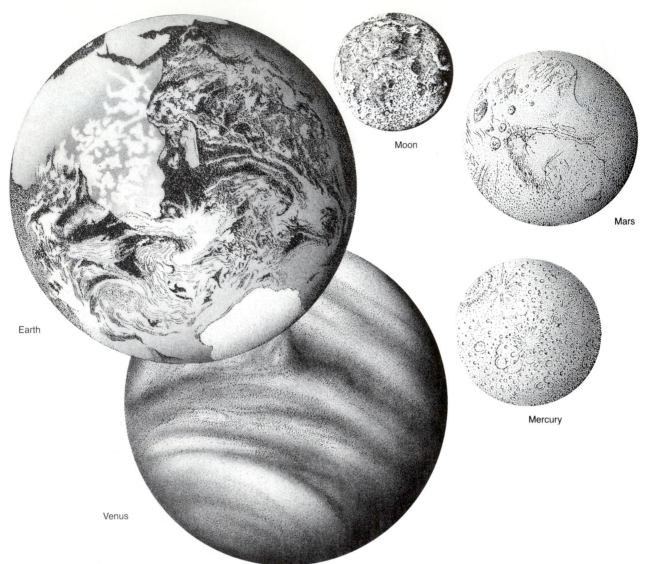

Moon

Mars

Earth

Mercury

Venus

Figure 1.9
Sketches of the terrestrial planets, shown here at the same scale, provide an insight into their planetary dynamics. The Moon and Mercury are relatively small planetary bodies and did not generate sufficient internal heat to sustain prolonged geologic activity. They thus remain primitive bodies dominated by impact structures. Mars is somewhat larger and has more internal heat and a thin atmosphere. Its original cratered surface has been modified by volcanic eruptions, huge canyons, and erosion by running water.

Earth and Venus are larger still and have more internal energy, which has continually changed their surfaces. Both have an atmosphere, although strikingly different in composition. Only Earth has large amounts of liquid water, which has influenced its unique development throughout its history.

SUMMARY

1. Space photography gives us both a new and important perspective of the features that make Earth unique and an insight into how it changes.

2. The atmosphere is a thin, fragile layer of gas that surrounds the planet. It is in constant motion and plays a critical role in the circulation of water across Earth's surface.

3. The hydrosphere is the total mass of water on the surface of our planet. It is in constant motion—evaporating from the ocean, moving with the at-mosphere, precipitating as rain or snow, and returning to the ocean through rivers, glaciers, and groundwater.

4. The biosphere is the complex web of life that covers much of Earth. It is a major geologic force operating at Earth's surface.

5. The lithosphere is the outermost solid shell of Earth. It is also in motion.

6. Earth is unique because of its planetary dynamics.

KEY TERMS

atmosphere (p. 2)   greenhouse effect (p. 10)   impact craters (p. 7)   planetary dynamics (p. 7)
biosphere (p. 5)   hydrosphere (p. 5)   lithosphere (p. 5)   terrestrial planets (p. 7)

REVIEW QUESTIONS

1. Why are the atmosphere and hydrosphere considered as much a part of Earth as solid rock?
2. Briefly describe Earth as seen from space.
3. How has the biosphere influenced the composition of our atmosphere?
4. Describe the paths in the movement of water over the surface of the planet Earth.
5. Compare and contrast Mercury, Mars, and Earth.
6. What sources of energy make Earth a dynamic planet?
7. How do the surfaces of Venus and Earth differ? Why is there a difference?
8. Explain why Earth has abundant liquid water and Venus, Mercury, and Mars do not?

Multiple-Choice Questions

1. Which of the following planets has high plateau-like continents?
   a. Mercury
   b. Moon
   c. Mars
   d. Venus
   e. Jupiter
2. On a space photograph of the whole Earth, such as Figure 1.4, most of the atmosphere would be in a layer approximately
   a. 3 in. thick.
   b. 2 in. thick.
   c. 1 in. thick.
   d. ½ in. thick.
   e. as thin as a sheet of paper.
3. Which of the following would *not* be considered part of the hydrosphere?
   a. water in rivers
   b. water in lakes
   c. water in the atmosphere
   d. water in the ocean
   e. water in small streams
4. Which of the following statements concerning the biosphere is *not* true?
   a. It extends as a thin continuous layer around the entire Earth.
   b. It includes all life on Earth.
   c. It changed the primitive atmosphere on Earth.
   d. It interacts with the atmosphere, hydrosphere, and lithosphere.
   e. A record of the biosphere is preserved in the rocks of Earth's crust.
5. Which of the following planetary bodies has the least number of craters on its surface?
   a. Mercury
   b. Mars
   c. The Moon
   d. Venus
   e. Earth

6. Which of the following statements concerning the terrestrial planets is *not* true?
   a. They were formed from the same types of material.
   b. They originated at the same time.
   c. They all have active volcanoes.
   d. They all were once saturated with craters.
7. Which of the following is least important in maintaining water in a liquid form on Earth?
   a. the size of Earth
   b. the distance from the sun
   c. composition of the atmosphere
   d. the biosphere
   e. the original composition of planetary material
8. Which of the following is *not* found on Mars?
   a. large volcanoes
   b. lava plains
   c. huge canyons
   d. stream valleys
   e. a hydrologic cycle
9. The outermost solid shell of Earth is called
   a. the mantle.
   b. the lithosphere.
   c. the asthenosphere.
   d. the atmosphere.
   e. the hydrosphere.
10. Which of the following makes Earth unique among the terrestrial planets?
    a. It has no impact craters.
    b. planetary dynamics
    c. It has an atmosphere.
    d. It has a lithosphere.
    e. its size

True/False Questions

1. *T F* Earth is the only terrestrial planet with stream valleys.
2. *T F* Earth is the only terrestrial planet with an atmosphere.
3. *T F* Earth is the only terrestrial planet with a hydrosphere.
4. *T F* Earth is the only terrestrial planet with a lithosphere.
5. *T F* Earth is the only terrestrial planet with a biosphere.

Fill in the Missing Term

1. The _____ is the outermost solid shell of Earth.
2. The _____ is the total mass of water on Earth's surface.
3. The _____ is the complex web of life that covers much of Earth.
4. _____ has an intense greenhouse effect.
5. _____ is noted for huge volcanoes and a great rift canyon.

2

THE STRUCTURE OF EARTH

We have not only seen Earth from space, we have also seen the ocean floor, mapped its topography and structure, and gained insight into its origin and history. Until recently, this part of Earth was as inaccessible as the planets in the outer solar system. We now know that the oceanic crust is completely different from the continental crust. It is much younger, composed of different material, and has had a separate history of its own.

We have also "X-rayed" the interior of Earth and recognized that it is a highly differentiated planet whose materials are segregated into layers according to density. From this new ability to see the internal structure of Earth, we have discovered how it functions as a dynamic system.

We now know that volcanoes, earthquakes, mountain building, and continental drift are manifestations of the Earth's internal heat. In our time the right tools, sophisticated technology, and decisive evidence have come together to allow us to understand how Earth works. In this chapter we will briefly describe the major geologic features of the planet Earth—features that make it unique in the solar system.

MAJOR CONCEPTS

1. Earth is a differentiated planet, with its materials segregated and concentrated into layers according to density.
2. The internal layers based on composition are (a) crust, (b) mantle, and (c) core.
3. The major internal layers based on physical properties are (a) lithosphere, (b) asthenosphere, (c) mesosphere, and (d) core.
4. Material within each of these units is in motion, making Earth a changing, dynamic planet.
5. Continents and ocean basins are the principal surface features of Earth.
6. The continents consist of three major components: (a) shields, (b) stable platforms, and (c) belts of folded mountains. All of these components show the mobility of the crust.
7. The ocean floor contains several major structural and topographic divisions: (a) the oceanic ridge, (b) the abyssal floor, (c) seamounts, (d) trenches, and (e) continental margins.

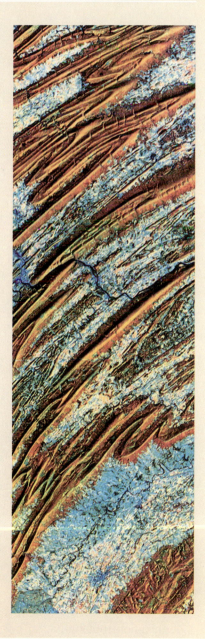

THE MAJOR STRUCTURAL UNITS OF EARTH

Which layers of Earth are most significant to the planet's dynamics?

Statement

Earth is a differentiated planet—that is, its constituent materials are separated and segregated into layers according to density. The denser materials are concentrated near the center, the less dense near the surface.

The internal layers are recognized on the basis of (a) composition and (b) physical properties. Compositional layers are

1. crust.
2. mantle.
3. core.

Layers based on physical properties are

1. lithosphere.
2. asthenosphere.
3. mesosphere.
4. core.

Discussion

Internal Layers Based on Composition

The nature of the atmosphere, the oceans, and the surface of the land is known in considerable detail because it can be studied by direct observation; but the internal structures of the planet present some of the most difficult problems faced by geologists. The deepest bore holes on Earth penetrate no deeper than 10 km, and structural deformation and erosion rarely expose rocks that formed more than 20–25 km below the surface. Volcanic eruptions provide small samples of material that come from greater depths, possibly as much as 200 km, but aside from these limited data we have no direct knowledge about the nature of Earth's interior. How then are we able to determine the structure and composition of Earth? The evidence comes largely from studies of the physical characteristics of the planet itself—its density, the way in which it transmits seismic (earthquake) waves, the nature of its magnetic field—and from comparisons with meteorites. Although these methods of study do not always provide absolute answers, they do indicate the limits of possibilities of what the interior of a planet may be.

The materials of Earth are separated into layers like an onion, but unlike an onion each layer has a different composition. There are three major compositional layers: the crust, mantle, and core (Fig. 2.1).

The Crust. The term *crust* is left over from the time when it was thought that the outermost layer of Earth formed as the hot molten planet cooled. Today the term designates the outer layer of Earth, extending from the solid surface down to the first major discontinuity in seismic wave velocity in the lithosphere. It heralds a compositional, *but not a structural change*. Moreover, the crust of the continents is distinctly different from the crust beneath the ocean basins (Figure 2.2). The continental crust is much thicker (as much as 50 km thick) and is composed of relatively light "granitic" rock that includes the oldest rock of the crust. By contrast, the oceanic crust is only about 8 km thick and is composed of dark, dense volcanic rocks (basalt) with densities much greater than granite. The oceanic crust is young and relatively undeformed by folding. The differences between the continental and oceanic crust, as we shall see, are of fundamental importance in understanding Earth.

The Mantle. The next major compositional layer of Earth, the mantle, surrounds or covers the core. This zone constitutes the great bulk of Earth (82 percent of its volume and 68 percent of its mass). The mantle is composed of iron and magnesium silicate rock, fragments of which have been brought to the surface by volcanic eruptions.

The upper boundary of the mantle is considered by tradition to be the first worldwide discontinuity in seismic wave velocity (the first zone below the Earth's surface where the velocity at which earthquake waves travel decreases significantly). This was discovered years ago by a Yugoslavian seismologist named Mohorovicic. The name has been shortened to Moho or M-discontinuity. The Moho is believed to be the result of a change in composition but is not a structural boundary.

The Core. The core of Earth is a central mass about 7,000 km in diameter. Its density increases with depth but averages about 10.78 g/cm^3. It constitutes only 16 percent of Earth's volume but accounts for 32 percent of Earth's mass. Indirect evidence indicates that the core is mostly iron.

Internal Layers Based on Physical Properties

Asthenosphere. During the last few decades, it has been recognized that there is a major zone within the upper mantle where temperature and pressure are at just the right balance so that part of the material melts. The rocks lose much of their strength and become soft plastic and flow like warm tar. This zone of easily deformed mantle is called the *asthenosphere* (meaning "weak sphere"). The asthenosphere is a distinctive zone in the upper mantle and is as much as 200 km thick. (Figure 2.2)

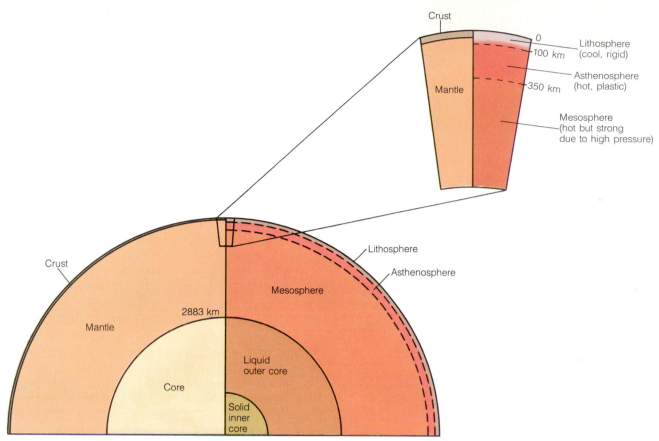

Figure 2.1

The internal structure of Earth is known from studies of its density, its magnetic field, and the way in which it transmits seismic waves. The left side of the diagram shows the layering based on composition; crust, mantle, and core. The right side shows the layering based upon physical properties. Note that the two do not coincide. The expanded view of the upper part of the diagram shows the nature of the outermost layers. The lithosphere is a hard, rigid shell surrounding the entire Earth. The asthenosphere is a zone where increased temperature makes the rock soft, weak, and capable of plastic flow. The mesosphere below is stronger and more rigid because increase in pressure offsets the high temperature that occurs with depth.

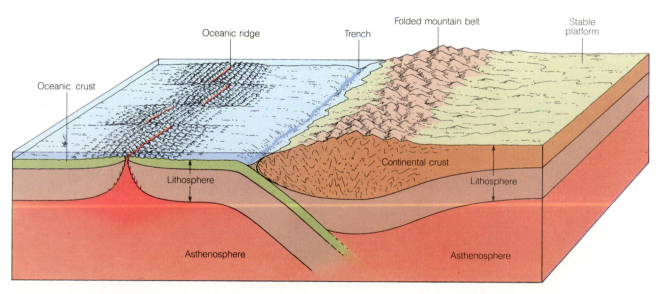

Figure 2.2

The outermost solid layers of Earth are the asthenosphere and the lithosphere. The asthenosphere is close to the melting point and is capable of flow. The lithosphere above it is rigid. It includes two types of crust: a thin, dense oceanic crust and a thick, lighter continental crust.

Lithosphere. The top of the asthenosphere is about 100 km below the surface. Above the asthenosphere the material is solid, strong, and rigid. This layer is called the *lithosphere* ("rock sphere"). The boundary between the lithosphere and the asthenosphere is distinct but does not correspond to a compositional change. The boundary is simply due to a major change in the physical properties of the rock. The lithosphere thus contains the continental crust and the uppermost part of the mantle (Figure 2.2).

Mesosphere. The rock below the asthenosphere is stronger and more rigid than the asthenosphere because the high pressure at this depth offsets the effect of high temperature. The region between the asthenosphere and the core-mantle boundary is called the *mesosphere.* The diagrams in Figure 2.1 shows how the major internal layers of Earth, based on composition and on physical properties, are related.

The Core. The core of the Earth marks a change in both physical properties and composition. It is composed mostly of iron and is therefore distinctly different from the silicate (rocky) material above. On the basis of physical properties, the core has two distinct parts—a solid inner core and a liquid outer core. Heat loss from the core and the rotation of the Earth probably causes the liquid outer core to circulate and its circulation generates the Earth's magnetic field.

CONTINENTS AND OCEAN BASINS

If the oceans were dry, would there by any fundamental difference between the continents and ocean basins? Why does Earth have continents and ocean basins instead of a cratered surface like the Moon, Mercury, and Mars?

Statement

Continents and ocean basins are the principle surface features of Earth. Continental and oceanic crust are distinctly different in composition, density, rock type, structure, and origin.

Discussion

If Earth had neither an atmosphere nor a hydrosphere, two principal regions—continents and ocean basins—would stand out as its dominant features. These major structural and topographic divisions of Earth differ markedly, not only in elevation, but also in rock types, density, chemical composition, age, and history. The ocean basins, which occupy about two-thirds of Earth's surface, are characterized by a spectacular topography, most of which originated from extensive volcanic activity and Earth movements that continue today. The continents rise above the ocean basins as large platforms, but the waters of the oceans more than fill their basins and flood a large part of the continental surface. The present shoreline, so important geographically and so carefully mapped, has no special structural significance with respect to the boundary between continents and ocean basins. Indeed, shorelines have fluctuated greatly throughout Earth's entire history.

From a geological viewpoint, the elevation of the continents, which rise high above the ocean basins, is much more significant than the position of the shore. The difference in elevation of continents and ocean basins is a clue to a fundamental difference in rock density. Continental rocks are less dense than the rocks of the ocean basins—that is, a given volume of continental rock weighs less than the same volume of oceanic rock. This difference causes the continental crust to float above the denser oceanic crust.

The elevation and area of the continents and ocean basins have been mapped with precision, and the data can be summarized in various forms. The data presented graphically in Figure 2.3 show that the average elevation of the continents is 0.84 km above sea level, and the average depth of the ocean is about 3.7 km below sea level. Only a relatively small percentage of the Earth's surface rises significantly above the average elevation of the continents (840 m) or drops below the average depth of the ocean (−3,700 m). If the continents did not rise quite so high above the ocean floor, the entire surface of Earth would be covered with water. Theoretically, a large rotating planet like Earth, with a strong gravitational field, would mold itself into a smooth spheroid covered with a layer of water approximately 2.4 km deep. Continents, as they are known today, have not been found on other planets in this solar system.

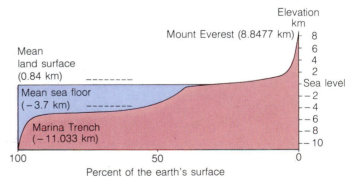

Figure 2.3
A graph of the elevation of the continents and ocean basins shows that the average height of the continents is only 0.84 km above sea level; the average depth of the ocean is 3.7 km below sea level. Only a small percentage of Earth's surface rises above the average elevation of the continents or drops below the average elevation of the ocean floor.

MAJOR FEATURES OF THE CONTINENTS

What are the fundamental structural features of continents?

Statement

Extensive geologic studies during the last 100 years have revealed several striking facts about the continents:

1. Most continents are roughly triangular in shape.
2. They are concentrated in the Northern Hemisphere.
3. Although each may seem unique, all continents have three basic components: (a) a shield, (b) a stable platform, and (c) folded mountain belts. Geologic differences among continents are mostly in the size, shape, and proportions of these components.
4. Continents consist of rock that is less dense than the rock in the ocean basins.
5. Many continental rocks are old, some as old as 3.96 billion years.
6. The climatic zones occupied by a continent usually determine the style and variety of landforms developed on it.

Discussion

The broad, flat continental masses, which rise above the ocean basins, present a great diversity of surface features, with an almost endless variety of hills and valleys, plains and plateaus, and mountains. From a regional perspective, however, the continents are remarkably flat. Most of their surface lies within a few hundred meters of sea level.

The extensive flat, stable regions of the continents in which complex crystalline rocks are exposed or buried beneath a relatively thin sedimentary cover are called **cratons.** These regions have been relatively undisturbed for more than a half billion years, except for broad, gentle warping. The cratons include the shields where large areas of highly deformed igneous and metamorphic rock, the basement complex, are exposed and the stable platforms, regions where the igneous and metamorphic rocks are covered with a relatively thin veneer of sedimentary rocks.

Shields

Without some firsthand knowledge of a shield, visualizing the nature and significance of this important part of the continental crust is difficult. Figure 2.4, showing part of the Canadian Shield of the North American continent, will help you to comprehend the

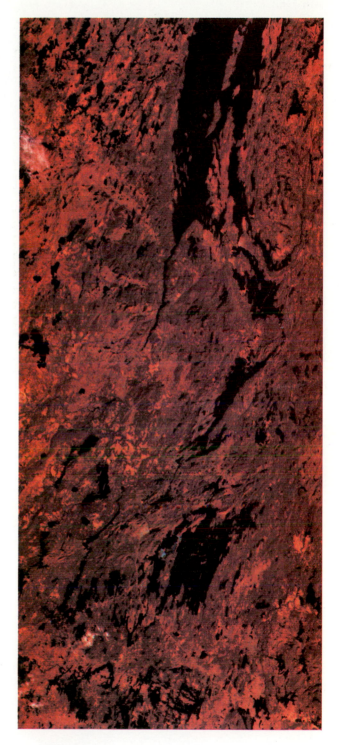

Figure 2.4
The Canadian Shield is a fundamental structural component of North America. It is composed of complexly deformed crystalline rock bodies, eroded down to an almost flat surface near sea level. Throughout much of the Canadian Shield, the topsoil has been removed by glaciers, and different rock bodies are etched out in relief by erosion. The resulting depressions commonly are filled with water, forming lakes and bogs that emphasize the structure of the rock bodies. Dark tones indicate areas of metamorphic rock. Light pink tones indicate areas of granitic rock.

extent, the complexity, and some of the typical features of shields. You should also study Figures 19.1 and 19.2.

Characteristically, a shield is a regional surface of low relief that generally has an elevation within a few hundred meters of sea level. The only features of relief are the resistant rocks that rise 50–100 m above their surroundings.

A second characteristic of shields is their complex structure and rock types. Many rock bodies were once liquid, and others have been compressed and extensively deformed. All of the rocks in the shields are complex and were formed several kilometers below the surface. They are now exposed only because the shields have been subjected to extensive erosion.

Stable Platforms

Large parts of the cratons are covered with a veneer of sedimentary rocks. These areas have been relatively stable throughout the last 600 million or 700 million years—that is, they have not been elevated a great distance above sea level or submerged far below it—and hence the term *stable platform*. In North America the stable platform lies between the Appalachian Mountains and the Rocky Mountains and extends from the Gulf Coast northward to the Lake Superior region and into western Canada. Throughout most of this area, the sedimentary rocks that cover the craton are nearly horizontal, but locally they have been warped into broad domes and basins (Figure 2.5).

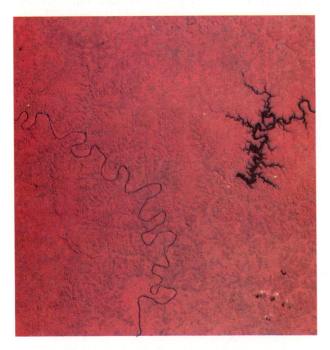

Figure 2.5
The stable platform in the vicinity of Nashville, Tennessee, consists of gently inclined layers of sedimentary rocks worked into a broad domel upwarp.

Folded Mountains

One of the most significant features of the continents is the young, folded mountain belts that typically occur along their margins. Figure 2.6 illustrates some of the characteristics of folded mountains and the extent to which the margins of continents have been deformed. The layers of rock shown in this photograph have been deformed by compression and are folded like wrinkles in a rug. Erosion has removed the upper part of the fold, so the layers that are resistant to erosion form zigzag patterns similar to those that would be produced if the crests of wrinkles in a rug were cut off.

Most people think of a mountain as simply a high, more or less rugged landform, in contrast to flat plains and lowlands. Mountains, however, are much more than high country. To a geologist, the term *mountain belt* means a long, linear zone in the Earth's crust where the rocks have been intensely deformed by compressive forces and generally intruded by molten rock material. The topography can be high and rugged, or it can be worn down to a surface of low relief. To a geologist, the topography of mountain belts is not as important as the way the rocks have been deformed. The great folds and fractures in mountain belts provide evidence that Earth's crust is, and has been, in motion.

We now know that the crusts of the Moon, Mars, and Mercury lack this type of deformation. All of their impact craters, regardless of age, retain their circular outline. If the crust of these planets had been compressed by horizontal forces, the circular craters would be deformed and wrinkled. This fact is proof that the crusts of these planets have not been compressed. Their crusts, unlike that of Earth, appear to have been essentially fixed and immovable throughout their history.

MAJOR FEATURES OF THE OCEAN FLOOR

What landforms and structural features characterize the ocean floor?

Statement

Profiles of the ocean floor show that submarine topography is as varied as that on the continents and, in some respects, more spectacular. Among the most significant facts we have learned about the oceanic crust are the following:

1. The oceanic crust is mostly basalt, a dense volcanic rock, and its major topographic features are somehow related to volcanic activity. The oceanic crust, therefore, is entirely different from the continental crust, whose major rocks are nonvolcanic in origin.

Figure 2.6
The Appalachian Mountains in the eastern United States show the typical style of deformation in a folded mountain belt. The folded strata are expressed by long, narrow ridges of resistant sandstone that rise about 300 m above the surrounding area. Erosion has removed the upper part of many of the folds, so that their limbs (which form the ridges) are exposed in elliptical or zigzag patterns. This type of deformation is an important structural feature of Earth's crust and clearly shows that the crust is mobile and has been in constant motion throughout most of geologic time.

2. The rocks of the ocean floor are young in a geologic time frame. Most are less then 150 million years old, whereas the ancient rocks of the shields are more than 700 million years old and commonly much older.
3. The rocks of the ocean floor have not been deformed by strong, horizontal compression. Their undeformed structure is in marked contrast to the complex deformation of rocks in the folded mountains and basement complex of the continents.
4. The major features of the ocean floor are (a) the oceanic ridge, (b) the abyssal floor, (c) seamounts, (d) trenches, and (e) continental margins.

Discussion

The ocean floor covers 69 percent of Earth's surface, so the oceans, not the continents, are the typical surface of the solid Earth. It is the ocean floor that holds the key to the evolution of Earth's crust, yet not until the 1960s did we obtain enough data about the ocean floor to get a clear picture of its regional characteristics. This new knowledge caused a revolution in geologists' ideas about the nature and evolution of the crust.

Before 1947 most geologists believed that the ocean floor was simply a submerged version of the continents, with huge areas of flat abyssal plains covered with sediment derived from the landmass. Although echo sounding was used to determine depths as early as 1922, the devices were primitive and limited to shallow water. In this process an observer with headphones timed the interval between the transmission of a sound pulse and the return of its echo. The time for the echo to return was a measure of the depth of the ocean floor. A major breakthrough occurred in 1953 with the development of a precision depth recorder that could automatically plot a continuous profile of the ocean floor in any depth of water. Since then, millions of kilometers of profiles have been made, and the information has been used to map the surface features of the ocean basins in remarkable detail.

Although most of the topography of the ocean floor can be seen only indirectly through profile records, some features are visible in satellite photographs, others have been photographed from deep submersible vessels, and some areas have been "seen" through radar imagery. However, physiographic maps that show the configurations of the landforms, such as the one shown on the book's inside covers and in Figure 2.7, provide the best visual reference for regional features of the ocean basins. You should refer to this map frequently as you study the following material.

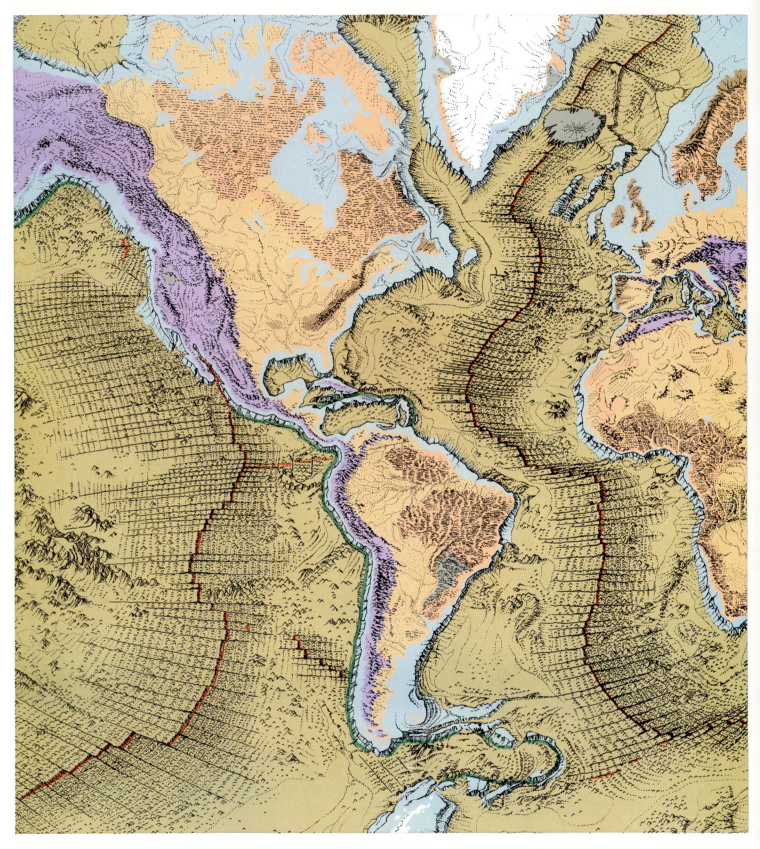

Figure 2.7
The major surface features of Earth reflect the structure of the crust. The continental crust rises above the ocean basins and forms continents that have as their major structural features the shields, stable platforms, and folded mountain belts. The oceanic crust, composed primarily of basalt, forms the ocean floor, the major features of which include the oceanic ridge, the abyssal floor, seamounts, and trenches.

Shield

Stable platform

Continental crust

Flood basalts

Young mountain belt (0.25 MY–0)

Older mountain belt (1.7–0.7 MY)

Oceanic crust

Trench

Rift zone

The Oceanic Ridge

The oceanic ridge is perhaps the most striking and important feature on the ocean floor. It extends continuously from the Arctic Basin down the center of the Atlantic Ocean, into the Indian Ocean, and across the South Pacific (Figure 2.8). The oceanic ridge is essentially a broad, fractured swell generally more than 1,400 km wide. Its higher peaks rise as much as 3,000 m above the ocean floor. A huge, cracklike valley, called the rift valley, runs along the axis of the ridge throughout most of its length. In addition, great systems of fractures, some as long as 4,000 km, extend across the ridge.

The Abyssal Floor

The oceanic ridge divides the Atlantic and Indian oceans roughly in half and cuts across the southern and eastern parts of the Pacific. On both sides of the ridge are vast areas of broad, relatively smooth, deep-ocean basins known as the abyssal floor. This surface extends from the flanks of the oceanic ridge to the continental margins and generally lies at depths of about 3,000 m.

The abyssal floor can be subdivided into two sections, the abyssal hills and the abyssal plains. The abyssal hills are relatively small hills, rising as much as 900 m above the surrounding ocean floor.

Near the continental margins, land-derived sediment completely covers the abyssal hills, forming flat, smooth abyssal plains.

Seamounts

Isolated peaks of submarine volcanoes are called seamounts. Some seamounts rise above sea level and form islands, but most are completely submerged and are known only from oceanographic soundings. Although many seem to occur at random, others, such as the Hawaiian Islands, form chains along well-defined lines. Islands and seamounts of various ages, all originating from volcanic activity, testify to the extensive and continuous volcanic activity that has occurred throughout the ocean basins.

Trenches

The deep-sea trenches are the lowest areas on Earth's surface. The Marianas Trench, in the Pacific Ocean, is the deepest part of the world's oceans—11,000 m below sea level—and many other trenches are more than 8,000 m deep. Trenches have attracted the attention of geologists for years, not only because of their depth, but also because they represent fundamental structural features of Earth's crust. As is illustrated in Figure 2.8, the trenches are invariably adjacent to island arcs or coastal mountain ranges of the

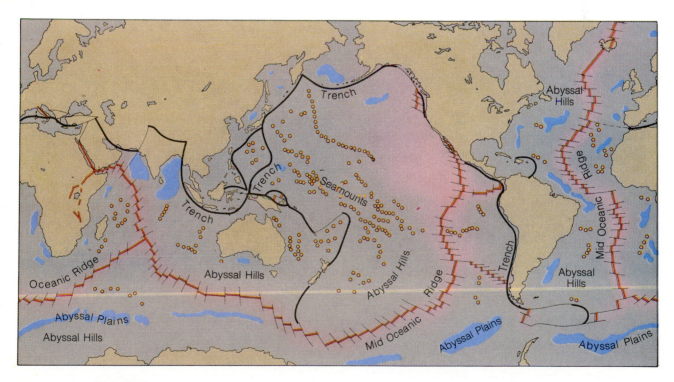

Figure 2.8
The major features of the ocean floor are related to plate boundaries. The oceanic ridge coincides with divergent plate margins. Trenches form where plates converge. The abyssal floor is the deep part of the ocean, off the flanks of the oceanic ridge.

continents. We shall see in subsequent chapters how the trenches are involved in the most intense volcanic and seismic (earthquake) activity on the planet.

Continental Margins

The zone of transition between a continental mass and an ocean basin is called a continental margin. The submerged part of a continent is referred to as a continental shelf. The shelf is the site of shallow seas, most of which have depths less than 200 m. Geologically it is part of the continent, not part of the ocean basin. The continental shelves presently constitute 11 percent of the continental surface, but at times in the geologic past, these shallow seas were much more extensive.

The seafloor descends in a long, continuous slope from the outer edge of the continental shelf to the deep-ocean basin. This continental slope marks the edge of the continental rock mass. Continental slopes are found around the margins of every continent and also around smaller fragments of continental crust, such as Madagascar and New Zealand. Look at the physiographic map inside this book's cover and study the continental slopes—especially those surrounding North America, South America, and Africa. You can see that they form one of the Earth's major topographic features. On a regional scale, continental slopes are by far the longest and highest slopes on Earth. Within this zone, from 20 to 40 km wide, the average relief above the sea floor is 4,000 m. Along the marginal trenches, relief is as great as 10,000 m. In contrast to the shorelines of the continents, the continental slopes are remarkably straight over distances of thousands of kilometers.

In many areas the continental slopes are cut by deep submarine canyons, which are remarkably similar to canyons that are cut by rivers into continental mountains and plateaus. As is shown on the physiographic map (inside covers), submarine canyons cut across the edge of the continental shelf and terminate on the deep abyssal floor, some 5,000 or 6,000 m below sea level.

SUMMARY

1. The major structural units of Earth, based on composition, are (a) crust, (b) mantle, and (c) core.
2. The internal layers of Earth, based on physical properties, are (a) lithosphere, (b) asthenosphere, (c) mesosphere, and (d) core.
3. The two major structural and topographic features of Earth are (a) the continents and (b) the ocean basins.
4. The continents consist of three major components: (a) shields, which are composed of complexly deformed rock eroded down to near sea level; (b) stable platforms, areas where the basement complex is covered with a veneer of horizontal sedimentary rocks; and (c) mountain belts, in which sedimentary rocks have been compressed into folds.
5. The major features on the ocean floor are (a) the oceanic ridge, (b) the abyssal floor, (c) seamounts, (d) trenches, and (e) continental margins.

KEY TERMS

abyssal floor (p. 25)
abyssal hill (p. 25)
abyssal plain (p. 25)
asthenosphere (p. 18)
continent (p. 20)
continental crust (p. 20)

continental margin (p. 26)
continental shelf (p. 26)
continental slope (p. 26)
core (p. 18)
craton (p. 21)

folded mountain belt (p. 22)
lithosphere (p. 20)
mantle (p. 18)
mountain belt (p. 22)
oceanic crust (p. 22)
oceanic ridge (p. 25)

rift valley (p. 25)
seamount (p. 25)
shield (p. 21)
stable platform (p. 22)
submarine canyon (p. 26)
trench (p. 25)

REVIEW QUESTIONS

1. Draw a diagram of the internal structure of Earth and briefly describe the core, mantle, asthenosphere, and lithosphere.
2. What are the major differences between the continents and the ocean basins?
3. Briefly describe the distinguishing features of continental shields, stable platforms, and folded mountain belts.
4. Briefly describe the distinguishing features of the oceanic ridge, the abyssal floor, seamounts, trenches, and continental margins.
5. Study the physiographic map of Earth (inside cover). Make a profile of the Earth's major structural features by tracing or sketching the continental crust, shields, stable platforms, and folded mountain belts, together with the oceanic ridge, the abyssal floor, and deep-sea trenches.

Multiple-Choice Questions

1. The core of Earth is
   a. nearly twice as dense as the mantle.
   b. composed mostly of iron and nickel.
   c. divided into a solid inner core and liquid outer core.
   d. all of the above.
   e. characterized by only (b) and (c).
2. Which of the following is *not* true of the asthenosphere?
   a. It is part of the crust.
   b. It is a "soft" layer below the lithosphere.
   c. It is located in the upper mantle.
   d. It is apparently partly molten.
   e. Volcanic processes and crustal deformation seem to be associated with its movement.
3. The lithosphere is
   a. the outermost shell of Earth.
   b. a rigid, solid layer about 65–100 km thick.
   c. is broken into a series of plates.
   d. all of the above.
   e. Only (a) and (b) are correct.
4. The continental crust
   a. is about the same thickness everywhere.
   b. is mostly basalt.
   c. ends at the shoreline of a continent.
   d. has a granitic layer underneath a basaltic layer.
   e. is much thicker than the oceanic crust.
5. Continental shields are composed of
   a. horizontal layers of rock.
   b. basalt.
   c. intensely deformed igneous and metamorphic rock.
   d. ancient rocks of the oceanic ridges.
   e. folded lava and sedimentary rock.
6. Which of the following is *not* true about young folded mountain belts?
   a. They typically occur near and along the trend of continental margins.
   b. They form as a result of horizontal stresses.
   c. They are long, linear zones made of rock that has been deformed and intruded by molten rock.
   d. They form as a result of vertical stress in the crust.
   e. They are commonly associated with volcanic activity during their early history.

7. The three major structural components of continents are
   a. shield, folded mountains, and plateaus.
   b. folded mountains, plateaus, and a stable platform.
   c. shield, folded mountains, and a stable platform.
   d. stable platform, shield, and plateaus.
   e. shield, maria, and a stable platform.
8. Oceanic trenches generally occur
   a. between a volcanic island arc and a continent.
   b. between the abyssal floor and a volcanic island arc.
   c. adjacent to the oceanic ridge system.
   d. along the east coast of continents.
   e. along the crest rift valleys.
9. The mid-oceanic ridge is
   a. a folded mountain belt.
   b. an up-arched sediment of the ocean floor split by rifting and transverse fractures.
   c. a zone where crustal plates collide and move back into the mantle.
   d. a zone where plates converge and descend into Earth's interior.
   e. isolated segments of continental crust.
10. Oceanic trenches are located
   a. along the margins of all ocean basins.
   b. near the midoceanic ridges.
   c. on the continental shelf areas.
   d. in the center of the ocean.
   e. parallel to chains of island arcs and/or young mountains.

True/False Questions

1. *T F* Continents and ocean basins do not differ markedly in rock type, density, or chemical composition.
2. *T F* The difference in elevation of continents and ocean basins represents a fundamental difference in rock density. Continental rocks are less dense than the rocks of the ocean basins.
3. *T F* Shields are covered with horizontal layers of sedimentary rock.
4. *T F* The rocks of the ocean floor are relatively old, in a geologic time frame.
5. *T F* The rocks of the ocean floor have been deformed by extensive compression.

Fill in the Missing Terms

1. The continents consist of three major components: _____, _____, and _____.
2. Those large areas of the continents in which highly deformed igneous and metamorphic rock of the basement complex are exposed are called _____.
3. Regions of the continents where the basement rocks are covered with a relatively thin veneer of sedimentary rock are called _____ _____.
4. The _____ is Earth's rigid outer layer.
5. The lowest areas on the surface of Earth are the deep-sea _____.

3

EARTH'S DYNAMIC SYSTEMS

We speak of Earth as a dynamic planet because the materials of the various layers are in motion. The most obvious motion is that of the surface fluids, air and water, which circulate in response to solar energy. The complex cycle by which water moves from the oceans to the atmosphere, to the land, and back to the oceans again is called the hydrologic system. Water precipitated on Earth furnishes a fluid medium for many processes that shape the surface features of Earth. Rivers erode the land as they flow to the sea. Groundwater percolating through the pore spaces in rocks dissolves and carries away soluble minerals. Where temperatures are low enough, glaciers form and spread over large parts of a continent, modifying its surface by erosion and deposition as the ice moves.

Earth's own internal energy also plays a major role in moving material inside of the Earth from place to place. Even though the Earth's crust appears to be fixed and stable, there is convincing evidence that it, too, is in constant motion. As you continue your study of geology, the great significance of the internal energy of Earth will become increasingly clear. Thermal energy is believed to be transferred from the asthenosphere to the Earth's surface by convection, a process by which material is moved as a result of density differences produced by heating. This movement of material produces what is known as the tectonic system. As material in the asthenosphere is heated, it rises and then moves laterally, causing the rigid lithosphere to arch upward and split apart. The lithosphere is thus broken into a series of large fragments, called plates, which are moved about by the flow in the asthenosphere. This motion is expressed by volcanic activity, earthquakes, mountain building, continental drift, and seafloor spreading.

MAJOR CONCEPTS

1. Earth's system of moving water—the hydrologic system—involves movement of water in rivers, as groundwater, in glaciers, in oceans, and in the atmosphere.
2. Heat from the sun is the source of energy for the Earth's hydrologic system.
3. As water moves, it erodes, transports, and deposits sediment and creates distinctive landforms.
4. The theory of plate tectonics explains the major structural features of Earth as a result of a series of moving lithospheric plates.
5. Where plates move apart, hot molten material from the mantle wells up to fill the void and creates new lithosphere. The major features formed where plates spread apart are continental rifts, oceanic ridges, basaltic volcanism, and new ocean basins.
6. Where plates converge, one slides beneath the other and plunges down into the mantle. The major features formed at converging plate margins are (a) folded mountain ranges, (b) volcanic arcs, and (c) deep-sea trenches.
7. Earth's lithosphere floats on the denser, plastic asthenosphere beneath, and it rises and sinks in attempts to maintain gravitational balance.

THE HYDROLOGIC SYSTEM

What are the major components of the hydrologic system, and how do they operate?

Statement

The complex cycle through which water moves from the oceans to the atmosphere, to the land, and back to the oceans again is called the hydrologic system. The energy source that drives the system is heat from the sun, which evaporates water from the oceans and causes the atmosphere to circulate. Water vapor moves with the circulating atmosphere and eventually condenses and falls as rain or snow. Acted on by the force of gravity, it then flows back to the oceans in several subsystems (rivers, groundwater, and glaciers). All of these subsystems involve gravity flow from higher to lower levels.

Water in the hydrologic system—moving as surface runoff, groundwater, glaciers, waves, and currents—erodes, transports, and deposits surface rock material as deltas, beaches, and other types of sedimentary accumulations. In this way, the surface material is in motion—motion that results in a continually changing landscape.

Discussion

This complex motion of Earth's surface water—the hydrologic system—operates on a global scale and unites all possible paths of water into a single, grand system of motion. The basic elements of the system can be seen from space (page 6 and Figure 1.4) and are diagrammed in Figure 3.1.

The hydrologic system is perhaps the most fundamental and significant geologic system operating on the Earth's surface. Its influence on the development of surface features is sometimes subtle, sometimes dramatic. Your challenge as a student, new to geology, will not be in understanding the process, but in conceiving the worldwide scope of the system and its influence on the surface features on our planet.

One of the best ways to gain an accurate conception of the magnitude of the hydrologic system is to study the space photography on pages 6 and 8. These photographs provide a view of the system in operation on a global scale. A traveler arriving from space would observe that the surface of the Earth is predominantly water (an obvious conclusion from Figure 1.4). The movement of water from the oceans to the atmosphere is expressed in the flow patterns of the clouds. The atmosphere and moving cloud cover are among the most distinctive features of Earth as it is viewed from space; the great river systems they produce

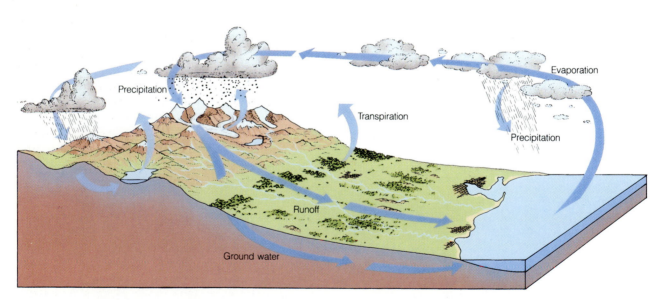

Figure 3.1
The circulation of water in the hydrologic system operates by solar energy. Water evaporates from the oceans, circulates with the atmosphere, and is eventually precipitated as rain or snow. Most of the water that falls on the land returns to the oceans by surface runoff and groundwater seepage. Variations in the major flow patterns of the system include the temporary storage of water in lakes and glaciers. Within this major system are many smaller cycles, or shortcuts, such as evaporation from lakes and transpiration from plants.

stand out in marked contrast to Mars, the Moon, and Mercury, where impact craters dominate the stark landscape. Without hydrologic systems the surfaces of these planets have remained largely unmodified for four billion years (see Figure 1.5).

In addition, the swirling, moisture-laden clouds carry an enormous amount of energy. For instance, the kinetic energy produced by a hurricane amounts to roughly 100 billion kilowatt-hours per day. That is considerably more than the energy used by all of the people of the world in one day.

Another way to grasp the magnitude of the hydrologic system is to consider the volume of water involved. From measurements of rainfall and stream discharge, together with measurements of heat and energy transfer in bodies of water, scientists have calculated that 400,000 km^3 of water evaporate each year (or 1,100 km^3 per day). Of this, about 336,000 km^3 are taken from the oceans, and 63,000 km^3 evaporate from bodies of water on the land. Of the total 400,000 km^3 of water evaporated annually, about 101,000 km^3 fall as rain or snow on the continents, and the rest falls back into the ocean. Most of the precipitation (from 60 to 80 percent) returns directly to the atmosphere by evaporation. About 38,000 km^3 of water flow back to the oceans over the surface as rivers or beneath as groundwater.

At these rates, if the hydrologic system were interrupted and water did not return to the oceans by precipitation and by surface runoff from the continents, sea level would drop 1 m per year. All of the ocean basins would be completely dry within 4,000 years. The "recent" Ice Age demonstrates this point clearly: the hydrologic system was partly interrupted, as much of the water that fell on the Northern Hemisphere froze and accumulated to form huge continental glaciers. This prevented the water from flowing immediately back to the oceans as surface runoff. Consequently, sea level dropped more than 100 m during the Ice Age.

As a final observation, consider the volume of water running off the surface of the Earth. The circulation of water in the hydrologic system provides a continuous supply of surface runoff over the land. At any given instant, about 1,260 km^3 of water are flowing in the world's rivers. Gauging stations on the major rivers of the world indicate that, if the ocean basins were empty, runoff from the continents would fill them completely in 40,000 years. (This estimate assumes no precipitation falling directly into the oceans.) Moreover, if all of the evaporated water returned to the oceans by runoff, and none of it returned to the atmosphere by immediate evaporation or transpiration (the return of water to the atmosphere by plants), then the ocean basins would be filled in slightly more than 3,000 years.

Gravity plays an important role in the hydrologic system. It is the force that causes rain to fall on the land and water to move downslope. It causes rivers to flow and to transport sediment from higher to lower elevations. Indeed, the movement of water in each part of the hydrologic cycle is a type of gravity flow system. The hydrologic system eroded the Grand Canyon, carved the fantastic Himalayan Mountains, and deposited the Mississippi Delta. It formed the glaciers of the last Ice Age, developed the Sahara Desert, and created the streams and valleys of your hometown area. There is little, if any, surface on Earth that is not affected in some way by the work of the hydrologic system. When you go for a walk in the country, it is very likely you will be walking over a surface that was formed by running water, a young surface that is still in the process of developing.

Major Subsystems of the Hydrologic System

The enormous energy of the hydrologic system is involved with each of the subsystems—rivers, glaciers, groundwater, oceans, and wind—all of which erode, transport, and deposit material and create new landforms in the process. We will explore the details of each major geologic process in subsequent chapters, but now let us examine some of the results of the movement of water over Earth.

River Systems. Most of the water precipitated on the land returns directly to the oceans through surface drainage systems. Actually, the volume of water in the rivers on Earth is relatively small; it is only about 0.0001 percent of the total water on Earth, or 0.005 percent of the water not in the oceans. But water flows through river systems very rapidly, at an average rate of 3 m^3 per sec. At this rate, water can travel through the entire length of the longest river in only twenty days. This means that although the volume of the water in rivers at any given time is small, the total volume passing through river systems in a given period can be enormous. As a result, most of Earth's landscape is dominated by features formed by running water. From viewpoints on the ground, we cannot appreciate the prevalence of stream channels on the surface of Earth. From space, however, we readily see that stream valleys are the most abundant landforms on the continents. In arid regions, where vegetation and soil cover do not obscure details, the intricate network of stream valleys is most impressive (Figure 3.2). Most of the surface of every continent is somehow related to the slope of a stream valley, which collects and funnels surface runoff toward the ocean.

Another important aspect of a river system is that it

Figure 3.2
Drainage systems are a clear record of how surface runoff has sculptured the land. They testify to the magnitude of Earth's hydrologic system, for few areas of the land are untouched by stream erosion. In this photograph of a desert region, details of the delicate network of tributaries are clearly shown.

provides the fluid medium that transports huge amounts of sand, silt, and mud to the oceans. This material forms the great deltas of the world, which are a record of the amount of material washed off the continents by rivers. The Mississippi Delta is a classic example (Figure 3.3). The Mississippi River is con-

fined to a single channel far downstream from New Orleans. It then splits into a series of branching channels, which build extensions of new land into the ocean. A large "cloud" of suspended sediment forms in the ocean where the river empties into the Gulf of Mexico and is actively building new land seaward.

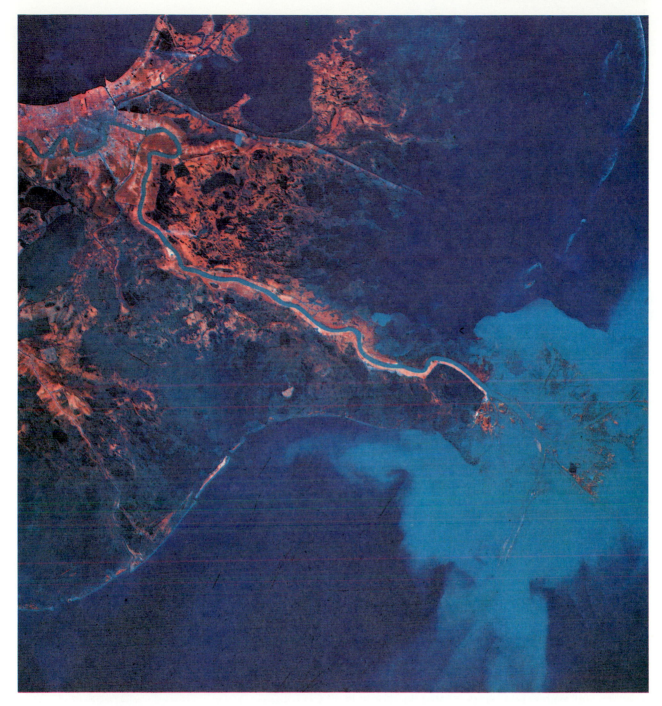

Figure 3.3
The Mississippi Delta, like deltas of other major rivers, is a record of erosion due to the hydrologic system. Sediment eroded from the land is transported by a river system and deposited in the sea. The dynamics of delta building are displayed vividly in this photograph. The cloud of mud and silt delivered to the ocean colors the water a lighter tone around the mouth of the river. This material is deposited as banks of mud, sand, and clay over the continental shelf as the delta grows seaward at a rate of nearly 20 km per 100 years. Measurements indicate that the Mississippi River pours more than a million metric tons of sediment into the Gulf of Mexico each day. In the process of deltaic growth, the river builds up a projection of new land into the ocean. Eventually, the river finds a shorter route to the ocean and abandons its active distributary channel for the shorter course. The abandoned distributary ceases to grow and is eroded back by wave action. Abandoned river channels and inactive subdeltas can be seen clearly on each side of the present river.

Ultimately, the main channel shifts its course to seek a more direct route to the ocean, and these extensions of land (subdeltas) are eroded back by waves and currents. Previous courses of the Mississippi can be seen on both sides of the present river as long, narrow lines of blue water still standing in the old river channel.

Glacial Systems. In cold climates, precipitation falls in the form of snow, most of which remains frozen and does not return immediately to the ocean as surface runoff. If more snow falls each year than melts during the summer months, huge bodies of ice build up to form glaciers. Figure 3.4 is an example of existing glaciers. Large valley glaciers originate from snowfall in the high country and slowly flow down valleys as rivers of ice. These glaciers greatly modify the normal hydrologic system because the water that falls upon the land does not return immediately to the ocean as surface runoff. It is not until the glaciers melt at their lower end that water returns to the hydrologic system as surface runoff.

At the present time, the continent of Antarctica is almost entirely covered with a continental glacier, a sheet of ice from 2.0 to 2.5 km thick. It covers an area of 13 million km^2—an area larger than the United States and Mexico combined. A large part of the Antarctic glacier can be seen on page 8. An ice sheet similar to that now on Antarctica covered a large part of North America and Europe during the last Ice Age and retreated only within the last 15,000 years. As the ice moved, it modified the landscape by creating numerous lakes and other landforms in Canada and the northern United States.

Water in the form of ice constitutes about 80 percent of the water not in the oceans, or about 2 percent of Earth's total water. Water in glaciers moves very slowly and may remain in a glacier for thousands or even millions of years. Present estimates suggest that water resides in glaciers on the average of 10,000 years.

Groundwater Systems. Another segment of the hydrologic system is the water that seeps into the ground and moves slowly through the pore spaces in soil and rocks. Surprisingly, about 20 percent of the water not in the oceans occurs as groundwater. As it slowly moves, groundwater dissolves soluble rocks and creates caverns and caves, which can enlarge and collapse to form surface depressions called sinkholes. This type of water solution-generated landform is common in Kentucky, Florida, and west Texas and is easily recognized from the air (Figure 3.5). Sinkholes commonly create a pockmarked surface somewhat resembling the cratered surface of the Moon. They may also become filled with water and form a series of circular lakes.

Shoreline Systems. The hydrologic system also operates along the shores of all continents, islands, and inland lakes by the unceasing work of waves. The oceans and lakes are bodies of mobile water subject to a variety of movements, waves, tides, and currents, all capable of eroding the coast and transporting vast

Figure 3.4
Valley glaciers, such as these in Alaska, occur where more snow accumulates each year than is melted during the summer months. Valley glaciers originate in the snowfields of high mountain ranges and flow as large tongues of ice down preexisting stream valleys. The moving ice is an effective agent of erosion and modifies the valleys in which it flows; thus, glaciers cause local modifications of the normal hydrologic system. The dark lines on the glaciers are rock debris derived from the valley walls.

quantities of sediment (Figure 3.6). The effects of shoreline processes are seen in wave-cut cliffs, shoreline terraces, deltas, beaches, bars, and lagoons.

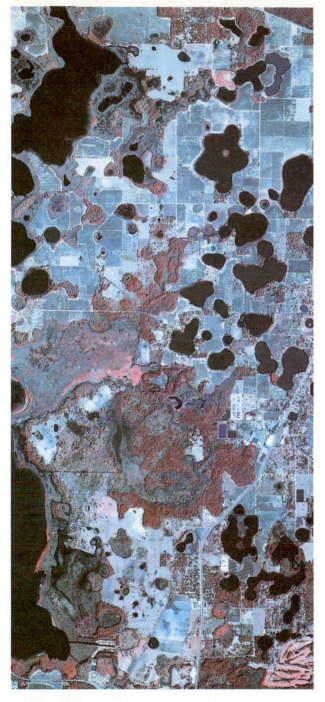

Figure 3.5
Groundwater is a largely invisible part of the hydrologic system because it occupies pore spaces in the soil and rocks beneath the surface. It can, however, dissolve soluble rocks, such as limestone, to form complex networks of caves and subterranean passageways. As the caverns enlarge, their roofs may collapse, so that circular depressions called sinkholes are formed. Sinkholes create a pockmarked surface called karst topography. The hundreds of lakes shown in this photograph of the area west of Cape Canaveral, Florida, occupy sinkholes and testify to the effectiveness of groundwater as a geologic agent.

Figure 3.6
The Atlantic coastline of the southern states is dominated by offshore bars and barrier islands. A considerable amount of sediment eroded from the continent is brought seaward by rivers and reworked by wave action to form long, narrow, sandy barrier beaches, which fringe the mainland. This photograph shows part of the shoreline of North Carolina. The lagoon of Pamlico Sound and the Great Dismal Swamp, protected by the barrier island from the vigorous wave action of the Atlantic, will eventually become filled with sediment, extending the mainland seaward to the present position of the barrier beach. The edge of the continental shelf is delineated by puffy clouds, which form where the cold surface water to the east meets the warmer surface water of the shallow nearshore zone. The edge of the shelf is only about 120 m deep.

Eolian (Wind) Systems. The hydrologic system also operates in the arid regions of the world. In many deserts, river valleys are still the dominant landform. There is no completely dry place on Earth. Even in the most arid regions some rain falls, and climatic patterns change over the years. River valleys can be obliterated, however, by the great seas of windblown sand that cover the desert landscape (Figure 3.7). Air circulating in the atmosphere constitutes the eolian system, which can transport enormous quantities of loose sand and dust, leaving a distinctive record of the wind's activity. In the broadest sense, the wind itself is part of the hydrologic system, a moving fluid on the planet's surface.

Figure 3.7
"Sand seas" of the great deserts of the Earth form in arid low-latitude regions where there is not enough precipitation for the hydrologic system to operate in its normal manner. The vast areas of migrating sand dunes in the world's deserts illustrate the effectiveness of the circulating atmosphere as a geologic agent, continually transporting enormous quantities of sediment over the surface of the Earth.

THE TECTONIC SYSTEM

What geologic features are explained by the theory of plate tectonics?

Statement

Plate tectonics is the theory of global dynamics in which the lithosphere is believed to be broken into individual plates that move in response to convection in the upper mantle. The margins of the plates are sites of considerable geologic activity, such as volcanoes, mountain building, earthquakes, and seafloor spreading.

Discussion

Geologists have long recognized that Earth has its own source of internal energy that is repeatedly manifested by earthquakes, volcanic activity, and folded mountain belts; but not until the middle 1960s was a unifying theory of the dynamics of Earth developed. This theory, known as plate tectonics, provided, for the first time, a master plan of Earth's crustal dynamics. The importance of the plate tectonic theory is that it provides a framework in which numerous observations of the Earth are united into a comprehensible whole. The great power of this theory is that it can explain, with equal ease, such diverse geologic phenomena as mountain building, evolution of ocean basins, the chemistry of lavas, rates of sedimentation, and the migration and extinction of plants and animals. Plate tectonics explains the dynamics of the planet Earth—the way it works—and is probably the most significant scientific breakthrough in the history of geology.

The basic elements of the plate tectonic theory are simple and can be easily understood by carefully studying the diagram in Figure 3.8. The lithosphere, which includes Earth's crust and part of the upper mantle, is rigid, but the underlying asthenosphere slowly flows. The fundamental idea of plate tectonics is that the segments, or plates, of the rigid lithosphere are in constant motion relative to one another and carry the lighter continents with them. The lithospheric plates form as hot mantle material rises along midoceanic ridges and are destroyed in subduction zones where one of the converging plates plunges down into the hotter mantle below (Figure 3.8). Their descent is marked by deep-sea trenches that border island arcs and some continents. Where plates slide by one another, large fracture systems occur. The movement and collision of plates account for most of Earth's earthquakes, volcanoes, and folded mountain belts, as well as for the drift of its continents.

The theory of plate tectonics largely grew out of the

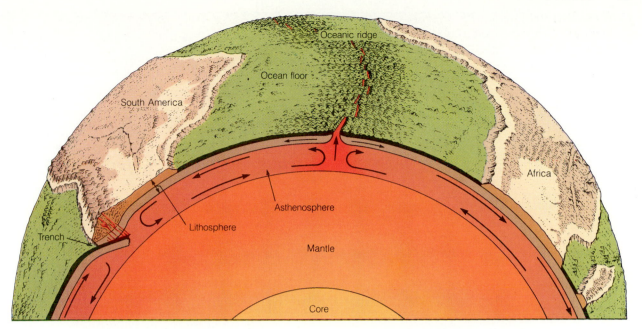

Figure 3.8
The Tectonic System operates from Earth's internal heat. The asthenosphere is more plastic than both the overlying lithosphere and the underlying lower mantle because of an optimum balance of temperature and pressure. Above the plastic asthenosphere, relatively cool and rigid lithospheric plates split and move apart as single mechanical units. Molten rock from the asthenosphere wells up to fill the void and creates new lithosphere. Convection circulation occurs in the asthenosphere where material fills the gap between the spreading plates and moves away to areas where plates descend into the lower mantle at oceanic trenches. Some plates contain blocks of thick, low-density continental crust, which cannot sink into the mantle. As a result, where a plate carrying continental crust collides with another plate, the continental margins are deformed into mountain ranges. The plate margins are the most active areas on Earth—the sites of the most intense volcanism, seismic activity, and crustal deformation. The depth of convection cells is uncertain.

new oceanographic technology that permits us to study the characteristics of the ocean floor, map its surface features, and measure its magnetic and seismic properties. Evidence for this revolutionary theory of crustal movement comes from many sources and includes data on the structure, topography, and magnetic patterns of the ocean floor; the locations of earthquakes; the patterns of heat flow in the crust; the locations of volcanic activity; the structure and geographic fit of the continents; and the nature and history of mountain belts (see Chapter 15 for details).

From the standpoint of the dynamics of Earth, the boundaries of the plates are where the action is. As seen in Figure 3.9, plate boundaries do not necessarily coincide with continental boundaries, although some do. There are seven very large plates and a dozen or more small plates (not all of which are shown in Figure 3.9). Each plate is about 100 km thick. Plates slide over the more mobile asthenosphere below, generally at rates between 1 and 10 cm per year, although rates of interactions between adjacent plates approach 20 cm per year. Because the plates are internally rigid, they interact mostly along their edges.

The plates diverge and move apart along midoceanic ridges (Figure 3.8). Hot molten material from the deeper mantle wells up to fill the void. Some of this material erupts at the surface as basaltic lava. Thus, new lithosphere is formed at the plate's trailing edge. The oceanic ridges stand high because their material is hot and, therefore, less dense than the colder adjacent oceanic crust.

The oceanic ridges are commonly displaced by large faults (fractures) that are also the sites of shallow earthquakes. It is here that plates horizontally slide past one another. The best-known example of this type of fault is the San Andreas Fault in California (see Figure 3.12).

Within the moving plate, the continents and oceanic crust experience little tectonic or volcanic activity as they passively move away from the midoceanic ridge. Hot spots in the deep mantle, however, may create upwarps and volcanic activity in the plates' interior. An excellent example is the Hawaiian Islands in the Pacific Ocean.

At convergent plate boundaries, the geologic activity is far more varied and complicated (Figure 3.8). We now know something about the mechanics of

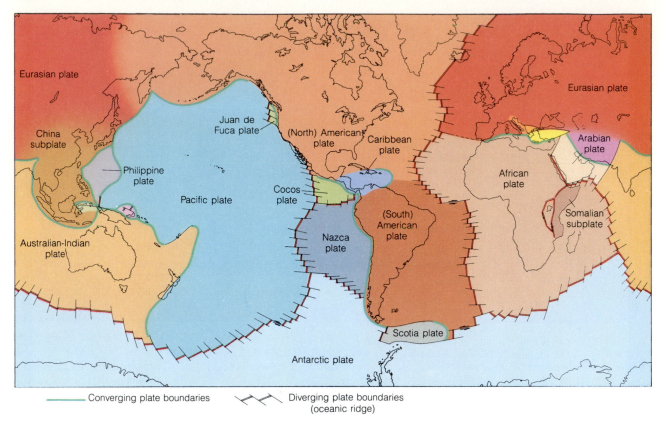

─── Converging plate boundaries   ⌐⌐⌐ Diverging plate boundaries (oceanic ridge)

Figure 3.9
A mosaic of plates forms Earth's lithosphere, or outer shell. The plates are rigid, and each moves as a single unit. There are three types of plate boundaries: (1) the axis of the oceanic ridge, where the plates are diverging and new oceanic crust is generated; (2) transform faults, where the plates slide past each other; and (3) subduction zones, where the plates are converging and one descends into the asthenosphere.

these junctions from geophysical studies. It is the processes at these junctions that ultimately result in folded mountain belts, alteration of the rock by heat and pressure, and the growth of continents. At the point where the plates converge, one tips down and slides beneath the other, a process known as subduction. The simplest form of the convergent junction is one where two plates with oceanic crust meet. Such plate junctions in the western and northern Pacific region lie along the volcanic islands of Tonga, the Marianas, and the Aleutians. A trench, normally 5–8 km deep, is formed where the undersliding plate plunges down into the mantle. Sediment on the ocean floor and slivers of oceanic crust are scraped off against the front of the overriding plate and stacked in a sequence against it as a chaotic mass of mud and basalt, commonly referred to as a *melange* (French, meaning "mixture"). Some large slices of oceanic crust and mantle may be incorporated into the melange. The Franciscan Formation in the area of San Francisco, California, is believed to have been plastered against the continental margin as sediment was stripped from a subducting plate. The subducting plate is heated as it descends into the mantle. Some of

the material rises and may be extruded to form a string of volcanic islands called island arcs, or they may produce a chain of volcanoes on a continent. It is a series of these volcanic arcs that make up the "ring of fire" around the Pacific Ocean.

The continental crust and material in the island arc are relatively buoyant and so resist subduction. The intense compression at plate margins is what forms mountain belts.

Although there are many unanswered questions, the basic source of energy for tectonic movement is believed to be the Earth's internal heat, which is transferred by convection. Hot material in the mantle rises to the base of the lithosphere, where it then moves laterally, cools, and descends to become reheated, beginning the cycle again. A familiar example of convection can be seen in the heating of a pot of soup (Figure 3.10). Heat applied to the base of the pot warms the soup at the bottom, which therefore expands and becomes less dense. The warm fluid rises to the top and is forced to move laterally. It then cools, becomes more dense, and sinks. The regular flow circuit of rising warm fluid and sinking cold fluid is called a convection current.

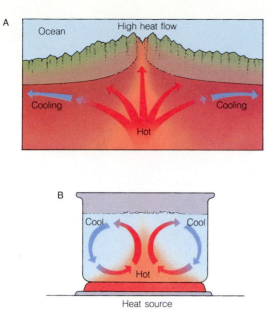

Figure 3.10
Convection in the asthenosphere can be compared to convection in a pot of soup. Heat from below causes the material to expand and become less dense. The warm material rises by convection and spreads laterally. It then cools, becomes more dense, and sinks. It is reheated as it descends, and the cycle is repeated.

Examples

In order to appreciate why the plate tectonic theory has had such an impact on the science of geology, let us consider some of the major geologic features of our planet and see how the theory explains them. Take a moment and study Figure 3.9 again. You will want to become very familiar with this map because it shows a new geography: the geography of plate tectonics.

Plates and Plate Motion

As can be seen on Figure 3.9, seven major lithospheric plates are recognized—the North American, South American, Pacific, Australian, African, Eurasian, and Antarctic plates—together with several smaller ones. The spreading axis, where the lithosphere is being pulled apart, is marked by the oceanic ridge, which extends from the Arctic down through the central Atlantic and into the Indian and the Pacific oceans. Movement of the plates is away from the crest of the oceanic ridge.

The North American and South American plates are moving westward and interacting with the Pacific, Juan de Fuca, Cocos, and Nazca plates along the west coast of the Americas. The Pacific plate is moving northwestward from the oceanic ridge toward the system of deep trenches in the western Pacific basin. The Australian plate includes Australia, India, and the

northeastern Indian Ocean. It is moving northward, causing India to collide with the rest of Asia and producing the Himalayan Mountains. The African plate includes the continent of Africa, plus the southeastern Atlantic and western Indian oceans. It is moving eastward and northward. The Eurasian plate, which consists of Europe and most of Asia, moves eastward. The Antarctic plate includes the continent of Antarctica, plus the Antarctic Ocean. It is unique in that it is nearly surrounded by oceanic ridges.

There are many other features of both the ocean basins and continents that can be nicely explained by the plate tectonic theory. We will consider many of them in some detail in subsequent chapters, but let us look briefly at the major geologic features of our planet and how they fit into the tectonic system.

Topography of the Ocean Floor

First look at the topography of the ocean floor (see the map on the inside of the book cover). One of the remarkable features of the seafloor is the great system of deep-sea trenches that almost surrounds the Pacific Ocean. They are long, narrow troughs and are the lowest features on the Earth. Invariably they are associated with volcanic activity and a zone of intense earthquakes, although the reason for this association was not always known. It is now obvious that the deep-sea trenches mark the zone where plates converge and one descends into the mantle. Their low topography, association with earthquakes, and lines of volcanic activity are clearly explained in light of plate tectonics. Similarly, the midoceanic ridge, the world's longest structural and topographic feature, has been known to exist for decades. But why is there a "mountain range" extending through the center of the world's ocean basins? Plate tectonics provides an answer. The midoceanic ridge marks the zone where the plate margins are uplifted and pulled apart.

Fractures and Rifts

Many geologic features on the continents that have been interpreted as independent phenomena are now explained in light of a single unifying theory of the Earth's dynamics. For example, the great rift of the Red Sea shown in Figure 3.11 is an extension of the spreading axis of the Indian Ocean, which splits the Sinai and Arabian peninsulas from Africa. Take the time to locate the area shown in this remarkable photograph on the physiographic map on the inside cover of this book. The structure of the area shown in Figure 3.11 is dominated by two long, linear fault valleys. One is occupied by the Gulf of Aqaba, the Dead Sea, and the Jordan River, and the other by the Gulf of Suez. These rifts express dramatically the tensional or shear stresses in the crust, and we now see the way these stresses affect the Earth's surface. The

Figure 3.11

The Red Sea rift, which separates Africa from Asia, is a major fracture system in the Earth's crust. This feature, an important part of the Earth's rift system, represents the incipient stages of crustal fragmentation and the movement of continental plates. The rift extends up the Red Sea and splits at its northern end, with one branch forming the Gulf of Suez and the other extending up the Gulf of Aqaba, into the Dead Sea, and up the Jordan Valley. Movement of the Arabian plate in a northeasterly direction, away from Africa, has created a new ocean basin, which is in the initial stage of its formation. As the rift widens, the edges of the continental block break off, forming a series of steps leading down toward the depression. These steplike blocks can be seen along the Gulf of Suez as distinct parallel lines in the bedrock and as linear trends in the offshore islands.

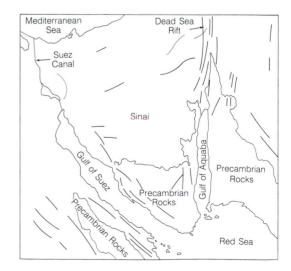

Figure 3.12
The San Andreas Fault is one of the most significant geologic structures in California. This view shows the intersection of the San Andreas Fault and the Garlock Fault, northwest of Los Angeles. These two major structural features intersect to form the boundary of the Mojave Desert, the light-colored triangular area in the upper-right-hand quarter of the photograph. At least a dozen major fault systems can be seen as linear trends in the landscape in this view. Most have been active during historical times. Movement along the San Andreas Fault is horizontal—that is, one block of the Earth's crust slides laterally past the other.

long straight lines extending through the Jordan River Valley and the Gulf of Aqaba are major fractures produced by the rift. Volcanic rocks extruded along the fracture zone are visible as a dark tone, especially evident near the Sea of Galilee, in the upper part of the photograph.

Another type of fracture in the crust results if plates slide past each other horizontally. The great San Andreas Fault in California, parts of which are clearly shown in Figure 3.12, is the result of this type of movement. The fault zone is marked by sharp linear landforms such as straight, narrow valleys, linear ridges, and displaced drainage patterns. The San Andreas fault system is an active boundary between the Pacific plate to the west and the North American plate to the east. The Pacific plate is moving at about 6 cm per year, relative to the North American plate. As stress builds up between the plates, the rock bodies are deformed until they break. The sudden release along the fault causes the rock bodies to snap back into place, which caused the recent earthquake in California.

Earthquakes and Volcanoes

Earthquakes and volcanic activity are closely related in both time and space and do not occur randomly. They are concentrated along specific zones, extending around the Pacific Ocean and through the Mediterranean Sea. More recently it has been discovered that a zone of shallow earthquakes extends along the crest of the midoceanic ridge. When these zones of earthquakes and volcanic activity are plotted on a map, they outline with dramatic clarity the plate boundaries (Figure 3.13). Plate tectonics can thus readily explain why Japan is tormented by repeated volcanic eruptions and earthquakes (Figure 3.14): it is created at the convergence of two tectonic plates. The same is true for the west coasts of Central and South America. It is equally clear that the earthquakes and volcanic eruptions in the Mediterranean area occur at plate margins. Most of the major earthquakes and volcanic eruptions you have read about in the past—and will likely hear of in the future—occur along the margins of tectonic plates.

The Origin of Mountains

These associations are enough in themselves to make the theory of plate tectonics attractive, but they are only the beginning. The great mountain belts of the world can be related to the moving plates. Where the moving plates converge, the crust is compressed and folded. The space photograph in Figure 3.15 shows the type of deformation that most vividly expresses motion in the crust. The complex system of ridges and valleys is produced by folded sedimentary rocks that were deformed by the collision between plates. The folded rocks now appear like wrinkles in a rug. A younger mountain belt extends from Alaska through the Rockies and the Andes of South America and is produced from the encounter of the American plates with the Pacific, Cocos, and Nazca plates. This is a young mountain system with many parts still being deformed as the plates continue to move. It is also apparent that the mountain systems of the Alps and Himalayas can be nicely explained by plate tectonics as the result of the collisions of the African and

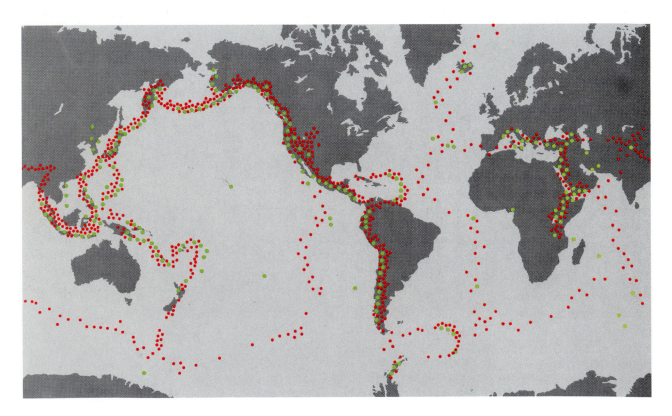

● Volcanoes   ● Earthquakes

Figure 3.13
Plate margins are outlined with remarkable fidelity by zones of earthquake activity and recent volcanism. Shallow earthquakes, submarine volcanic eruptions, and tensional fractures occur along the oceanic ridges, where plates are moving apart. Deep earthquakes, volcanic eruptions, and folded mountain belts occur along margins where plates converge.

Figure 3.14
Volcanoes of the Japanese Island arc are formed by volcanic activity produced where two oceanic plates converge, and one is thrust under the other and assimilated into the mantle. As the descending plate melts, the molten rock rises to form a chain of volcanoes. If the upper plate contains oceanic crust, the volcanic activity produces an arc of volcanic islands. This area in the vicinity of Sapporo, Japan, is dotted with both active and extinct volcanic cones built up from the seafloor. If the upper plate contains continental crust, volcanic material is extruded on a folded mountain belt. An excellent example of this type of volcanic activity occurs in the Andes Mountains of South America.

Australian-Indian plates with Eurasia. These are young, active mountain belts, still growing, still being deformed by the present motion of plates. Older mountain belts, such as the Appalachian Mountains in the eastern United States and the Ural Mountains of Russia, mark zones where older plates collided long ago and were "welded" together.

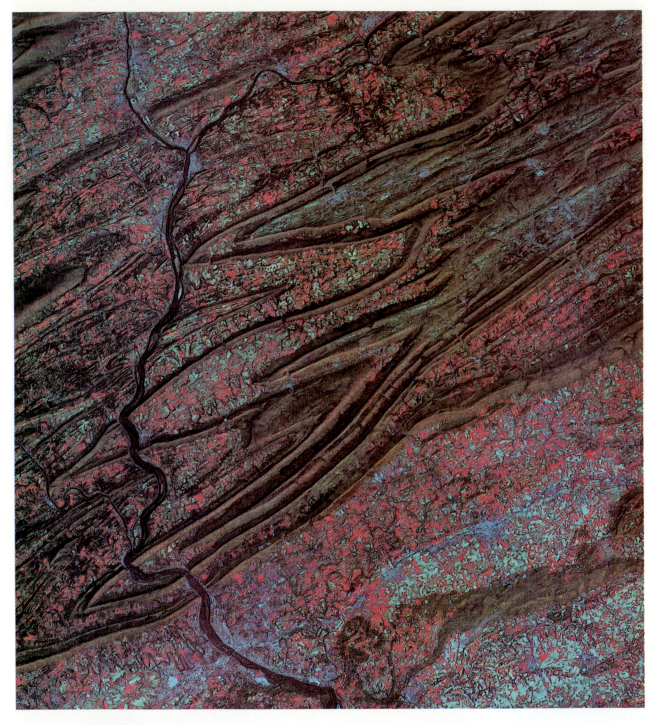

Figure 3.15
The Folded strata of the Appalachian Mountains were formed approximately 250 million years ago at converging plate margins. Formations of sedimentary rock, which were originally horizontal, have been compressed into tight folds that have been subsequently eroded. The resistant layers of sandstone appear as ridges, forming a zigzag pattern. Folded mountain belts like the Appalachians are one of the most significant results of converging plates.

GRAVITY AND ISOSTASY

How is gravity involved in the Earth's processes?

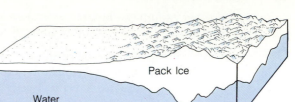

Figure 3.16
Isostasy is the universal tendency of segments of the Earth's crust to establish a condition of gravitational balance. Differences in both density and thickness can cause isostatic adjustments in the Earth's crust. The thickness of pack ice in polar regions illustrates the concept of isostasy very well. Measurements made by airborne radar and laser profilometers and by submarine upward-directed sonar give us a complete picture of the upper and lower extent of an ice body. Where major bodies collide, pressure ridges form that stand as much as 8 m above sea level, and corresponding pressure roots may extend more than 20 m below the surface. The block diagram, which shows a reconstruction based on numerous measurements, indicates the isostatic adjustments of pack ice are similar to isostatic adjustments of the continental crust of the Earth.

Statement

Gravity plays a fundamental role in Earth's dynamics, and all parts of the crust adjust to establish a gravitational balance.

Discussion

In addition to changes brought about by the hydrologic and tectonic systems, the Earth's lithosphere continually responds to the force of gravity as it tries to maintain a gravitational balance, or *isostasy* (Greek *isos,* "equal"; *stasis,* "standing"). Isostasy occurs because the crust is buoyed up by the more dense mantle beneath it, and each portion of the lithosphere displaces the mantle according to its bulk and density (Figure 3.16). Denser crustal material sinks deeper into the mantle than less dense crustal material.

Isostatic adjustment in the Earth's crust can be compared to adjustments in a sheet of ice floating on a lake as you skate on it. The layer of ice bends down beneath you, displacing a volume of water with a weight equal to your weight. As you move ahead, the ice rebounds behind you, and the displaced water flows back.

As a result of isostatic adjustments, high mountains and plateaus having a great vertical thickness sink deeper into the mantle than areas of low elevation do. Any change in the thickness in an area of the crust—such as removal of material by erosion or addition of material by sedimentation, volcanic extrusion, or accumulations of large continental glaciers—causes an isostatic adjustment.

The concept of isostasy, therefore, is fundamental to studies of the major features of the crust, such as continents, ocean basins, and mountain ranges, and also to understanding the response of the crust to erosion, sedimentation, glaciation, and the tectonic system.

Examples

How do we know that isostatic adjustments occur?

The construction of the Hoover Dam, on the Colorado River, provides an excellent, well-documented illustration of isostatic adjustment. Before the dam was built, the land had adjusted to some degree of isostatic balance. The water that accumulated behind the dam to form Lake Mead upset this balance. In a matter of years, this added weight caused the crust to subside in the vicinity of the lake as much as 1.7 m.

Continental glaciers are another clear example of isostatic adjustment of the crust. The weight of an ice sheet several thousand meters thick disrupts the crustal balance and causes the crust beneath to be depressed. In both Antarctica and Greenland, the weight of the ice has depressed the central part of the landmasses below sea level. A similar isostatic adjustment occurred in Europe and North America during the last Ice Age, when continental glaciers existed there. Parts of both continents, such as the Baltic Sea and Hudson Bay, are still below sea level. Now that the ice is gone, however, the crust is rebounding at a rate of 5–10 m per 1,000 years.

Tilted shorelines of ancient lakes provide another means of documenting isostatic rebound. Lake Bonneville, for example, was a large lake in Utah and Nevada during the Ice Age but has since dried up into such small remnants as Utah Lake and the Great Salt Lake. As the lake basin filled with water, the surface was depressed under the additional weight of the water. When the lake was full, its shorelines were level. As the lake dried up, the weight of the water was removed and the crust rebounded, which caused the shorelines to tilt. The lake was only 305 m deep and covered a much smaller area than a continental glacier, but this relatively small weight was sufficient to depress the crust. Since it has been removed, the shorelines near the deepest part of the lake have rebounded nearly 60 m.

These and other examples illustrate several important facts about gravity and isostatic adjustments.

1. Gravity is the driving force for all isostatic adjustments. All types of loading and unloading therefore cause vertical isostatic movements. Isostasy is involved in all of the processes that shift material on the Earth's surface. Some of the more obvious isostatic adjustments occur in
   a. mountains and highlands: as erosion removes material, the crust should rebound.
   b. deltaic areas: where sediment is deposited, the added weight causes the crust to subside.
   c. areas of volcanic activity: the added weight of extrusions causes the crust to subside.
   d. regions of continental glaciation: the thick ice sheets cause the crust to subside; as the ice is removed, the crust rebounds.
2. Very small loads, such as water a few hundred meters deep, are also sufficient to cause isostatic adjustments.
3. Isostatic adjustments can occur very rapidly in a geologic time frame (60 m in fewer than 20,000 years).

SUMMARY

1. The hydrologic system includes all possible paths by which water moves from the oceans through the atmosphere, over the land, and below the ground.
2. Rivers, glaciers, groundwater, shorelines, and wind are the major subsystems of the hydrologic system, and each produces distinctive landscapes.
3. The theory of plate tectonics explains Earth's crustal dynamics by the movement of rigid lithospheric plates. New crust is created and heat is brought to the surface, where mantle material from below moves into the gap between the spreading lithospheric plates. At a subduction zone, cool lithosphere descends and is ultimately consumed in the hotter mantle below.
4. The major structural features of Earth are formed along plate boundaries.
5. At divergent plate boundaries the lithosphere is upwarped and spreads apart. The major features produced in this area are (a) oceanic ridges, (b) continental rifts, (c) volcanic activity, and (d) generation of new lithosphere.
6. At converging plate margins, one plate slides under the other and descends into the mantle. The major features produced in this area are (a) folded mountain belts, (b) deep-sea trenches and island arcs, (c) volcanism, and (d) earthquakes.
7. Isostasy plays an important role in the Earth's dynamics.

KEY TERMS

convergent plate
  boundary (p. 37)
divergent plate
  boundary (p. 37)
eolian system (p. 36)

glacier (p. 34)
groundwater (p. 34)
hydrologic system (p. 30)
isostasy (p. 45)
Mid-Atlantic Ridge (p. 39)

mountain belt (p. 42)
plate tectonics (p. 36)
pore spaces (p. 34)
residence time (p. 31)
river system (p. 31)

runoff (p. 31)
shoreline system (p. 34)
tectonic plate (p. 38)
tectonics (p. 36)

REVIEW QUESTIONS

1. Diagram the paths by which water circulates in the hydrologic system.
2. What is the source of energy for the hydrologic system?
3. Describe the major landforms resulting from (a) rivers, (b) groundwater, (c) glaciers, and (d) wind.
4. Draw a diagram (cross section) showing (a) converging plates and (b) diverging plates.
5. Show the plate boundaries on a map.
6. What surface features mark plate boundaries?
7. Explain how the Andes Mountains, the midoceanic ridge, deep-sea trenches, island arcs, and volcanoes are related to plate tectonics.
8. Explain convection in the Earth's interior.
9. Explain the concept of isostasy in the Earth's system and give two examples of isostatic adjustment of the Earth's crust in recent geologic time.

Multiple-Choice Questions

1. Which of the following is an *inaccurate* statement about isostasy?
   a. Gravity is the ultimate force for all isostatic adjustments.
   b. Any major loading and unloading may result in vertical movements of the crust.
   c. A body of water a few hundred meters deep is sufficient to cause isostatic adjustments.
   d. Isostatic adjustments require at least one million years to take effect.
   e. Gravity is a force in all geologic systems.

2. Which is *not* the result of the hydrologic system?
   a. surface form of mountains
   b. internal structure of mountains
   c. canyons
   d. shorelines
   e. glaciers

3. The source of energy that makes the hydrologic system operate is
   a. radiogenic heat from Earth.
   b. solar energy.
   c. geothermal energy.
   d. heat from the core.
   e. wind.

4. If the hydrologic system were interrupted and water were prevented from returning to the ocean, the ocean basins would be dry in
   a. four million years.
   b. one million years.
   c. 500,000 years.
   d. 4,000 years.
   e. 400 years.

5. Which of the following subsystems is the most important in terms of modifying the surface of most continents?
   a. glacial systems
   b. river systems
   c. groundwater systems
   d. eolian systems
   e. shoreline systems

6. At a transform boundary, tectonic plates
   a. slide past each other.
   b. move away from each other.
   c. move toward each other.
   d. move over spreading centers.
   e. move vertically.

7. Which of the following features develop where plates converge and one descends into the mantle?
   a. a midoceanic rift
   b. a seamount chain
   c. an abyssal plain
   d. a trench
   e. a fracture system

8. Earth's hydrologic system includes
   a. river systems.
   b. glacial systems.
   c. groundwater systems.
   d. shoreline systems.
   e. all of the above.

9. The effects of gravity on the Earth's dynamics can be seen in
   a. isostatic adjustment of the crust.
   b. differentiation of the interior of Earth.
   c. the preservation of continental crust on the lithosphere.
   d. the geologic subsystems of the hydrologic system.
   e. all of the above.

10. At a convergent boundary, tectonic plates
   a. slide past each other.
   b. move away from each other.
   c. move toward each other.
   d. move over spreading centers.
   e. do not move.

True/False Questions

1. *T F* If precipitation and runoff did not return water to the ocean, all of the ocean basins would be completely dry within 4,000 years.
2. *T F* The fundamental idea of the plate tectonic theory is that the asthenosphere moves in response to flow in the lithosphere.
3. *T F* Plates containing continental crust move down into the asthenosphere at the deep-sea trenches and are consumed.
4. *T F* The boundaries between lithospheric plates are delineated with dramatic clarity by belts of active earthquakes and volcanoes.
5. *T F* The mountain systems of the Alps and the Himalayas resulted from the collisions of the African plate and the Australian plate with the Eurasian plate.

Fill in the Missing Terms

1. The source of energy for Earth's hydrologic system is heat from the _____.
2. The source of energy driving the tectonic system is thought to be heat from the _____.
3. The zone where the Pacific plate descends into the mantle is marked by a system of deep-sea _____.
4. The greatest volcanic activity on Earth occurs under water, along the _____ _____.
5. The driving force for all isostatic adjustment is _____ _____.

4

MINERALS

All of the Earth's dynamic processes involve the growth and destruction of minerals—matter changing from one state to another. Minerals grow, melt, dissolve, or are broken and modified by physical forces. As the Earth's surface weathers and erodes, some minerals are destroyed and others grow in their place. As sediments accumulate in the oceans, minerals grow from solution. In volcanic eruptions, minerals grow as lava cools. Deep below the surface, high pressure and temperature cause the removal of atoms from the crystal structure of some minerals and then recombine them into new structures, to form minerals that are more stable deep in the crust. As tectonic plates move and continents drift, minerals are created and destroyed by a variety of chemical and physical processes. Some knowledge of the Earth's major minerals, therefore, is essential to an understanding of the Earth's dynamics.

In this chapter we survey the general characteristics of minerals and the physical properties that identify them. We then explore the silicate mineral group (the major rock-forming minerals) in preparation for a study of the major rock types in Chapters 5, 6, and 7.

MAJOR CONCEPTS

1. An atom is the smallest unit of an element that possesses the properties of the element. It consists of a nucleus of protons and neutrons and a surrounding "cloud" of electrons.
2. An atom of a given element is distinguished by the number of protons in its nucleus. Isotopes are varieties of an element, distinguished by the different numbers of neutrons in their nuclei.
3. Ions are electrically charged atoms, produced by a gain or loss of electrons.
4. Matter exists in three states: (a) solid, (b) liquid, and (c) gas. The differences among the three states are related to the degree of ordering of the atoms.
5. A mineral is a natural, inorganic solid possessing a specific internal atomic structure and a chemical composition that varies only within certain limits.
6. Minerals grow when atoms are added to the crystal structure as matter changes from the gaseous or the liquid state to the solid state. Minerals dissolve or melt when atoms are removed from the crystal structure.
7. All specimens of a mineral have well-defined physical and chemical properties (such as crystal structure, cleavage, fracture, hardness, and specific gravity).
8. Silicate minerals form more than 95 percent of the Earth's crust.
9. The most important rock-forming minerals are feldspars, micas, olivines, pyroxenes, amphiboles, quartz, clay minerals, and calcite.

MATTER

What is the structure of an atom?
What are the distinguishing characteristics of an isotope? Of an ion?

Statement

A simplified description of modern atomic theory includes the following fundamental concepts:

1. An atom is the smallest fraction of an element that can exist and still show the characteristics of that element.
2. The main building blocks of an atom are protons, neutrons, and electrons, although many other subatomic particles have been identified in recent years.
3. A typical atom consists of a nucleus of protons and neutrons and a cloud of electrons surrounding the nucleus.
4. The distinguishing feature of an atom of a given element is the number of protons in the nucleus. The number of electrons and neutrons in an atom of a given element can vary within limits, but the number of protons is constant.
5. Normally atoms are electrically neutral because they have one negatively charged electron for each positively charged proton.
6. Electrically charged atoms, called ions, are produced by the gain or loss of electrons.
7. Isotopes, which are varieties of a given atom (element), are the result of differences in the number of neutrons in the nucleus.
8. Atoms combine, mostly through ionic or covalent bonding, to form minerals.

Discussion

Atoms

To understand the dynamics of the Earth and how rocks and minerals are formed and changed through time, you must have some knowledge of the fundamental structure of matter and how it behaves under various conditions. The solid materials that make up Earth's outer layers are called rocks. Most rock bodies are mixtures, or aggregates, of minerals. A mineral is a naturally occurring inorganic solid compound with a definite chemical formula and a specific internal structure. Because minerals, in turn, are composed of atoms, we must understand something about atoms and the ways in which they combine.

Atoms have been measured and described by models constructed from mathematical formulas. Recently, however, images of atoms have been made. An example is shown in Figure 4.1. In its simplest form,

an atom is characterized by a relatively small nucleus of tightly packed protons and neutrons, with a surrounding cloud of electrons. Each proton carries a positive electrical charge, and the mass of a proton is taken as the unit of atomic mass. The neutron, as its name indicates, is electrically neutral and has approximately the same mass as the proton. The electron is a much smaller particle, with a mass approximately $1/1,850$ of the proton. It carries a negative electrical charge equal in intensity to the positive charge of the proton. Because the electron is so small, for practical purposes the entire mass of the atom is considered to be concentrated in the protons and neutrons of the nucleus.

Hydrogen is the simplest of all elements. It consists of one proton in the nucleus and one orbiting electron (Figure 4.2). The next heavier atom is helium, with two protons, two neutrons, and two electrons. Each subsequently heavier element contains more protons, neutrons, and electrons. The distinguishing feature of an element is the number of protons in the nucleus of each of its atoms. The number of electrons and neutrons in an atom of a given element can vary within limits, but the number of protons is constant.

Atoms normally have the same number of electrons as protons and thus do not carry an electrical charge. As the number of protons increases in the succeedingly heavier atoms, the number of electrons also increases. The electrons fill a series of energy-level shells around the nucleus, each shell having a maximum capacity.

Isotopes

Although the number of protons in each atom of a given element is constant, the number of neutrons can vary within certain limits. This means that atoms of a given element are not necessarily all alike. Iron atoms, for example, have 26 protons but can have 28, 30, 31, or 32 neutrons. These varieties of iron are examples of isotopes. They all have the properties of iron and differ from one another only in number of neutrons. Most common elements exist in nature as a mixture of isotopes. Some isotopes are unstable, emitting atomic particles as radioactive energy.

Ions

Atoms that have as many electrons as protons are electrically neutral, but atoms of all elements can gain or lose electrons in their outermost shells. If this occurs, an atom loses its electrical neutrality and becomes charged. These electrically charged atoms are called ions. The loss of an electron results in a positively charged ion because the number of protons then exceeds the number of negatively charged electrons. If an electron is gained, the ion has a negative charge. The electrical charges of ions are important

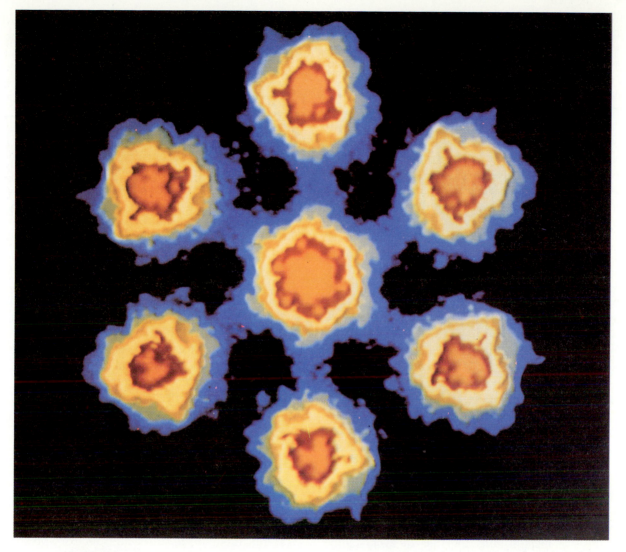

Figure 4.1
Photographs of atoms have been made by M. Isaacson, Cornell University, and M. Ohtsuki, University of Chicago. This illustration shows a uranyl acetate cluster on a very thin carbon substrate. The individual uranium atoms are the colored spots with the red-orange centers. The carbon substrate appears black. Separation between the uranium atoms is 0.34 nm.

because the attraction between positive ions and negative ions is the bonding force that sometimes holds matter together.

Bonding

Atoms are most stable if their outermost shells are filled to capacity with electrons. Except for the inner shell, which can hold no more than 2 electrons, each shell can hold 8 electrons up to argon (atomic number 18). Shells can then have 18 electrons per shell, and later 32. Neon, for example, has 10 protons in the nucleus and 10 electrons, of which 2 are the first shell and 8 are in the outermost shell. A neon atom does not have an electrical charge. Its two electron shells are complete because the second shell has a limit of eight electrons. As a result, neon is stable: it does not interact chemically with other atoms. Argon and the other noble gases also have eight electrons in their

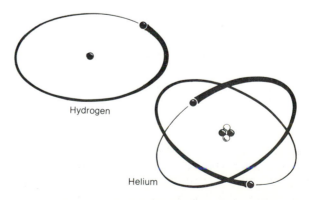

Hydrogen

Helium

Figure 4.2
The atomic structures of hydrogen and helium illustrate the major particles of an atom. Hydrogen has one proton and one orbiting electron. Helium has two protons, two neutrons, and two orbiting electrons.

outermost shells, and they normally do not combine with other elements. Most elements, however, have incomplete outer shells. Their atoms readily lose, gain, or share electrons to achieve a structure like that of argon, neon, and the other noble gases, with eight electrons in the outermost shell.

For example, an atom of sodium has only one electron in its outer shell but eight in the shell beneath (Figure 4.3). If it could lose the lone outer electron, the sodium atom would have a stable configuration like that of the noble gas neon. The chlorine atom, in contrast, has seven electrons in its outer shell, and if it could gain an electron, it too would attain a stable configuration. Whenever possible, therefore, sodium gives up an electron and chlorine gains one. The sodium atom thus becomes a positively charged sodium ion, and the chlorine atom becomes a negatively charged chloride ion. With opposite electrical charges, the sodium ions and chloride ions attract each other and bond to form the compound sodium chloride (common salt, also known as the mineral halite). This type of bond, between ions of opposite electrical charge, is called an ionic bond.

Atoms can also attain the electronic arrangement of a noble gas, and thus attain stability, by sharing electrons. No electrons are lost or gained, and no ions are formed. Instead, an electron "cloud" surrounds both nuclei. This type of bond is called a covalent bond.

A third type of bond is called a metallic bond. In a metal, each atom contributes one or more outer electrons that are not attached to a specific pair but move relatively freely throughout the entire aggregate of ions. This sea of negatively charged electrons holds the positive metallic ions together in a crystalline structure and is responsible for the special characteristics of metals.

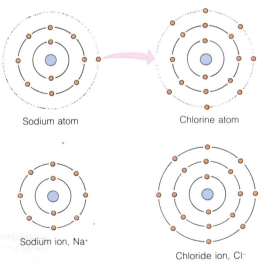

Sodium atom

Chlorine atom

Sodium ion, Na+

Chloride ion, Cl−

Figure 4.3
The formation of sodium and chlorine ions by transfer of an electron from the outermost shell of a sodium atom to the outermost shell of a chlorine atom results in a stable outer shell for each ion.

States of Matter

Why do solids, liquids, and gaseous forms of a substance have such different physical properties and still have the same composition?

The principal differences between solids, liquids, and gases involve the degree of ordering of their atoms. In the typical solid, atoms are arranged in a rigid framework. The arrangement in crystalline solids is a regular, repeating, three-dimensional pattern called a crystal structure. Each atom in a solid occupies a more or less fixed position but has a vibrating motion. As the temperature rises, the vibration increases, and the atoms move farther apart. Eventually they become free and are able to glide past one another, and the solid matter passes into the liquid state.

In a liquid, the atoms are in random motion, but they are packed closely together. They slip and glide past one another or collide and rebound, but they are held together by forces of attraction greater than those in gases but less than solids. This explains why density generally decreases and compressibility increases as matter changes from solid to liquid to gas. If a liquid is heated, the motion of the particles increases, and individual atoms, or molecules, become separated as they move about at high speeds.

In a gas, the particles are in rapid motion and travel in straight lines until their direction changes when they collide with another atom. The individual atoms, or molecules, are separated by empty spaces and are comparatively far apart. This is why gases can be markedly compressed and can exert pressure. Gases have the ability to expand indefinitely, and the continuous, rapid motion of the particles results in rapid diffusion.

Water is undoubtedly the most familiar example of matter that changes through the three basic states. At pressures prevailing on the Earth's surface, water changes from a solid to a liquid to a gas in a temperature range of only 100° C. Other forms of matter in the solid Earth are capable of similar changes, but usually their transitions from solid to liquid to gas occur at comparatively high temperatures. At normal room temperature and pressure, 93 of the 106 elements are solids, 2 are liquids, and 11 are gases.

Most people are familiar with the effects of temperature changes on the state of matter because of their experience with water as it freezes, melts, and boils. Increase in temperature causes matter to change from solid to liquid to gas. Fewer people are familiar with the effects of pressure. In general, an increase in pressure can cause matter to pass from a gas to a liquid to a solid. Under great pressure, water will remain liquid at temperatures as high as 371° C (superheated water).

THE NATURE OF MINERALS

What is a mineral?

Statement

Minerals are the major solid constituents of the Earth. Their distinguishing characteristics are as follows:

1. They occur naturally as inorganic solids.
2. They have a specific internal structure; that is, their atoms are precisely arranged into a crystalline solid.
3. They have a chemical composition that varies within definite limits and can be expressed by a chemical formula.
4. They have definite set of physical properties (hardness, cleavage, crystal form, etc.) that result from their crystalline structure and composition.

The differences among minerals arise from the kinds of atoms they contain and the ways the atoms are arranged in a crystalline structure.

Discussion

Inorganic Solids

Many people commonly think of minerals only as exotic crystals in museums or as valuable gems and metals. But grains of sand, snowflakes, and salt particles are also minerals, and they have much in common with gold and diamonds.

By definition only naturally occurring inorganic solids are minerals—that is, natural elements or inorganic compounds in a solid state. Synthetic products, such as artificial diamonds, are therefore not minerals in the strictest sense, nor are they organic compounds, such as coal and petroleum, which are organic materials but not crystalline solids.

Minerals can consist of a single element, such as gold, silver, copper, diamond, and sulfur. However, because the abundant elements in the Earth have a strong tendency to combine, most are compounds of two or more elements.

The Structure of Minerals

The key words in the definition of mineral are *internal structure*. The atoms of a mineral have a specific arrangement in a definite geometric pattern. Every specimen of the same mineral has the same internal structure, regardless of when, where, and how it was formed. Mineralogists had long suspected this property of minerals because they had observed the many expressions of order in crystals.

With modern methods of X-ray diffraction (the apparent reflection of X-rays as they pass through a crystalline substance), we can determine precisely a mineral's internal structure and learn much about its atomic arrangement. The technique is illustrated in Figure 4.4. If a thin beam of X-rays is passed through a mineral, it is diffracted (or dispersed) by the framework of atoms. The dispersed rays produce an array of dots on photographic film placed behind the crystal. From measurements of the relationships among the dots, the systematic orientation of planes of atoms within the crystal can be deduced and expressed in mathematical formulas. Detailed models of crystal structures can thus be constructed and analyzed. The X-ray diffraction instrument is now the most basic device for determining the internal structure of minerals, and geologists use it extensively to precisely identify and analyze minerals.

To understand the importance of structure in a mineral, consider the characteristics of diamond and graphite. These two minerals are identical in chemical composition. Both consist of a single element, carbon (C). Their crystal structures and physical properties, however, are very different. In a diamond, which forms only under high pressure, the carbon atoms are packed closely together, and the bonds between the atoms are very strong. This explains why diamonds are extremely hard—the hardest substance known. In graphite, the carbon atoms are loosely bound in a layered structure. The layers separate easily, which accounts for graphite's slippery, flaky property. Because of its softness and slipperiness, graphite is used as a lubricant and is also the main constituent of common "lead" pencils. The important point to note is that different structural arrangements of the same element produce different minerals with different properties. This ability of a specific chemical substance to crystallize with more than one type of structure is known as polymorphism.

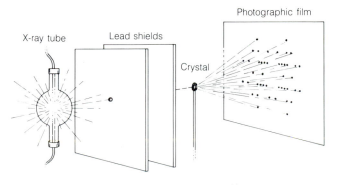

Figure 4.4
X-ray diffraction patterns are made by passing a thin beam of X-rays through a mineral. The X-rays are diffracted (dispersed) by the framework of atoms in the crystal, forming an orderly arrangement of dots on a photographic film placed behind the specimen.

The Composition of Minerals

A mineral has a definite chemical composition, in which specific elements occur in definite proportions. Thus, a precise chemical formula can be written to express the chemical composition—i.e., SiO_2, $CaCO_3$, etc. The chemical composition of some minerals can vary, but only within specific limits. In such minerals, two or more kinds of ions can substitute for each other in the mineral structure, a process called ionic substitution. Ionic substitution results in a chemical change in the mineral without a change in the crystal structure, so substitution can occur only within definite limits. The composition of such a mineral can be expressed by a chemical formula that specifies ionic substitution and how the composition can change.

The suitability of one ion to substitute for another is determined by several factors, the most important being the size and the electrical charge of the ions in question (Figure 4.5). Ions can readily substitute for one another if their ionic radii differ by less than 15 percent. If a substituting ion differs in charge from the ion for which it is substituted, the charge difference must be compensated for by other substitutions in the same structure.

Ionic substitution is somewhat analogous to substituting different types of equal-sized bricks in a wall. The substitute brick may be composed of glass, plastic, or whatever, but because it is the same size as the original brick, the structure of the wall is not affected. An important change in composition has, however, occurred, and as a result there are changes in physical properties. In minerals, ionic substitution causes changes in hardness, color, etc., without changing the internal structure.

Ionic substitution is common in rock-forming minerals and is responsible for mineral groups whose members have the same structure but varying composition. For example, in the olivine group, with the formula $(Mg,Fe)_2 SiO_4$, ions of iron (Fe) and magnesium (Mg) can substitute freely for one another. The total number of Fe and Mg ions is constant in relation to the number of silicon (Si) and oxygen (O) atoms in the olivine, but the ratio of Fe to Mg may vary in different samples. The common minerals feldspar, pyroxene, amphibole, and mica each constitute a group of related minerals in which atomic substitution produces a range of chemical composition and a range of physical properties within certain limits.

Physical Properties of Minerals

Because a mineral has a definite chemical composition and a definite internal crystalline structure, all specimens of a given mineral, regardless of when or where they were formed, have the same physical and chemical properties. If ionic substitution occurs, variation in physical properties also occurs; but because ionic substitution can occur only within specific limits, the range in physical properties also can occur only within specific limits. This means that one piece of quartz, for example, is as hard as any other piece, that it has the same specific gravity (the ratio of the density of a substance to the density of water), and that it breaks in the same manner.

The more significant and readily observable physical properties of minerals are crystal form, cleavage, hardness, specific gravity, color, and streak.

Crystal Form.   If a crystal is allowed to grow in an unrestricted environment, it develops natural crystal faces and assumes a specific geometric form. The shape of a crystal is a reflection of its internal structure and is an identifying characteristic for many mineral specimens. If the atoms are arranged in a long chain, the crystal may be long and slender, like a needle. If the atoms are arranged in a boxlike network, the crystal will likely be in the form of a cube (Figure 4.6). Quartz, for example, typically forms long, hexagonal crystals (Figure 4.7). If the space for growth is restricted, however, smooth crystal faces cannot develop.

Most crystals are rather small, measuring from a few tenths of a millimeter to several centimeters in diameter. Some are so small they can only be seen enlarged thousands of times with an electron microscope (Figure 4.8). In an unrestricted environment,

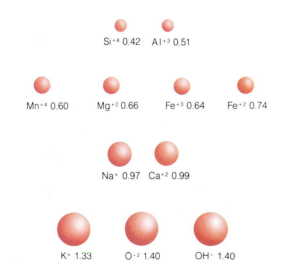

Si^{+4} 0.42 Al^{+3} 0.51

Mn^{+4} 0.60 Mg^{+2} 0.66 Fe^{+3} 0.64 Fe^{+2} 0.74

Na^+ 0.97 Ca^{+2} 0.99

K^+ 1.33 O^{-2} 1.40 OH^- 1.40

Figure 4.5
The relative size and electrical charge of ions are important factors governing the suitability of one ion to substitute for another in a crystal structure. Iron can be replaced by magnesium, sodium by calcium, and silicon by aluminum. If the substituting ions do not have exactly the same electrical charge, the difference in charge must be balanced or compensated for by other substitutions in the structure.

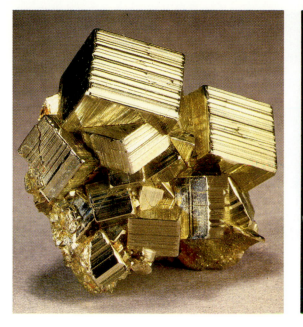

Figure 4.6
Crystals of pyrite have a cubic form because the constituent atoms are arranged in a cubic pattern.

Figure 4.7
Quartz crystals form prisms in which some of the crystal faces meet at 120_o, regardless of the gross shape and size of the crystals or when and where they were formed. The constancy of interfacial angles is an expression of the atomic structure of the mineral. Every mineral has a characteristic crystal form as a result of its crystalline structure.

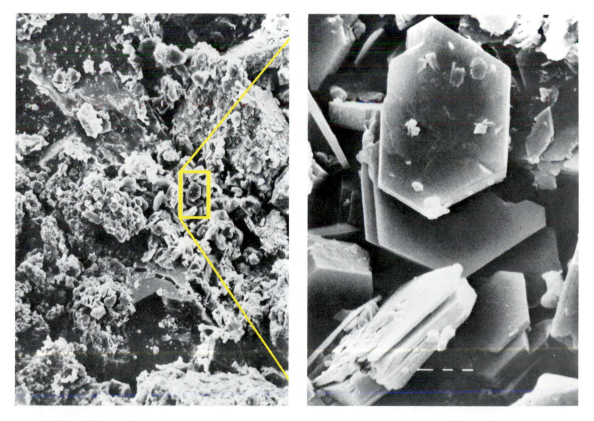

Figure 4.8
Submicroscopic crystals growing in the pore spaces between sand grains can be seen through an electron microscope. Each crystal contains all of the physical and chemical properties of the mineral.

Figure 4.10
Cleavage in the mineral halite occurs in three directions at right angles. The resulting fragments have a characteristically cubic form.

Figure 4.9
Large crystals can form where there is ample space for growth, such as in caves. These crystals of gypsum are more than 1 m long.

however, crystals can grow to enormous sizes (Figure 4.9). A single crystal, regardless of size, contains all the physical and chemical properties of that mineral.

Cleavage. Cleavage is the tendency of a crystalline substance to split or break along smooth planes parallel to zones of weak bonding in the crystal structure (Figures 4.10–4.12). If the bonds are especially weak in a given direction, as in mica or halite, perfect cleavage occurs with ease, and breaking the mineral in any direction other than along a cleavage plane is difficult (Figure 4.11). In other minerals, the differences in bond strength are not great, so cleavage is poor or imperfect. Cleavage can occur in more than one direction; however, the number and direction of cleavage planes in a given mineral species are always the same.

Some minerals have no weak planes in their crystalline structure. As a result, they do not have cleavage but break along various types of fracture surfaces. Quartz, for example, characteristically breaks by conchoidal fracture—that is, along curved surfaces, like the curved surfaces of chipped glass.

Hardness. Hardness is a measure of a mineral's resistance to abrasion. This property is easily determined and is used widely for field identification of minerals. More than a century ago, Friedrich Mohs (1773–1839), a German mineralogist, assigned arbi-

Figure 4.11
Perfect cleavage in one direction is illustrated by the mineral mica.

Figure 4.12
Cleavage of calcite occurs in three planes that do not intersect at right angles. This results in a form called a rhombohedron. Each cleavage fragment is bounded by similar sets of cleavage planes.

Table 4.1 The Mohs Hardness Scale

Hardness	Mineral	Test
1	Talc	Fingernail (2.5)
2	Gypsum	
3	Calcite	Copper coin (3)
4	Fluorite	Knife blade (5.5)
5	Apatite	glass plate (5.5 +)
6	K-feldspar	
7	Quartz	streak plate (7)
8	Topaz	
9	Corundum	
10	Diamond	

Figure 4.13
Gypsum has a hardness of 2 on the Mohs hardness scale. It is a very soft mineral and can easily be scratched with a fingernail.

trary relative numbers to ten common minerals in order of their hardness. He assigned the number 10 to diamond, the hardest mineral known. Softer minerals were ranked in descending order, with talc, the softest mineral, assigned the number 1. The Mohs hardness scale (Table 4.1) provides a standard for testing minerals for preliminary identification. Gypsum, for example, has a hardness of 2 and can be scratched by a fingernail (Figure 4.13).

Specific Gravity. Specific gravity is the ratio of the weight of a given volume of a substance to the weight of an equal volume of water. For example, a liter of solid lead weighs a little over eleven times more than a liter of water, and thus the specific gravity of lead is 11.

Specific gravity is one of the more precisely defined properties of a mineral. It depends on the kinds of atoms making up the mineral and how closely they are packed in the crystal structure. Clearly, the more numerous and compact the atoms, the higher the specific gravity. Most common rock-forming minerals have specific gravities from 2.65 (for quartz) to about 3.37 (for olivine).

With a little experience, you will be able to estimate the relative specific gravity of a mineral merely by lifting a specimen. Most metallic minerals feel heavy, whereas most common nonmetallic minerals seem relatively light.

Color. Color is one of the more obvious properties of a mineral. Unfortunately, it is usually not a good diagnostic tool to use in identifying minerals. Most minerals are found in various hues, depending on such factors as subtle variations in composition and the presence of impurities. Quartz, for example, ranges through the spectrum from clear, colorless crystals to purple, red, white, yellow, gray, and jet black.

Streak. The color of a mineral in powder form, referred to as streak, is usually a better diagnostic tool than the color of a large specimen. For example, the mineral pyrite (fool's gold) has a gold color but a black streak, whereas real gold has a gold streak—the same color as that of its larger grains. To test for streak, a mineral is rubbed vigorously against the surface of an unglazed piece of white porcelain. Minerals softer than the porcelain leave a streak, or line, of fine powder. For minerals harder than porcelain, a fine powder can be made by crushing a mineral fragment. The powder is then examined against a white background.

Growth and Destruction of Minerals

Minerals are characteristically susceptible to chemical change. They grow as matter changes from a gaseous or liquid state to a solid state, and they break down as the solid changes back to a liquid or gas. All minerals came into being because of specific physical and chemical conditions, and all are subject to change as physical and chemical conditions change. Minerals, therefore, provide an important means of interpreting the changes that have occurred in the Earth throughout its history.

The growth and breakdown of minerals are of paramount importance in geologic processes because rocks originate and change as minerals are formed and destroyed. Minerals grow when lava cools and solidifies. They grow from solution in the sea. They grow at the Earth's surface, where minerals in common rocks react chemically with elements in the atmosphere. They also form and are destroyed down deep within Earth's crust, where heat and pressure cause some crystal structures to collapse and new minerals with a more dense, compact atomic structure to form in their place.

SILICATE MINERALS

What are silicate minerals?
Why are they important in geology?

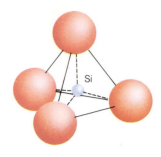

Figure 4.14
The silicon-oxygen tetrahedron is the basic building block of the silicate minerals. In this figure the diagram to the right is expanded to show the position of the small silicon atom. Four large oxygen atoms are arranged in the form of a pyramid (tetrahedron) with a small silicon atom fitted into the central space between them. This is the most important building block in geology because it is the basic unit for 95 percent of the minerals in the Earth's crust.

Statement

Silicate minerals are complex in both chemistry and crystal structure, but all contain a basic building block called the silicon-oxygen tetrahedron. This is a complex ion, SiO^4, in which four large oxygen ions, O^{-2}, are arranged to form a four-sided pyramid with a smaller silicon ion, SiO^{-4}, fitted into the cavity between them (Figure 4.14). This geometric shape is known as a tetrahedron. The major groups of silicate minerals differ mainly in the arrangement of such silica tetrahedra in their crystal structure.

Discussion

Although more than 2,000 minerals have been identified, 95 percent of the Earth's crust is composed of a group of minerals called the silicates. This should not be surprising because silicon and oxygen constitute nearly 75 percent of the mass of the Earth's crust and therefore must predominate in most rock-forming minerals. Perhaps the best way to understand the unifying characteristics of the silicates, as well as the reasons for their differences, is to study the models shown in Figure 4.15. These were constructed on the basis of X-ray studies of silicate crystals. Silicon-oxygen tetrahedra combine to form minerals in two ways. In the simplest combination, the oxygen ions of the tetrahedra form bonds with other elements, such as iron or magnesium. Olivine is an example. Most silicate minerals, however, are formed by the sharing of an oxygen ion between two adjacent tetrahedra. In this way, the tetrahedra form a larger ionic unit, just as beads are joined to form a necklace. The sharing of oxygen ions by the silicon ions results in several fundamental configurations of tetrahedral groups: single chains, double chains, two-dimensional sheets, and three-dimensional frameworks (Figure 4.15). These structures define the major silicate mineral groups.

1. Single chains—pyroxenes
2. Double chains—amphiboles
3. Two-dimensional sheets—micas, chlorites, and clay minerals
4. Three-dimensional frameworks—feldspars and quartz

The unmatched electrons are balanced by various metal ions, such as ions of calcium, sodium, potassium, magnesium, and iron. The silicate minerals thus contain silica tetrahedra linked in various patterns by metal ions.

   Considerable ionic substitution can occur in the

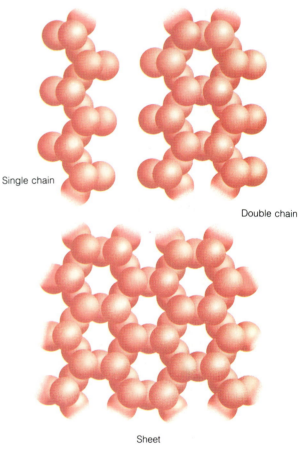
Single chain

Double chain

Sheet

Figure 4.15
Silicon-oxygen tetrahedral groups can form single chains, double chains, and sheets by the sharing of oxygen ions among silicon ions.

basic crystal structure. For example, sodium can substitute for calcium, or iron can substitute for magnesium. Minerals of a major silicate group can thus differ chemically among themselves but have a common silicate structure.

ROCK-FORMING MINERALS

What are the distinguishing characteristics of the major minerals in igneous rocks, metamorphic rocks, and sedimentary rocks?

Statement

Fewer than twenty kinds of minerals account for the great bulk of the Earth's crust and most of the upper mantle. The most important minerals in each of the major rock types are as follows:

1. In igneous rocks: feldspars, micas, amphiboles, pyroxenes, olivines, and quartz
2. In sedimentary rocks: quartz, calcite, dolomite, clays, halite, gypsum, and feldspars
3. In metamorphic rocks: quartz, feldspars, amphiboles, pyroxenes, micas, garnet, and chlorites

Discussion

Rocks

Rocks are aggregates of minerals, and as a result they exhibit a wide range in composition and physical properties. The chemical composition of a rock is not as definite as that in a mineral, and one type of rock may grade imperceptibly into another. Some may be so soft as to crumble between the fingers, whereas others are extremely hard and resistant. But rocks are much more than mineral aggregates; they are also documents that record events far back in the Earth's history. It is the minerals within the rocks that contain many of the important clues as to events that occurred in eons past because rocks are formed, destroyed, and remade by geologic processes involving the growth and destruction of minerals. The rocks in the visible part of the Earth are divided into three great classes: igneous, sedimentary, and metamorphic. These are shown in Figure 4.16.

Figure 4.16
The three major rock types are shown in this photograph of the inner gorge of the Grand Canyon. Igneous rocks are the pink granite that cooled from the molten liquid that was squeezed into fractures of the dark metamorphic rock and cooled. The metamorphic rock forms the jagged inner gorge and was formed in the deep roots of an ancient mountain belt. The heat and pressure in the mountain root modified and obliterated the characteristics of the original rocks. Sedimentary rocks form the horizontal layers that overlie the igneous and metamorphic rocks.

Igneous Rocks.   Igneous rocks (Latin *ignis,* "fire") are formed by the cooling of molten liquids. A well-known example is basalt, a dark, dense rock formed by the "freezing" of molten lava. However, the most abundant igneous rock is granite, which cools slowly and solidifies far below the surface. Practically all varieties of igneous rock consist of aggregates of interlocking crystals.

Sedimentary Rocks.   Sedimentary rocks form at the surface through the interaction of the atmosphere and water with preexisting rocks. Most are composed of fragments of other rocks and minerals, but some are composed of organic matter, and others are composed of minerals precipitated from waters. The distinctive characteristic of sedimentary rocks is that they occur in layers, or beds, and are thus distinctly stratified (Figure 4.16).

Metamorphic Rocks.   Metamorphic rocks are igneous and sedimentary rocks which have been subjected to considerable heat and pressure and the chemical actions of subsurface fluids so that the component minerals have been recrystallized. The diagnostic features of the original igneous or sedimentary rocks are modified or obliterated. The new rock is an interlocking aggregate of new minerals with new textures and structures (Figure 4.16).

The identification of minerals in rocks presents some special problems because the mineral grains are usually small and rarely have well-developed crystal faces. You will find, however, that the distinguishing properties of the rock-forming minerals are easy to learn if you examine a specimen while you study the written description.

A careful examination of the minerals that make up granite (a common igneous rock) is a good beginning. The polished surface of a granite rock in Figure 4.17 shows that the rock is composed of a myriad of mineral grains of different sizes, shapes, and colors. Although the different mineral grains interlock to form a tight, coherent mass, each one has distinguishing properties.

Feldspar

Granite consists largely of a pink porcelainlike mineral that has a rectangular shape and a milky-white mineral that is somewhat smaller but similarly shaped. These are feldspars (German, "field crystals"), the most abundant mineral group composing about 50 percent of the Earth's crust. The feldspars have good cleavage in two directions, a porcelain luster, and a hardness of about 6 on the Mohs hardness scale.

The crystal structure permits considerable ionic substitution, giving rise to two major types of feld-

spars—potassium feldspar (K-feldspar) and plagioclase feldspar.

Potassium feldspar ($KAlSi_3O_8$) is commonly pink in granitic rocks. Plagioclase feldspar (shown in white in Figure 4.17) permits complete substitution of sodium (Na) for calcium (Ca) in the crystal structure, which gives rise to a compositional range from $NaAlSi_3O_8$ to $CaAl2Si_2O_8$. White plagioclase in granite is rich in sodium.

Feldspars are common in most igneous rocks, in many metamorphic rocks, and in some sandstones.

Micas

The tiny, black shiny grains in Figure 4.17 are mica. This group of minerals is readily recognized by its perfect one-directional cleavage, which permits breakage into thin, elastic flakes. Mica is a complex silicate with a sheet structure, which is responsible for its perfect cleavage. Two common varieties occur in rocks: biotite, which is black mica, and muscovite, which is white or colorless. Mica is abundant in granites and in many metamorphic rocks and is also a significant component of many sandstones.

Quartz

The glassy, irregularly shaped grains in Figure 4.17 are quartz. It is usually the last mineral to form in a granite and is thus forced to grow in the spaces between the earlier formed feldspars and micas. As a result, quartz in granite typically lacks well-developed crystal faces.

Quartz is abundant in all three major rock types. It has the simple composition SiO_2 and is distinguished by its hardness (7), its conchoidal fracture, and its glassy luster. Pure quartz crystals are colorless, but slight impurities produce a variety of colors. When quartz crystals are able to grow freely, their form is elongated, has six sides, and terminates in a point (Figure 4.7), but well-formed crystals are rarely found in rocks. In sandstone, quartz is abraded into rounded sand grains. Quartz is stable both physically (it is very hard and lacks cleavage) and chemically (it does not react with elements at or near the Earth's surface). It is therefore a difficult mineral to alter or break down once it has formed.

Another category of silicate minerals is the ferromagnesian minerals, so named because they contain appreciable amounts of iron and magnesium. These minerals generally range from dark green to black in color and have a high specific gravity. Biotite is classified in this general group, together with the olivines, pyroxenes, and amphiboles. Biotite is common in granite, but the other ferromagnesian minerals are rare or absent. The ferromagnesian minerals are common, however, in basalt (volcanic rock) (Figure 4.18).

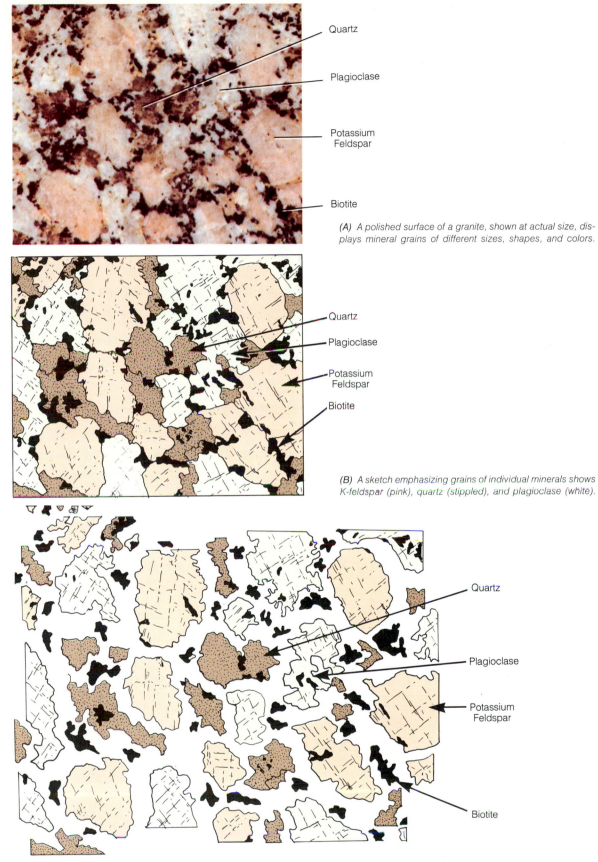

(A) *A polished surface of a granite, shown at actual size, displays mineral grains of different sizes, shapes, and colors.*

Quartz

Plagioclase

Potassium Feldspar

Biotite

Quartz

Plagioclase

Potassium Feldspar

Biotite

(B) *A sketch emphasizing grains of individual minerals shows K-feldspar (pink), quartz (stippled), and plagioclase (white).*

Quartz

Plagioclase

Potassium Feldspar

Biotite

(C) *An exploded diagram of B shows the size and shape of individual mineral grains.*

Figure 4.17
Mineral grains in a granite form a tight, interlocking texture because each mineral is forced to compete for space as it grows.

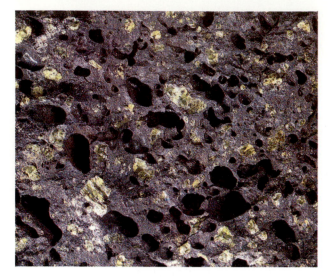

(A) In a hand specimen, only a few large grains can be seen.

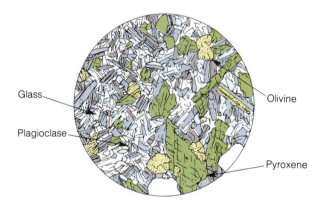

Glass

Plagioclase

Olivine

Pyroxene

(B) Viewed under a microscope, the grains can be seen forming an interlocking texture. Plagioclase crystals typically form lathlike grains.

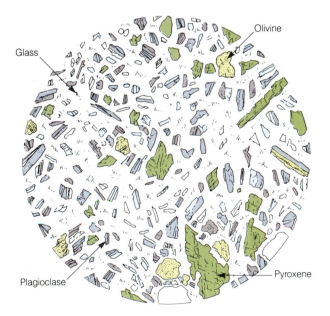

Glass

Olivine

Plagioclase

Pyroxene

(C) An exploded diagram of B shows the size and shape of individual mineral grains.

Figure 4.18
Mineral grains in basalt are microscopic in size.

Olivines

The only mineral clearly visible in the hand specimen in Figure 4.18 is the green, glassy mineral called olivine. The olivine family is a group of silicates in which iron and magnesium substitute freely in the crystal structure. The composition is expressed as $(Mg,Fe)_2SiO_4$. This hard mineral is characterized by an olive-green color (if magnesium is abundant) and a glassy luster. In rocks, it rarely forms crystals larger than a millimeter in diameter. Olivine crystallizes at high temperatures and is common in basalt. It is probably a major constituent of the material beneath the Earth's crust, the mantle.

Pyroxenes

Pyroxenes are high-temperature minerals found in many igneous and metamorphic rocks. In Figure 4.18, pyroxene occurs as microscopic crystals, but some basalt samples contain larger grains of this mineral, which typically range from dark green to black in color. Their internal structure shows single chains of linked tetrahedron (Figure 4.15).

Amphiboles

Figure 4.19 shows a rock that is similar to granite but has less quartz and contains appreciable amounts of the mineral amphibole. Amphiboles have much in common with the pyroxenes. Their chemical compositions are similar, except that amphibole contains hydroxyl ions (OH) and pyroxene does not. Their internal structure consists of double chains of tetrahedra (Figure 4.15). The minerals differ in structure, however. The amphiboles produce elongate crystals that cleave perfectly in two directions, which are not at right angles. The color of amphibole ranges from green to black. This mineral is common in rocks closely related to granite and may be more abundant in that group than biotite. It is especially common in the metamorphic rock known as amphibolite. Hornblende is the most common variety of amphibole.

Clay Minerals

The clay minerals constitute a major part of the soil and are thus encountered more frequently than other minerals in everyday experience. Clay minerals form at the Earth's surface where air and water interact with the various silicate minerals, breaking them down to form clay and other products. Like the micas, the clay minerals are sheet silicates, but their crystals are microscopic and usually can be detected only with an electron microscope (Figure 4.8). More than a dozen clay minerals can be distinguished on the basis of their crystal structures and variations in composition.

Figure 4.19
Amphibole crystals are among the first to crystallize in "granitic" rocks and therefore have well-developed crystal faces.

Calcite

Calcite is composed of calcium carbonate ($CaCO_3$), the principal mineral in limestone. It can be precipitated directly from seawater and is removed from it by organisms as they use it to make their shells. Calcite is dissolved by groundwater and reprecipitated as new crystals in caves and fractures in rock. It is usually transparent or white in color, but the aggregates of calcite crystals that form limestone contain various impurities, which give them gray or brown hues. Calcite is common at the Earth's surface and is easy to identify. It is soft enough to scratch with a knife (hardness of 3), and it effervesces (bubbles) in dilute hydrochloric acid. It has perfect cleavage in three directions, which are not at right angles, so that cleaved fragments look like a distorted cube (Figure 4.12). Besides being the major component of limestone, calcite is the major mineral in the metamorphic rock, marble.

Dolomite

Dolomite is composed of calcium and magnesium and carbonate (CO_2). Large crystals form rhombohedra, but most dolomite occurs as granular masses of small crystals. Dolomite is widespread in sedimentary rocks, forming when calcite reacts with solutions of magnesium carbonate in seawater or groundwater. Dolomite can be distinguished from calcite because it effervesces in dilute hydrochloric acid only if it is in powdered form.

Halite and Gypsum

Halite and gypsum are the two most common minerals formed by the evaporation of seawater or saline lake water. Halite, common salt (NaCl), is easily identified by its taste. It also has one of the simplest of all crystal structures—the sodium and chloride ions are united in a cubical array. Most physical properties of halite are related to this structure. Halite crystals cleave in three directions, at right angles, to form cubic or rectangular fragments (Figure 4.10). Salt, of course, is very soluble and readily dissolves in water.

Gypsum is composed of calcium sulfate and water ($CaSo_42H_2O$). It forms crystals that are generally colorless, with a glassy or silky luster. It is a very soft mineral and can be scratched easily with a fingernail. It cleaves perfectly in one direction to form thin, non-elastic plates (Figure 4.13). Gypsum occurs as single crystals, as aggregates of crystals in compact masses (alabaster), and as a fibrous form (satin spar).

SUMMARY

1. All change requires the interaction of energy and matter.
2. The atomic theory is the basis for our present-day understanding of matter. A simplified description of the fundamental ideas in this theory follows.
   a. An atom is the smallest fraction of an element that can exist and still show the characteristics of that element.
   b. Although many subatomic particles have been identified, the basic building blocks of atoms are (1) protons, (2) neutrons, and (3) electrons.
   c. An atom consists of a nucleus of protons and neutrons and a surrounding cloud of electrons.
   d. The atoms of each element have a unique number of protons in the nucleus. The number of electrons and neutrons may vary from atom to atom of the same element.
   e. Isotopes are varieties of an atom (element) produced by variations in the number of neutrons in the nucleus.
   f. Normal atoms are electrically neutral because they have one negatively charged electron for each positively charged proton.
   g. Electrically charged atoms are called ions. They are produced by the gain or loss of electrons.
   h. Atoms combine to form minerals by ionic bonding, covalent bonding, and metallic bonding.
3. There are three states of matter: solid, liquid, and gas. The three states are distinguished by completely different physical properties, but matter retains the same chemical composition in each state. The dynamic processes in the Earth's dynamics mostly involve matter changing from one state to another.
4. A mineral is a natural inorganic solid with a specific internal structure and a chemical composition that varies only within specific limits.
5. All specimens of a given mineral, regardless of where, when, or how they were formed, share certain physical properties. Some of the more readily observable physical properties are cleavage, crystal form, hardness, specific gravity, color, and streak, all of which aid in mineral identification.
6. Minerals are the building blocks of rocks. They grow or are broken down by chemical reaction as matter changes to and from the solid state. Minerals, therefore, provide an important means of interpreting the changes that have occurred in the Earth throughout its history.
7. More than 95 percent of Earth's crust is composed of the silicate minerals, a group of minerals containing silicon and oxygen linked together in tetrahedral units, with four oxygen atoms to one silicon atom. Several fundamental configurations of tetrahedral groupings—single chains, double chains, two-dimensional sheets, and three-dimensional frameworks—result from the sharing of oxygen ions among silicon ions.
8. Fewer than twenty minerals form the great bulk of the Earth's crust. The most important minerals in igneous rocks are feldspars, micas, amphiboles, pyroxenes, olivines, and quartz. The most important in the sedimentary rocks are quartz, calcite, dolomite, clay minerals, and feldspars. Quartz, feldspars, amphiboles, pyroxenes, micas, and garnet are the predominant minerals in metamorphic rocks.

KEY TERMS

amphibole (p. 62)
atom (p. 50)
calcite (p. 63)
clay mineral (p. 62)
cleavage (p. 56)
conchoidal
   fracture (p. 56)
covalent bond (p. 52)
crystal (p. 53)
crystal face (p. 54)

crystal form (p. 54)
crystal structure (p. 53)
crystallization (p. 57)
dolomite (p. 63)
feldspar (p. 60)
gas (p. 52)
gypsum (p. 63)
halite (p. 63)
hardness (p. 56)
hornblende (p. 62)

ion (p. 50)
ionic bond (p. 52)
ionic substitution (p. 54)
isotope (p. 50)
liquid (p. 52)
metallic bond (p. 52)
mica (p. 60)
mineral (p. 63)
olivine (p. 62)
plagioclase (p. 60)

polymorphism (p. 53)
pyroxene (p. 62)
quartz (p. 60)
silicate (p. 58)
silicon-oxygen (p. 58)
tetrahedron (p. 58)
solid (p. 52)
specific gravity (p. 57)
streak (p. 57)
X-ray diffraction (p. 53)

REVIEW QUESTIONS

1. Give a brief but adequate definition of a mineral.
2. Explain the meaning of "the internal structure of a mineral."
3. Why does a mineral have a definite chemical composition?
4. How do geologists identify minerals too small to be seen in a hand specimen?
5. Briefly explain how minerals grow and are destroyed.
6. Explain the origin of cleavage in minerals.
7. Describe the silicon-oxygen tetrahedron. Why is it important in the study of minerals?
8. Why does mica have excellent cleavage, whereas quartz and olivine lack cleavage?
9. What are the silicate minerals? List the silicate minerals that are important in the study of rocks.
10. Give three examples of how silicon-oxygen tetrahedra are arranged in silicate minerals.
11. Study Figure 4.17 and explain why most of the mineral grains in a granite have an irregular shape.

Multiple-Choice Questions

1. Which of the following is *not* a mineral in the strict definition of the term?
   a. ice
   b. coal
   c. graphite
   d. diamond
   e. salt
2. In the study of minerals, X-ray diffraction measures
   a. chemical composition.
   b. internal structure.
   c. hardness.
   d. specific gravity.
   e. cleavage.
3. The range in chemical composition in a mineral typically results from
   a. impurities.
   b. time and place of origin.
   c. change in atomic structure.
   d. ionic substitution.
   e. changes in physical properties.
4. Cleavage is most conspicuous in
   a. quartz.
   b. mica.
   c. chalk.
   d. chert.
   e. granite.
5. The color of a finely powdered mineral is called its
   a. fracture.
   b. streak.
   c. cleavage.
   d. crystal form.
   e. luster.

6. Most rock-forming minerals are
   a. oxides.
   b. carbonates.
   c. silicates.
   d. sulfides.
   e. none of the above.
7. The major mineral found in limestone is
   a. quartz.
   b. calcite.
   c. halite.
   d. feldspar.
   e. agate.
8. Minerals are defined on the basis of their
   a. physical properties.
   b. color.
   c. atomic structure and chemical composition.
   d. chemical composition, but they are classified on the basis of their origin.
   e. origin.
9. Which of the following is *not* a physical property of a mineral?
   a. specific gravity
   b. color
   c. chemical composition
   d. crystal form
   e. hardness
10. Most rock-forming minerals are
   a. oxides.
   b. carbonates.
   c. silicates.
   d. sulfides.
   e. none of the above.

True/False Questions

1. *T F* A mineral occurs naturally as an inorganic solid.
2. *T F* A mineral has a specific internal structure.
3. *T F* Each crystal of a given mineral contains all of the physical and chemical properties of the mineral.
4. *T F* The clay minerals constitute a major part of the soil.
5. *T F* Halite and gypsum are the two most common minerals formed by the evaporation of seawater or saline lake water.

Fill in the Missing Terms

1. The _____ diffraction instrument is now the basic device for determining the internal structure of minerals.
2. _____ is usually the last mineral to form in a granite.
3. The _____ minerals constitute a major part of igneous rocks.
4. _____ is the principal mineral in limestone.
5. The mineral _____ is soft enough to scratch with a knife and effervesces in dilute hydrochloric acid.

5

IGNEOUS ROCKS

Igneous rocks are records of the thermal history of the Earth. Their origin is closely associated with the movement of tectonic plates, and they play an important role in seafloor spreading, the origin of mountains, and the evolution of continents. The best-known examples of igneous activity are volcanic eruptions, where liquid rock material works its way to the surface and erupts from volcanic vents and fissures. Less spectacular, though just as important, are the tremendous volumes of liquid rock that never reach the surface, but remain trapped in the crust where they cool and solidify. Granite is the most common variety of this type of igneous rock and is commonly exposed in eroded mountain belts and in the roots of ancient mountain systems now preserved in the shields.

In this chapter, we study the major types of igneous rocks and what they reveal about the thermal activity of the Earth. We pay particular attention to the distinctive texture of igneous rocks and how we can read, from the interlocking network of grains, a fascinating history of how the hot liquid became part of the solid crust.

MAJOR CONCEPTS

1. Magma is molten rock that is capable of penetrating into, or through, the Earth's crust.
2. Magma originates from the partial melting of the lower crust and the upper mantle, usually at depths between 50 and 200 km below the surface.
3. Two major types of magma are recognized: (a) basaltic magma and (b) granitic magma. Intermediate types also occur.
4. The texture of a rock provides important insight into the cooling history of the magma.
5. The major textures of igneous rocks are (a) glassy, (b) aphanitic, (c) phaneritic, (d) porphyritic, and (e) pyroclastic.
6. High-silica magmas produce rocks of the granite-rhyolite family, which are composed of quartz and K-feldspar with minor amounts of Na-plagioclase, biotite, and amphibole.
7. Low-silica magmas produce rocks of the basalt-gabbro family, which are composed of Ca-plagioclase and pyroxene, with minor amounts of olivine or amphibole.
8. Magmas with composition intermediate between high and low silica produce rocks of the diorite-andesite family.
9. Basaltic magma is generated by the partial melting of the mantle along rift zones. Granitic magma is produced at converging plate margins by the partial melting of the oceanic and lower continental crust.

NATURE OF MAGMA

What are the chemical and physical characteristics of molten rock?

Statement

Observations of volcanic activity, together with observations of igneous rocks formed beneath the surface but now exposed by erosion, indicate that the following are the essential characteristics of magma:

1. Magma is molten silicate material, including early formed crystals and dissolved gases. It is a complex mixture of liquid, solid, and gas.
2. The principal elements of magma are oxygen (O), silicon (Si), aluminum (Al), calcium (Ca), sodium (Na), potassium (K), iron (Fe), and magnesium (Mg).
3. Silica (SiO_2) is the principal constituent, comprising from 37 to 75 percent of magma.
4. Dissolved gases can constitute a small percentage of magma. They are important in determining certain properties, such as viscosity (resistance to flow) and the explosive characteristics of volcanic eruptions. More than 90 percent of the gas emitted from volcanoes is water (H_2O) and carbon dioxide (CO_2).

Discussion

The term *magma* comes from the Greek work that means "kneaded mixture," like a dough or paste. In its geologic application, it refers to any hot, partly molten, mobile material within the Earth that is capable of penetrating into or through the rocks of the crust. This definition may be quite different from the concept most people have, which is derived largely from illustrations of spectacular volcanic eruptions. Popular articles about volcanism usually emphasize the sensational, and readers are impressed with the idea that all magma is a very fluid, red-hot liquid closely resembling molten steel (Figure 5.1).

Most magmas, in reality, are not entirely liquid but are a combination of liquid, solid, and gas. Early formed crystals (minerals that form at high temperature) may make up a large portion of the mass, so a magma could be thought of more accurately as a slush, a liquid melt mixed with a mass of mineral crystals. Such a mixture has a consistency similar to that of freshly mixed concrete, slushy snow, or thick oatmeal. Movement of these magmas is slow and sluggish.

Water vapor and carbon dioxide are the principal gases in a magma, and they can constitute as much as 14 percent of its volume. These volatiles (materials that are readily vaporized) are important because

Figure 5.1
Magma is molten rock material. Much of it cools deep in the Earth's crust, but some of it works its way to the surface and is extruded as lava. Lava eruptions such as this permit us to study the nature of magma—its composition and physical characteristics—and give us an insight into the origin of igneous rocks.

they strongly influence the mobility and melting point of a magma and the types of volcanic activity that can be produced. Volatiles tend to increase the fluidity of a magma. Magmas rich in volatiles also tend to erupt more violently.

Chemical analyses of many igneous rocks distinguish two principal kinds of magma: (1) basaltic magmas, which contain about 50 percent SiO_2 and have temperatures ranging from 900 to 1200°C, and (2) granitic (silicic) magmas, which contain at least 60 or 70 percent SiO_2 and generally have temperatures lower than 800°C. Basaltic magmas are characteristically fluid, whereas granitic magmas are thick and viscous. This is because granitic magmas have lower temperatures and greater amounts of SiO_2. Viscosity is the tendency of a body to resist flow. If a fluid has a high viscosity, it is thick and pasty. If it has low viscosity, it is highly fluid and flows readily. The viscosity in a magma is influenced by SiO_2 content because silica tetrahedra link together even before crystallization occurs, and the linkages offer resistance to flow. The higher the silica content, therefore, the greater the magma's viscosity.

Like most fluids, magma is less dense than the solid material it forms, and it tends to migrate upward through the mantle and crust. Magma can intrude into the overlying rock by forceful injection into fractures, or it can melt and assimilate the rock it invades. As it rises, it cools and eventually solidifies to form an igneous rock. When magma reaches the surface and flows out over the landscape as lava, it forms extrusive rock bodies. Magma that solidifies below the surface forms intrusive rock bodies.

ROCK TEXTURES

What does the texture of a rock indicate about the cooling history of a magma?

Statement

Igneous rocks have distinctive textures, characterized mostly by the interlocking grains that grow from cooling magma.

The major textures of igneous rocks are

1. Phaneritic: most of the grains large enough to be seen without a microscope.
2. Aphanitic: most of the grains too small to be seen without a microscope.
3. Porphyritic: two or more distinctly different crystal sizes.
4. Glassy: no grains.
5. Pyroclastic: volcanic fragments.

Discussion

Texture refers to the size, shape, and boundary relationships of adjacent minerals in a rock mass. In igneous rocks, texture develops primarily in response to the composition and the rate of cooling of the magma. Magmas located deep in the Earth's crust cool slowly. Individual grains grow to a more or less uniform size and may be more than an inch in diameter. In contrast, a lava extruded at the Earth's surface cools rapidly, so the mineral grains have only a short time in which to grow. The grains from such a magma are typically so small that they cannot be seen without the aid of a microscope, and the resultant rock appears massive and structureless. Regardless of the size of the grains, the texture of most igneous rocks is distinguished by a network of interlocking grains.

Igneous rock textures are divided into the following types: (1) phaneritic, (2) aphanitic, (3) glassy, (4) porphyritic-phaneritic, (5) porphyritic-aphanitic, and (6) pyroclastic (Figure 5.2).

Phaneritic Texture

In phaneritic textures, the individual grains are large enough to be plainly visible to the naked eye (Figure 5.2a). The grains in each specimen are approximately equal in size and form an interlocking mosaic. The size of grains in a phaneritic texture can range from those barely visible to crystals of more than an inch in length. Phaneritic texture develops from magmas that cool slowly, commonly in intrusive igneous bodies. Very coarse phaneritic rocks, in which the crystals are several feet long, almost invariably are found in large fissures or veins.

Aphanitic Texture

In aphanitic texture, individual crystals are so small that they cannot be detected without the aid of a microscope (Figure 5.2b). Rocks with this texture therefore appear to be massive and structureless. When a thin *section* (a thin, transparent slice of rock) of an aphanitic rock is viewed under a microscope, however, the crystalline structure is readily apparent. The rock is seen to be composed of numerous small crystals and, frequently, some glass.

Glassy Texture

A glassy texture in an igneous rock is similar to that of ordinary glass. It may occur in massive units (Figure 5.2c) or in a threadlike mesh similar to spun glass. Crystals cannot be discerned in a glassy texture, even when the specimen is viewed under high magnification.

Porphyritic-Phaneritic Texture

A porphyritic-phaneritic texture is characterized by two distinct crystal sizes, both of which can be seen with the naked eye. The smaller crystals constitute a matrix, or groundmass, that surrounds the larger crystals, called phenocrysts (Figure 5.2d).

Porphyritic-Aphanitic Texture

A porphyritic-aphanitic texture is defined as a rock with an aphanitic matrix in which embedded phenocrysts make up more than 10 percent of the total rock volume (Figure 5.2e). The phenocrysts are visible to the unaided eye. When phenocrysts are abundant, the rock may at first glance appear to be phaneritic. Careful study of the area between the phenocrysts will indicate whether the matrix is aphanitic or phaneritic.

Fragmental Texture (Pyroclastic)

Fragmental textures consist of broken, angular fragments of rock material (Figure 5.2f). In pyroclastic rocks, the fragmental material is composed of pumice, glass, and broken crystals. Some sorting and stratification are generally present. Rocks composed of material finer than 4 mm are called tuff. Fragmental rocks made up of material larger than 4 mm are referred to as volcanic breccia. If the fragments are exceptionally hot when deposited, they may fuse or weld together to form a dense mass.

(A) *A phaneritic texture* consists of grains large enough to be seen with the unaided eye. All grains are roughly the same size and interlock to form a tight mass. The large crystals suggest a relatively slow rate of cooling.

(D) *A porphyritic-phaneritic* texture results from two stages of cooling, both of which are relatively slow, so all of the crystals are relatively large and can be seen without a microscope. In this specimen, the large pink crystals of K-feldspar developed during the first stage of cooling. The second stage of cooling was more rapid, so the remaining crystals are smaller.

(B) *An aphanitic texture* consists of mineral grains too small to be seen without a microscope. Only a few grains are large enough to be seen; most are microscopic. Aphanitic texture results from rapid cooling.

(E) *A porphyritic-aphanitic* texture results from two separate rates of cooling. In the specimen, the larger dark phenocrysts formed first, when the magma cooled relatively slowly. The microscopic crystals (aphanitic matrix) formed later, during a period of more rapid cooling. The rock is a porphyritic rhyolite.

(C) *A glassy texture* develops when molten rock material cools so rapidly that the migration of ions to form crystal grains is inhibited. Glassy texture typically forms on the crust of lava flows and in viscous magma. The sample shown here is obsidian.

(F) *A pyroclastic texture* forms when crystals, fragments of rock, and glass are blown out of a volcano as ash. Generally the particles fall to the Earth and accumulate like sediment. Ash flows are very hot and move as a body close to the ground and the individual fragments become fused in a dense mass. In this specimen, the black glass material was originally frothy pumice but was so hot that the fragments collapsed into dense glass.

Figure 5.2
Textures of igneous rocks.

KINDS OF IGNEOUS ROCKS

What are the major types of igneous rocks?
What is the basis for this classification?

Statement

Igneous rocks are classified on the basis of texture and composition. Texture provides important information on the cooling history of the magma, and composition provides insight into the nature and origin of the magma. Rocks that cool below the surface are called intrusive, and those that cool at the surface are called extrusive.

Discussion

A simple chart of the major types of igneous rocks is shown in Figure 5.3. Variations in composition are arranged horizontally, and variations in texture are arranged vertically. The rock names are printed in bold type, the size of which is roughly proportionate to the relative abundance of the rock. Rocks in the same vertical column have the same composition but different textures. Rocks in the same horizontal row have the same texture but different compositions. The chart shows that granite, for example, has a phaneritic texture and is composed predominantly of quartz and K-feldspar. The size of the type its name is printed in indicates that it is the most abundant intrusive igneous rock. Rhyolite has the same composition as granite but is aphanitic. Basalt has an aphanitic texture and is composed predominantly of Ca-rich plagioclase and pyroxene. It has the same composition as gabbro but is much more abundant.

Rocks with Aphanitic Textures

Basalt. Basalt is the most common aphanitic rock. It forms most of the oceanic crust and the volcanic islands, and is common in continental rift zones. Basalt is a very fine-grained, usually dark-colored rock that originates from the cooling lava flows. The mineral grains are commonly so small that they can rarely be seen without a microscope. If a thin section is viewed through a microscope, the individual minerals can then be seen and studied.

Basalt is composed predominantly of calcium-rich plagioclase and pyroxene, with smaller amounts of olivine or amphibole. The plagioclase occurs as a mesh of elongate, lathlike crystals surrounding the more equidimensional pyroxene and olivine grains. In some cases, large crystals of olivine or pyroxene form phenocrysts, resulting in a porphyritic texture. Many basalts have some glass, especially near the tops of flows.

Andesite. Andesite is an aphanitic rock composed of Na-rich plagioclase, pyroxene, and amphibole. It usually contains little or no quartz and has the same composition as diorite. The texture of andesite is generally porphyritic, with phenocrysts of feldspar and ferromagnesian minerals. The rock is named after the Andes Mountains, where volcanic eruptions have produced it in great abundance. Andesite is the next most abundant lava type after basalt and occurs most frequently along continental margins. It is rare in ocean basins and is not abundant in continental interiors. The origin of andesite along continental margins is probably related to partial melting of the oceanic crust and the lower continental crust at convergent plate boundaries.

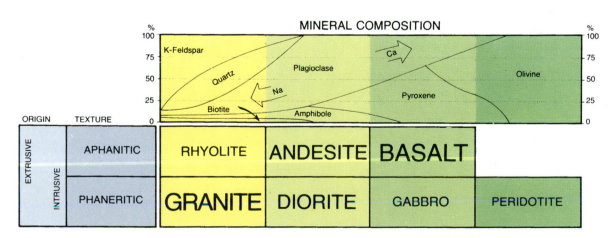

Figure 5.3
The classification of igneous rocks is based on texture (shown vertically on the chart) and composition (shown horizontally). The size of type in which the names of the rocks are printed is roughly proportional to their abundance.

Rhyolite. Rhyolite is an aphanitic rock with the same composition as that of granite. It usually contains a few phenocrysts of feldspar, quartz, or mica, but not enough to be considered porphyritic. Rhyolite flows are viscous. Instead of spreading in a linear flow, rhyolite typically piles up in large, bulbous domes. Many rhyolites have a significant glass content.

Rocks with Fragmental (Pyroclastic) Textures

Tuff. Volcanic eruptions of rhyolitic and andesitic lavas commonly produce large volumes of fragmental material. The fragments range from dust-size pieces to large blocks more than a meter in diameter. The rock resulting from the accumulation of ash falls is referred to as tuff. Although of volcanic origin, tuff has many of the characteristics of sedimentary rock because the fragments composing tuff settle out from suspension in the air and are stratified like sedimentary rocks.

Ash-Flow Tuff. Ash-flow tuff is rock composed of fragments of volcanic glass, broken fragments of crystal, rock fragments, and pieces of solidified lava—all fused in a tight, coherent mass. Typically, many of the fragments are flattened or bent out of shape. This unique texture indicates that at the time they were extruded, the ash fragments were hot enough to fuse in a rock mass.

Rocks with Phaneritic Textures

Granite. Granite is a coarse-grained igneous rock composed predominantly of feldspar and quartz (Figure 5.3). K-feldspar is the most abundant mineral, and usually it is easily recognized by its pink color. Na-Ca-plagioclase is present in moderate amounts, usually distinguished by its white color and its porcelainlike appearance. Mica is conspicuous as black or bronze-colored flakes, usually distributed evenly throughout the rock. The texture of granite, together with laboratory experiments, indicates that Na-plagioclase, biotite, and amphibole are the first minerals to crystallize from a granitic magma. A seemingly insignificant, but very important, property of granite is its relatively low specific gravity, about 2.7. In contrast to basalt and related rocks, which have a specific gravity of 3.2, granite is light. This fact is important in considering the nature of continents and the contrast between continental crust and oceanic crust. Granite and related rocks make up the great bulk of continental crust, whereas most of the oceanic crust is composed of basalt.

Diorite. Diorite is similar to granite in texture, but it differs in composition. Na-plagioclase feldspar is the dominant mineral, and quartz and K-feldspar are minor constituents. Amphibole is an important constituent, and some pyroxene may be present. In composition, diorite is intermediate between granite and basalt.

Gabbro. Gabbro is not commonly exposed at the Earth's surface, but it is a major constituent of the lower part of the oceanic crust and is present in some ancient parts of the continents. It has a coarse-grained texture similar to that of granite, but it is composed almost entirely of pyroxene and calcium-rich plagioclase, with minor amounts of olivine. Gabbro is dark green, dark gray, or almost black because of the predominance of dark-colored minerals.

Peridotite. Peridotite is composed almost entirely of two minerals, olivine and pyroxene (Figure 5.4). It is not common at the Earth's surface or within the continental crust, but it is a major constituent of the mantle.

Figure 5.4
Peridotite is a phaneritic rock composed mostly of olivine, but minor amounts of pyroxene and Ca plagioclase may be present. This specimen is composed almost entirely of olivine. It is believed that the mantle is composed of peridotite and closely related rocks.

EXTRUSIVE ROCK BODIES

*What are the major types of volcanic eruptions?
How does each originate?*

Statement

The two major types of magma, basaltic and granitic, produce contrasting types of eruptions and form different types of extrusive rocks.

1. Basaltic magmas are low in silica and are relatively fluid. Dissolved gases escape readily, so the lava is typically extruded quietly from fissures and fractures. Fissure eruptions produce a succession of thin lava flows that cover large areas.
2. Silicic magmas are thick and viscous. The escape of gas is thus retarded, and pressure builds up within the magma. These eruptions are typically violent, and the lava is extruded as thick flows, bulbous domes, or ash flows.

Discussion

Basaltic Eruptions

Basaltic eruptions are probably the most common type of volcanic activity on Earth. The lava is gener-ally extruded from fractures or fissures in the crust. After the lava is extruded, it tends to flow freely downslope and spreads out to fill valleys and topographic depressions (Figure 5.5). This type of eruption occurs along fractures in the oceanic ridge to form new oceanic crust where the tectonic plates move apart; it is the major type of eruption in the volcanic plains and plateaus of the continents.

There are two common types of basaltic flows, referred to by the Hawaiian terms *aa* (pronounced ah'ah') and *pahoehoe* (pronounced pa ho'e ho'e) (Figures 5.6a and 5.6b). An aa flow contains relatively little gas and is slow moving, a flow 3 to 10 m thick. The surface of the flow cools and forms a crust, while the interior remains molten. The flow may move only a few meters per hour. As it continues to move, the hardened crust is broken into a jumbled mass of angular blocks and clinkers (Figure 5.6a). Gas within the fluid interior of the flow migrates toward the top, but it may remain trapped beneath the crust. These "fossil gas bubbles," called vesicles, make the rock light and porous.

Compared with aa flows, pahoehoe flows contain much more gas and are thus more fluid. They are not as thick (usually less than 1 m thick) and move much faster because of their lower viscosity. As a pahoehoe flow moves, it develops a thin, glassy crust, which is

Figure 5.5
Fissure eruptions are the most common type of volcanic eruption on the Earth. Lava is simply extruded through cracks or fissures in the crust. This type of eruption is typical of fluid basaltic magma and is the dominant eruptive style along the oceanic ridge. This photograph shows a recent fissure eruption on the island of Hawaii.

(A) **The surface of an aa flow** *consists of a jumbled mass of angular blocks, which form when the congealed crust is broken by the slow-moving flow. Aa flows are viscous and much thicker than pahoehoe flows. This photograph shows a recent aa flow in Craters of the Moon National Monument, Idaho.*

(B) **The surfaces of pahoehoe flows** *are commonly twisted, ropy structures. Pahoehoe flows form on fluid lava and typically are very thin.*

(C) **Hexagonal columnar joints** *commonly form by contraction when a lava cools. The long axis of the column is approximately perpendicular to the cooling surface.*

Figure 5.6
A variety of surfaces develop on a lava flow and reflect the manner of flow, rates of cooling, amounts of dissolved gases, and viscosity.

molded into billowy forms or surfaces that can resemble coils of rope. A variety of flow features (such as those shown in Figure 5.6b) can develop on the surface of the flow.

Under certain circumstances, a flow will contract as it cools and may develop a system of polygonal cracks called columnar joints (Figure 5.6c).

Droplets and blobs of lava blown out from a volcanic vent usually cool by the time they fall back to the ground. This material forms volcanic ash and dust, collectively known as tephra (Figure 5.7a-b), and larger fragments called volcanic bombs (Figure

5.7c). As the tephra travels through the air, it is sorted according to size. The larger particles accumulate close to the vent, or volcanic neck, to form a cinder cone, and the finer, dust-size particles are transported great distances by the wind. Cinder cones, which are generally less than 200 m high and 2 km in diameter, are relatively small features, compared with the large shield volcanoes and stratovolcanoes (composite volcanoes).

If the extrusion of large quantities of fluid basaltic lava continues, a broad cone called a shield volcano may form around the vent (Figure 5.8). With each

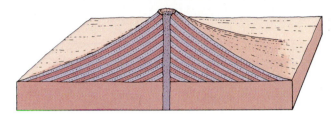

(A) The form and internal structure of a cinder cone are relatively simple. Ash, dust, and coarser pyroclastic material are blown out from a central vent and accumulate as a smooth, conical hill. The internal structure consists of layers of tephra inclined away from the summit crater. The vent, or volcanic neck, is commonly filled with solidified lava and volcanic fragmental debris.

(B) Tephra is a general term referring to all pyroclastic material ejected from a volcano. It includes ash, dust, bombs, and rock fragments. It is commonly stratified.

Figure 5.7
A cinder cone is a distinctive type of volcano built up almost exclusively of pyroclastic material.

(C) Volcanic bombs are fragments of lava ejected in a liquid or plastic state. As they move through the air they twist and turn and form spindle-shaped masses.

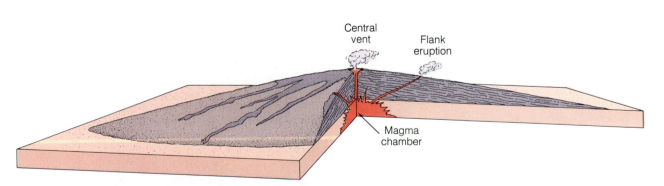

Figure 5.8
Shield volcanoes are the largest volcanoes on Earth. They are composed of innumerable thin, basaltic flows with relatively little ash. Some of the best examples are the islands and seamounts of the Pacific. The island of Hawaii, rising approximately 10,000 m above the seafloor, is the largest volcano on the Earth.

eruption, the fluid basaltic lava flows freely for some distance, spreading into a thin sheet, or tongue, before congealing. Shield volcanoes, therefore, have a wide base (commonly over 100 km in diameter) and gentle slopes (generally less than 10 degrees). Their internal structure consists of innumerable thin basalt flows with comparatively little ash. The Hawaiian Islands are excellent examples of shield volcanoes. They are enormous mounds of basaltic lava, rising as high as 10,000 m above the seafloor. The younger volcanoes typically have summit craters as much as 3 km wide and several hundred meters deep, which resulted from subsidence following the eruption of magma from below.

Silicic Eruptions

Silica-rich magmas produce granitic and associated igneous rocks. They are relatively cool, and the mechanics of eruption and flow of their lavas is quite different from that of basaltic lavas.

Some silicic magmas are so thick and viscous that small volumes hardly flow at all, but instead form massive plugs or bulbous domes over the volcanic vent (Figure 5.9). The high viscosity of silicic magmas inhibits the escape of dissolved gas, so tremendous pressure can build up. Consequently, when eruptions occur, they may be highly explosive and violent and commonly produce large quantities of tephra. The alternating layers of tephra and thick, viscous lava typically produce a composite volcano, or stratovolcano, characterized by a high, steep-sided cone around the vent (Figure 5.10). This is probably the most familiar form of continental volcano, with such famous examples as mounts Shasta, Fuji, Vesuvius, Etna, and Stromboli. A depression at the summit, the crater, usually marks the position of the vent.

The explosive eruptions of silicic volcanoes can blow out large volumes of ash and magma. As a result, the summit area sometimes collapses, forming a large basin-shaped depression known as a caldera. Crater Lake, Oregon (Figure 5.11), for example, formed when a volcano's summit collapsed after its last major eruption and the resulting caldera filled with water. Wizard Island, a small cinder cone in the caldera, resulted from subsequent minor eruptions. Krakatoa is another well-known example. The 1883 eruption was one of the largest explosions in recent history. The cone was demolished, and great quantities of volcanic ash were blown high into the atmosphere. The explosion and subsequent subsidence produced a caldera 6 m in diameter, which completely altered the configuration of the island.

A spectacular type of eruption associated with silicic magmas is the lateral-flowing movement of large masses of ash and lava particles. This phenomenon is not a liquid lava flow or an ash fall (in which particles settle independently), but a flow consisting of gas and suspended fragments of hot mineral grains, droplets of lava, and pieces of rock suspended like a dense dust cloud that moves rapidly, close to the surface. This type of eruption is therefore called an ash flow (Figure 5.12). As a magma works its way to the surface, confining pressure is released and trapped gas bubbles rapidly expand. Near the surface, the magma can literally explode, ejecting pieces of lava, bits of solid rock, ash, early formed crystals, and gas. This material is very hot, sometimes incandescent. Being denser than the air, it flows across the ground surface as a thick, dense cloud of hot ash. Ash flows can reach velocities greater than 250 km per hr because the expanding gas continually forces the cooling lava particles apart. Some have moved as much as 400 km from the vent. When an ash flow comes to rest, the particles of hot crystal fragments, glass, and ash fuse to form welded tuff (Figure 5.13). Ash flows can be very large. Some form flow units more than 100 m thick and cover thousands of square miles. As it cools, the contracting mass can develop columnar jointing.

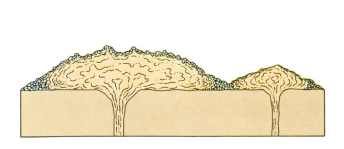

Figure 5.9
Domes of silicic lava form because silica-rich lava is viscous and resists flow. It therefore tends to pile up over the vent to form large, bulbous domes.

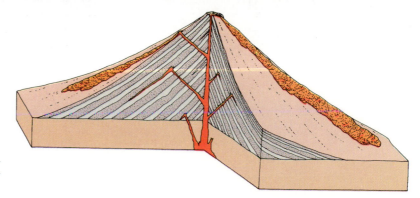

Figure 5.10
Composite volcanoes are built up of alternating layers of ash and lava flows, which characteristically form high, steep-sided cones.

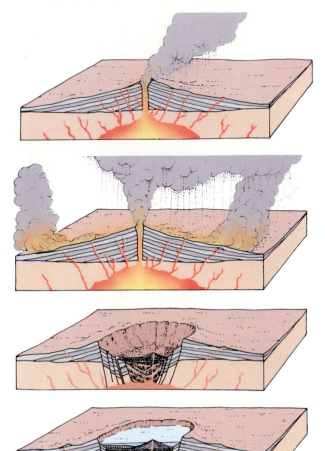

(A) Early eruptions formed the prehistoric volcano called Mount Mazma.

(B) Great eruptions of ash flows emptied the magma, causing the top of the volcano to collapse.

(C) The collapse of the summit into the magma chamber formed the caldera.

(D) A lake formed in the caldera, and subsequent minor eruptions produced small volcanic islands in the lake.

Figure 5.11
The evolution of the caldera at Crater Lake, Oregon, involved a series of great eruptions followed by the collapse of the summit into the magma chamber.

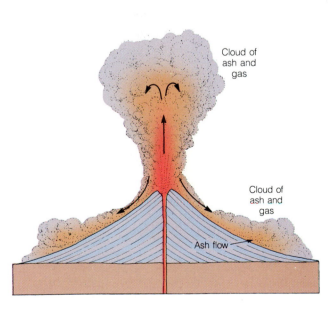

Cloud of ash and gas

Cloud of ash and gas

Ash flow

Figure 5.12
An ash flow is a hot mixture of highly mobile gas and ash that moves rapidly over the surface of the ground away from the vent.

Figure 5.13
Devastation caused by the 1951 ash flow of Mount Lamington, New Guinea, illustrates the force of hot flowing ash. Entire buildings were ripped from their foundations and automobiles were picked up and thrown to the top of trees.

INTRUSIVE ROCK BODIES

What are the characteristics of the major types of intrusions?
How do intrusions originate?

Statement

Intrusions usually are classified according to their sizes, shapes, and relationships to the older rocks that surround them. The major intrusive bodies are batholiths, stocks, dikes, sills, and laccoliths (Figure 5.14).

Discussion

Batholiths

Batholiths, which are masses of coarsely crystalline rock, generally of granitic composition, are the largest rock bodies in the Earth's crust (Figures 5.14 and 5.15). They form almost exclusively in continents and large island arcs and do not occur in oceanic islands. Many cover several thousand square kilometers. The Idaho batholith, for example, is a huge body of granite, exposed over an area of nearly 41,000 km^2 (Figure 5.15). The British Columbia batholith, to the north, is over 2,000 km long and 300 km wide and at least 30 km thick. The true three-dimensional form of batholiths is difficult to determine because of uncertainty about their shape and how deep they extend below the surface. Evidence from gravity measurements and from seismic studies showing the layered nature of the crust and the mantle suggests that batholiths are limited in depth to the thickness of the crust and do not extend down into the mantle. They must therefore be less than 60 km thick. The nature of the base of a batholith remains a matter of conjecture. Indirect evidence suggests that the size of batholiths may increase with depth for some distance and then taper off at still greater depth, somewhat like the root of a tooth.

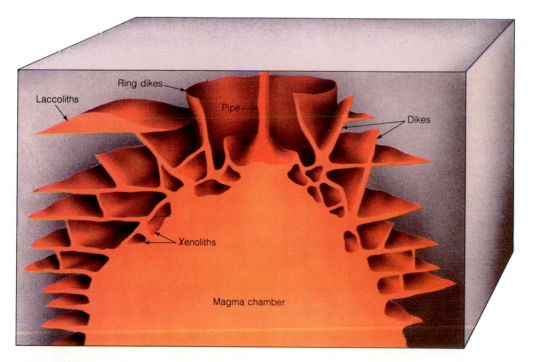

Figure 5.14
Magmatic intrusions may assume a variety of forms. Batholiths are large masses of coarsely crystalline rock that cools in the major magma chamber. Stocks are smaller masses and may be protrusions from a batholith. Dikes are narrow, tabular bodies formed as magma is squeezed into fractures and cools. Many dikes are related to conduits leading to volcanoes. Some radiate out from the volcanic neck; others form a circular pattern above a stock and are called ring dikes. Sills are layers of igneous rock squeezed in between sedimentary strata. Laccoliths, dome-shaped bodies with a flat floor, are formed where magma is able to arch up the overlying strata. Inclusions of the surrounding rock in the magma are called xenoliths. A pipe is a tabular passageway through which magma migrates upward.

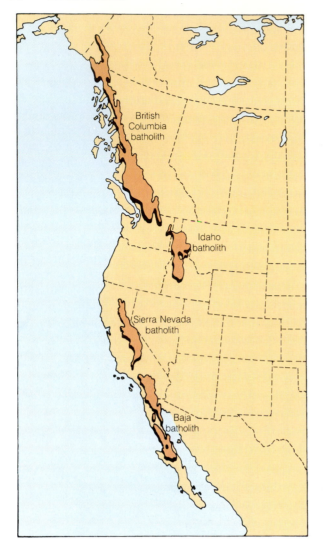

Figure 5.15
The shape of batholiths can be determined by mapping the surface exposure and by studies of seismic wave velocities. Some well-known batholiths in North America, shown on the map, are huge tabular bodies with an irregular elliptical shape.

Batholiths typically form in the deeper zones of mountain belts and are exposed only after considerable uplift and erosion. Some of the highest peaks of mountain ranges, such as the Sierra Nevada and the Coast Ranges of western Canada, are carved in the granite of batholiths. These rocks originally cooled thousands of feet below the surface. The trend of a batholith usually parallels the axis of the mountain range, although the intrusion can cut locally across folds within the range.

Extensive exposures of batholiths also are found in the shield areas of the continents (Figure 5.16). These batholith exposures are considered to be the roots of ancient mountain ranges, which have long since been eroded to lowlands.

Stocks

A stock is an intrusive body with an outcrop area (the area exposed at the surface) of less than 100 km^2. Some stocks are known to be small protrusions rising from the main batholith, but the downward extent of others is unknown. Stocks are generally composed of granitic rock, and many are porphyritic with a fine-grained groundmass. Many deposits of silver, gold, lead, zinc, and copper are deposited in fractures and form veins extending from a stock into the surrounding rock.

Dikes

One of the most familiar signs of ancient igneous activity is a narrow, tabular body of igneous rock called a dike (Figure 5.17). A dike forms where magma squeezes into fractures that cut across layers of the surrounding rock and then cools. The width of a dike can range from a fraction of a centimeter to hundreds of meters. The largest known example is the Great Dike of Zimbabwe, Africa, which is 600 km long and has an average width of 10 km.

Sills

Rising magma follows the path of least resistance. If this path includes a bedding plane, which separates layers of sedimentary rock, magma will be injected between those layers to form a sill—a tabular intrusive body parallel to the layering (Figures 5.14 and 5.17). Sills range from a few centimeters to hundreds of meters thick and can extend laterally for several kilometers.

Sills can form as local offshoots from dikes, or they can be connected directly to a stock or a batholith. Characteristically, they are formed from highly fluid basaltic magma, which can be injected between the older rocks without deforming them. The more viscous granitic magma rarely forms sills.

Laccoliths

When viscous magma is injected between layers of sedimentary rock, it tends to arch up the overlying strata. The resulting intrusive body, a laccolith, is lens shaped, with a flat floor and an arched roof (Figures 5.14 and 5.18). Laccoliths usually occur in blisterlike groups in areas of flat-lying strata. They can be several kilometers in diameter and thousands of meters thick. Typically they are porphyritic.

Laccoliths were first discovered in the Colorado Plateau, where they occur along the margins of a central stock and appear as an inflated sill. Laccoliths can also be fed through a passageway from below.

Figure 5.16
Batholiths in the shield of western Australia are shown on Landsat images as large, elliptical, yellowish-white masses. These granitic rocks have intruded and arched up the older, greenish metamorphic rock, so that the older layered structures are nearly vertical.

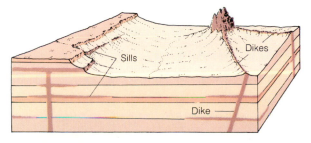

Figure 5.17
Dikes and sills are tabular intrusive bodies. Dikes cut across layers of the surrounding rock, and sills are injected between strata.

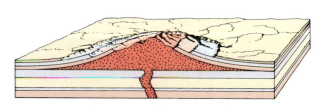

Figure 5.18
Laccoliths are masses of igneous rock injected between layers of the surrounding rock. They tend to deform the overlying strata in a domelike structure.

THE ORIGIN OF MAGMA

How is magma produced in the Earth's tectonic system?

Statement

From the global distribution of volcanoes, batholiths, and mountain belts, we know that igneous activity is related to processes operating at active plate margins. The two fundamental types of magma apparently form in separate tectonic settings.

1. Basaltic magma is generated by the partial melting of the mantle that wells up along the rift zones, where plates move apart. Basaltic volcanism dominates the igneous activity of the ocean basins.
2. Granitic magma is generated in the subduction zone by the partial melting of the oceanic crust and by the partial melting of the lower continental crust in the deeper roots of an active mountain belt (in the zone where the plates collide and temperature and pressure are high).

Discussion

Magma that produces igneous rocks in the Earth's crust is not generated in the Earth's lower mantle or core. It originates in the lower crust and upper mantle. Figure 5.19 is a diagram summarizing our present understanding of the origin of magma. Seismic evidence shows that the crust and the mantle are solid and that no permanent, worldwide reservoirs of magma exist from the surface to a depth of nearly 2,900 km. The Earth's core responds to seismic waves as if the core were a liquid, but gravity measurements indicate that the core is made of material far denser than any magma, so the core material could not rise to the surface. If it did reach the surface, core material would produce rocks several times denser than any known igneous rock. Contrary to popular belief, therefore, it is very unlikely that magma comes from the liquid core or even from the middle and lower parts of the mantle. Magma must originate between 50 and 200 km below the surface from the partial melting of the upper mantle and lower crust.

Understanding the concept of the partial melting of rocks in the upper mantle and lower crust is critical to the understanding of the origin of magma. You must bear in mind that most rocks are composed of more than one mineral, and thus rocks do not have a specific melting temperature. Each component mineral in the rock melts at a different temperature. Hence, a magma solidifies over a range of temperatures. The following are the major factors that influence melting:

1. Temperature
2. Pressure
3. Amount of water and other volatiles (material that readily vaporizes) present
4. Composition of the rock that is melting

A rock will begin to melt because of either an increase

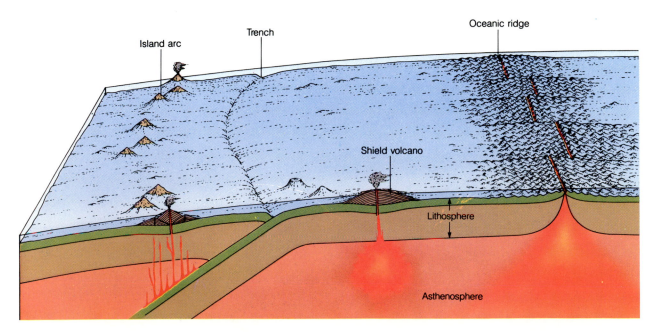

Figure 5.19
The origin of magma is involved with the dynamics at plate margins. Basaltic magma originates by partial melting of the upper mantle at diverging plate margins. As mantle material moves upward in a convection cell, the peridotite (major rock in the mantle) begins to melt because of a decrease in pressure. The material that melts first produces a magma of basaltic composition. Granitic magma is generated at

in temperature or a decrease in pressure. If the temperature or the pressure (or both) changes in a direction to cause melting, the component minerals melt in a definite sequence. At a given temperature and pressure, a rock body may thus be only partly melted, and the liquid (or melt) may have a composition quite different from that of the original rock. This process is known as partial melting. Conversely, as a magma cools, the various minerals crystallize in a definite sequence. When partial crystallization occurs, the remaining liquid can be separated from the crystals, so that it forms a magma quite different from the parent material.

Generation of Basaltic Magma

The asthenosphere is believed to be composed of peridotite, a rock consisting of the minerals olivine and pyroxene. The balance between temperature and pressure in the asthenosphere is just about right for a slight degree of melting to occur. At greater depths in the mantle, however, pressure is too great for partial melting. The balance of temperature and pressure to produce a small amount of melting explains why the asthenosphere is soft and weak. As the soft asthenosphere slowly moves upward along the rift zone between plates, the decrease in pressure causes additional partial melting. The partial melting of peridotite produces a basaltic magma because the first minerals to melt yield basaltic constituents. Laboratory experiments on the melting of peridotite at high pressures indicate that basaltic magma is produced if a portion of the peridotite (ranging from less than 10 percent up to 30 percent) melts. The basaltic magma, being less dense than peridotite, rises along the oceanic ridge and is extruded as new crust in the spreading ocean basins (Figure 5.19).

Generation of Granitic Magma

At a subduction zone, the basaltic ocean crust and its veneer of marine sediment descend into the mantle, where they are heated (Figure 5.19). There, melting is much more complex. In addition to basalt, the oceanic crust contains oceanic sediments with water-rich clay and silica-rich material derived from the erosion of the continents. Partial melting of this material produces a magma with a high silica content, which forms andesitic or granitic rock. Where continental collisions and mountain building occur in subduction zones, a greater variety of silica-rich rocks develops from the partial melting of the metamorphic rocks in the roots of the mountain systems. The lighter magma rises and collects in large bodies, producing the great granitic batholiths that characteristically form in mountain belts.

The simple model of plate tectonics discussed in Chapter 2 thus explains the major facts about most igneous rocks on the Earth—facts derived from field observations, studies of composition and texture of rock types, and laboratory experiments on synthetic magma.

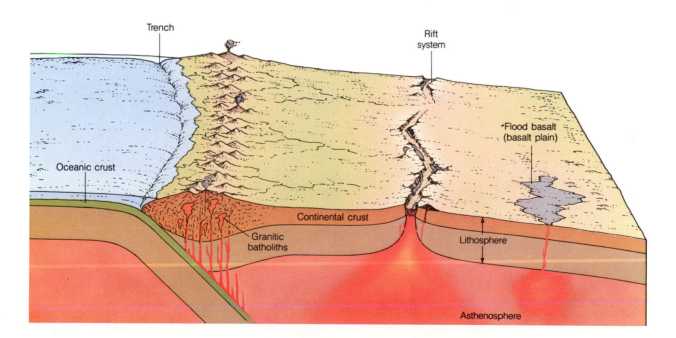

subduction zones by the partial melting of an oceanic plate and of lower continental crust. As the oceanic crust (containing basalt, oceanic sediments, and water) descends into the mantle, it is heated and begins to melt. The resulting liquid has a granitic composition and, being lighter than the surrounding rock, rises to form granitic intrusions or andesitic volcanoes.

SUMMARY

1. Magma is molten rock material that can penetrate into or through rocks of the crust. It can include crystals that formed early in the cooling history (solids), silicate melt (liquid), and water and other volatiles (gas).

2. Two major types of magma are common: (a) basaltic magma is typically very hot (from 900 to 1,200° C) and highly fluid and contains about 50 percent silica (SiO^2); (b) granitic (silicic) magma is cooler (less than 800° C), and highly viscous and contains between 60 and 70 percent silica (SiO^2).

3. The texture of a rock refers to the size, shape, and arrangement of the constituent mineral grains.

4. The major textures in igneous rocks are (a) glassy, (b) aphanitic, (c) phaneritic, (d) porphyritic, and (e) pyroclastic.

5. The classification of igneous rocks is based on texture and composition. High-silica magmas produce rocks of the granite-rhyolite family, characterized by quartz, K-feldspar, and Na-plagioclase. Ferromagnesian and low-silica magmas produce rocks of the gabbro-basalt family, characterized by Ca-plagioclase, amphibole, and pyroxene with little or no quartz. Magmas with an intermediate composition produce rocks of the diorite-andesite family, with a composition intermediate between that of granite and that of basalt.

6. The two major types of magma—basaltic and granitic—produce contrasting types of eruptions.

7. Basaltic magmas are more fluid than granitic magmas. They are commonly extruded as quiet fissure eruptions, producing a succession of thin flows that cover broad areas on the ocean floor, and as flood basalt on continents. Basaltic flows commonly develop columnar jointing, cinder cones, and shield volcanoes.

8. Granitic magmas are more viscous than basaltic magmas and contain large amounts of trapped gas. They therefore explode violently and/or extrude either lava flows in a thick, pasty mass or ash flows. Extrusion of granitic magma commonly produces composite volcanoes.

9. Masses of igneous rocks formed by the cooling of magma beneath the surface are called intrusions. They are classified according to their size and shape and their relationship with the older rocks that surround them. The most important types of intrusions are batholiths, stocks, dikes, sills, and laccoliths.

10. Magma probably originates by the partial melting of the lower crust and the upper mantle, usually at depths from 50 to 200 km below the surface. It is not generated in the Earth's core.

11. Basaltic magma is formed by the partial melting of the mantle along rift zones as tectonic plates move apart.

12. Granitic magmas are generated along subduction zones by the partial melting of oceanic crust descending into the hotter mantle and by the partial melting associated with metamorphism in the continental crust.

KEY TERMS

aa flow (p. 73)
andesite (p. 71)
aphanitic texture (p. 69)
ash-flow tuff (p. 72)
basalt (p. 71)
batholith (p. 79)
caldera (p. 76)
cinder cone (p. 73)
columnar joint (p. 74)
composite volcano (p. 76)
crater (p. 76)
dike (p. 80)
diorite (p. 72)

extrusive rock (p. 73)
fissure (p. 73)
gabbro (p. 72)
glass (p. 69)
glassy texture (p. 69)
granite (p. 72)
groundmass (p. 69)
igneous rock (p. 69)
intrusion (p. 79)
intrusive rock (p. 79)
laccolith (p. 80)
lava (p. 73)
magma (p. 68)

matrix (p. 69)
pahoehoe flow (p. 73)
partial melting (p. 82)
peridotite (p. 72)
phaneritic texture (p. 69)
phenocryst (p. 69)
porphyritic
  texture (p. 69)
pyroclastic texture (p. 69)
rhyolite (p. 72)
shield volcano (p. 72)
sill (p. 80)
stock (p. 80)

stratovolcano (p. 72)
subduction zone (p. 83)
tephra (p. 74)
texture (p. 69)
thin section (p. 71)
tuff (p. 72)
volatiles (p. 68)
volcanic ash (p. 74)
volcanic bomb (p. 74)
volcanic neck (p. 74)
volcanism (p. 68)
welded tuff (p. 76)

REVIEW QUESTIONS

1. Define the term *magma*.
2. What are the principal gases (volatiles) in magma?
3. What are the two principal types of magma?
4. List the major types of igneous rock textures.
5. List the major types of igneous rocks and briefly describe their texture and composition.
6. Describe some of the common surface features of basaltic flows.
7. Describe the extrusion of an ash flow.
8. Describe and illustrate the major types of igneous intrusions. What is the textural difference between intrusive rocks and extrusive rocks?
9. Explain how an igneous rock can be produced from magma that does not have the same composition as the rock.
10. Draw a simple diagram and explain how basaltic magma may originate from the partial melting of the asthenosphere.
11. Draw a simple diagram and explain how granitic and andesitic magma originate from the partial melting of the lithosphere as it descends into a subduction zone.

Multiple-Choice Questions

1. Molten rock on the Earth's surface is called
   a. lava.
   b. obsidian.
   c. magma.
   d. till.
   e. tephra.
2. The texture of an igneous rock
   a. is controlled by its composition.
   b. is determined by the kind of feldspar present.
   c. records its cooling history.
   d. determines the color of the rock.
   e. is caused by the settling of mineral grains during cooling.
3. Volcanic glass is produced by
   a. slow cooling.
   b. rapid cooling.
   c. slow cooling followed by rapid cooling.
   d. any two different rates of cooling.
   e. rapid cooling followed by slow cooling.
4. A porphyritic texture is produced by
   a. explosive eruptions that blow ash and glass into the air as a mixture of hot fragments.
   b. an initial stage of rapid cooling followed by a later slow cooling stage.
   c. gas bubbles being trapped in the lava as it is solidifying.
   d. an initial stage of relatively slow cooling and a later stage of rapid cooling.
5. Composite volcanoes are composed largely of
   a. basaltic lava flows and breccia.
   b. basic lava and acidic lava.
   c. andesitic lava and tephra.
   d. dikes and sills.
   e. granitic rock and ash.
6. Batholiths
   a. are large masses of coarsely crystalline rock, generally of granitic composition.
   b. typically form in deeper zones of folded mountain belts.
   c. generally cut across the country rock into which they intrude.
   d. all of the above.
   e. only a and c are correct.

7. Granitic magma is typically generated
   a. at spreading zones between plates.
   b. where plates slide past each other horizontally.
   c. in subduction zones.
   d. at hot spots in the mantle.
   e. in the central part of plates.
8. Granite, rhyolite, and the volcanic glass obsidian have the same
   a. texture.
   b. cooling history.
   c. composition.
   d. structural deformation.
   e. postcooling history.
9. The surface of a lava flow characterized by a jumbled mass of angular blocks is referred to as a(n)
   a. pahoehoe flow.
   b. laccolith.
   c. aa flow.
   d. pressure ridge.
   e. spatter joint.
10. The origin of basaltic magma is believed to be from
    a. partial melting of a granitic crust.
    b. complete melting of the upper mantle.
    c. partial melting of the upper mantle.
    d. complete melting in the liquid core.
    e. complete melting of the lithosphere.

True/False Questions

1. *T F* Crystalline rock indicates a faster rate of cooling than occurs in glassy rock.
2. *T F* Igneous rocks are classified on the basis of texture and composition.
3. *T F* Rocks that cool below the surface are called intrusive, and those that cool at the surface are called extrusive.
4. *T F* Andesite has the same composition as granite but has an aphanitic texture.
5. *T F* Magma that produces igneous rocks in the Earth's crust is generated in the Earth's lower mantle or core.

Fill in the Missing Terms

1. _____ magmas are characteristically fluid, whereas granitic magmas are thick and viscous.
2. A _____ texture usually indicates two stages of cooling.
3. _____ are the largest rock bodies in the Earth's crust.
4. _____ magma is generated by the partial melting of mantle material that rises along rift zones, where plates are moving apart.
5. _____ volcanoes are the largest volcanoes on the Earth.

6

SEDIMENTARY ROCKS

The geologic processes operating on the Earth's surface produce only subtle changes in the landscape during a human lifetime, but over a period of tens of thousands, or millions, of years, the effect of these processes is considerable. Given enough time, the erosive power of the hydrologic system can reduce an entire mountain range to a featureless lowland. The eroded debris is transported by rivers and deposited as new sedimentary formations. Sedimentary rock sequences preserve the record of erosion through time. Each layer is a remnant of what was once the surface of the Earth. Each rock layer, together with its internal structure, is the product of a previous period of erosion.

To interpret the sedimentary record correctly, you must first understand something about modern sedimentary environments, such as deltas, beaches, and rivers. The study of how sediment is deposited in these areas provides insight into how ancient sedimentary rocks formed. From the sedimentary record, geologists can find the trends of ancient shorelines, map the positions of former mountain ranges, and determine the drainage patterns of ancient river systems. With this information, they are able to make maps of ancient landscapes, which allow us to look back through time and see the planet as it was millions of years ago.

MAJOR CONCEPTS

1. Sedimentary rocks are formed at the Earth's surface by the hydrologic system. Their origin involves the weathering (decomposition) of preexisting rock, the transporting of the material away from the original site, and the depositing of the eroded material in the sea or in some other sedimentary environment.

2. Two main types of sedimentary rocks are recognized: (a) clastic rocks, consisting of rock and mineral fragments, and (b) chemical rocks and organic rocks, consisting of chemical precipitates or organic material.

3. The most significant sedimentary structures are stratification, cross-bedding, graded bedding, ripple marks, and mud cracks.

4. Sedimentary processes result in sedimentary differentiation, in which material is sorted, segregated, and concentrated according to grain size, composition, and density.

5. The major sedimentary environments include (a) fluvial, (b) alluvial-fan, (c) eolian, (d) glacial, (e) delta, (f) shoreline, (g) organic-reef, (h) shallow-marine, and (i) deep-marine systems.

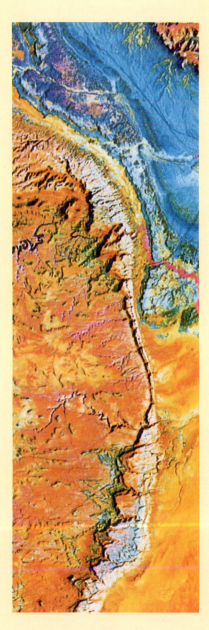

NATURE OF SEDIMENTARY ROCKS

What are the distinguishing features of sedimentary rocks?

Statement

Sedimentary rocks are sediments that have been compacted or cemented to form solid rock bodies. The original sediment can be various substances:

1. Fragments of other rocks and minerals, such as gravel in a river channel, sand on a beach, and mud in the ocean
2. Chemical precipitates, such as salt in a saline lake and calcium carbonate in a shallow sea
3. Organic materials, such as coral reefs and vegetation in a swamp

All are sediment, and all can become sedimentary rocks.

Discussion

The single most distinctive feature of sedimentary rocks is stratification. The sediments are laid down in a series of individual beds, one on top of another, with each bedding plane representing a preserved segment of a former surface of the Earth (such as a river bed, a flood plain, or the sea floor). Layers of sedimentary rock blanket most of the continents, and major layers or groups of layers, called formations, can be traced over areas of thousands of square kilometers. Sedimentary rocks are important because they preserve a

Figure 6.1
The sequence of horizontal sedimentary rocks exposed in the Grand Canyon, Arizona, is almost 2,000 m thick and was deposited over a period of 300 million years. Each major rock unit erodes into a distinctive landform. Formations that are resistant to weathering and erosion (such as sandstone and limestone) erode into vertical cliffs. Rocks that weather easily (such as shale) form slopes or terraces.

record of ancient landscapes, climates, and mountain ranges, as well as the history of the erosion of the Earth. In addition, fossils are found in abundance in sedimentary rocks younger than 600 million years and provide evidence of changing plant and animal communities. The Earth's geologic time scale was worked out using this record of sedimentary rocks and fossils.

Sedimentary rocks are probably more familiar to most people than the other major rock types because they cover approximately 75% of the surface of the continents and therefore form most of the landscape. Few people, however, are aware of the true nature and extent of sedimentary rock bodies.

Consider the rocks in the Grand Canyon area, where many features of sedimentary rocks are well exposed (Figure 6.1). Their most obvious characteristic is that they occur in distinct layers, or *strata,* many of which are more than 100 m thick. Rock types that are resistant to weathering and erosion form cliffs, and nonresistant rocks erode into gentle slopes. Figure 6.1 is a view looking west (downstream) and Figure 6.2 is a cross section of the same area showing graphically the rock layers as if a large slice were cut through the layers from the area of the river channel to the canyon rim. Note that the rim of the canyon is formed on the resistant Kaibab and Torowcap Limestones. Other resistant major rock layers which form high cliffs are the Coconino Sandstone, the Redwall Limestone, and the Tapeats Sandstone. The V-shaped inner gorge of the canyon is cut into igneous and metamorphic rocks which do not have distinctive layering. The major formations exposed in the Grand Canyon can be traced across much of northern Arizona. In fact, they cover an area of more than 250,000 km². A close view of sedimentary rocks in the canyon (Figure 6.1) reveals that each formation has a distinctive texture, composition, and internal structure. The major layers of the sandstone in 6.2 actually consist of smaller units separated by *bedding planes* (planes separating successive layers of rock). These bedding planes are marked by some change in composition, grain size, color, or other physical features (Figure 6.3a). Animal and plant *fossils* are common in most of the rock units and can be preserved in great detail. The term *fossil* is generally used to refer to any evidence of former life (plant or animal). It may be direct evidence, such as shells, bones, or teeth; or it may be indirect, such as tracks and burrows produced by organic activity. The texture of most sedimentary rocks consists of mineral grains or rock fragments that show evidence of abrasion (Figure 6.3b) or consists of interlocking grains of the mineral calcite. In addition, many layers show ripple marks (Figure 6.3c), mud cracks (Figure 6.3d), and other evidence of water deposition preserved in the bedding planes. All of these features indicate that sedimentary rocks form at the Earth's surface in environments similar to present-day deltas, streams, beaches, tidal flats, lagoons, and shallow seas.

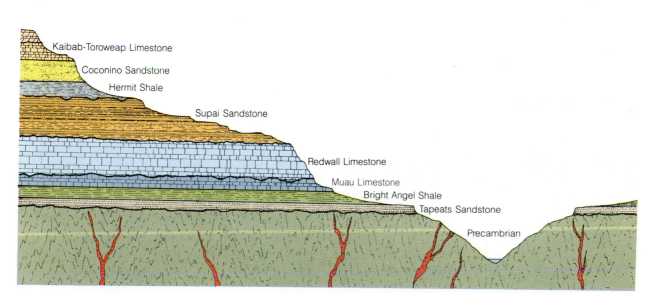

Figure 6.2
A cross section of the Grand Canyon graphically illustrates the major sedimentary formations. The sedimentary strata are essentially horizontal and were deposited on older igneous and metamorphic rocks.

(A) Stratification in sedimentary rocks consists of numerous layers from 0.5 to 1.0 m thick. Each major layer contains cross bedding or other types of layering.

(B) A microscopic view of sand grains in sedimentary rocks shows the effects of transportation by running water. The grains are rounded and sorted to approximately the same size.

(C) Ripple marks preserved in a sandstone indicate that the sediment was deposited by the current action of wind or water.

(D) Mud cracks form where sediment dries while it is temporarily exposed to the air. This structure is common on tidal flats, in shallow lakebeds, and on stream banks. Mud cracks in a rock layer provide important clues to the environment in which the sediment was deposited.

Figure 6.3
A variety of features in sedimentary rocks indicate their origin at Earth's surface as a result of the hydrologic system.

TYPES OF SEDIMENTARY ROCKS

How are the different types of sedimentary rock distinguished and classified?

Statement

Sedimentary rocks are classified on the basis of the size, shape, and composition of their constituent particles. Two main groups are recognized:

1. Clastic rocks, formed from fragments of other rocks.
2. Chemical rocks and organic rocks, formed by chemical or biological processes. Some examples are shown in Figure 6.4.

Discussion

Clastic Rocks

Generally, clastic rocks are subdivided according to the grain size of the component materials. From the largest grain size to the smallest, the types of clastic rocks are conglomerate, sandstone, siltstone, and shale.

Conglomerates. A conglomerate consists of consolidated deposits of gravel, with variable amounts of sand and mud in the spaces between the larger grains (Figure 6.4a). The cobbles and pebbles usually are well-rounded fragments over 2 mm in diameter. Most conglomerates show a crude stratification and include beds and lenses of sandstone. Conglomerates are accumulating today at the base of many mountain ranges, in stream channels, and on beaches.

Sandstones. Sandstone is probably the most familiar sedimentary rock (Figure 6.4b). The sand grains range from $\frac{1}{16}$ mm to 2 mm in diameter and can be composed of almost any material, so that sandstones can be various colors. Quartz grains are usually the most abundant mineral because quartz is a common constituent in many other rock types and because it is not easily broken down by abrasion or chemical action. The particles of sand in most sandstones are cemented by calcite, silica, or iron oxide.

Siltstones. Siltstone is a fine-grained clastic rock in which most of the material is $\frac{1}{16}$ to $\frac{1}{256}$ mm in diameter (finer than sand but coarser than mud). Siltstones commonly contain very thin layers, called laminae, and may show evidence of burrowing by organisms. Silt is a material frequently carried in suspension by rivers and deposited in floodplains and deltas.

Shales (Mudstone). Solidified deposits of mud and clay are known as shale. The particles that make up the rock are less than $\frac{1}{256}$ mm in diameter and in many cases are too small to be clearly seen and identified even under a microscope (Figure 6.4c). Shale is the most abundant sedimentary rock. It usually is soft and weathers rapidly into a slope, so relatively few fresh, unweathered exposures are found. Most shale is well stratified, with thin laminae. Black shales are rich in organic material and accumulate in a variety of quiet-water environments, such as lagoons, restricted shallow seas, and tidal flats. Red shales are colored with iron oxide and indicate oxidizing conditions in the environments in which they accumulate, such as stream channels, floodplains, and tidal flats.

Chemical and Organic Rocks

Nonclastic rocks are formed from chemical precipitation and from biological activity.

Limestone. By far the most abundant nonclastic rock is limestone. It is composed principally of the mineral calcite—calcium carbonate ($CaCO_3$)—and originates by both chemical and organic processes. Limestones have a great variety of rock textures, and many different types have been classified. Some of the major groups are skeletal limestone, oolitic limestone, and microcrystalline limestone.

Many plants and invertebrate animals extract calcium and carbonate (limestone) from water in their life processes and use it to construct their shells and hard parts of calcite. When these organisms die, their shells accumulate on the sea floor. Over a long period of time, the shells build up a deposit of limestone with a texture consisting of shells and shell fragments (Figure 6.4d). This type of limestone, composed mostly of skeletal debris, can be several hundred meters thick and extends over thousands of square kilometers. Chalk, for example, is a skeletal limestone in which the skeletal fragments are remains of microscopic plants and animals.

Some limestones are composed of small, spherical concretions of calcium carbonate called oolites. The individual grains, about the size of a grain of sand, can presently be observed forming in the shallow waters off the Bahamas, where currents and waves are active. Evaporation and increased temperatures in the seawater raise the concentration of $CaCO_3$ until it is precipitated. Small fragments of shells and tiny grains of $CaCO_3$ become coated with successive layers of $CaCO_3$ as they are rolled along the seafloor by waves and currents.

In quiet water, calcium carbonate is precipitated as tiny, needlelike crystals, which accumulate on the bottom as limy mud. Soon after they are deposited, the grains commonly are modified as they are com-

(A) Conglomerate is a coarse-grained, clastic rock in which most of the particles are larger than 2 mm in diameter. It is, in essence, gravel that has been cemented and consolidated into a solid rock body. The fragments in a conglomerate are mostly rounded pebbles and cobbles, but generally there are also considerable amounts of sand and mud filling the space between pebbles. Most conglomerates are deposited by streams.

(B) Sandstone is a fine-grained, clastic rock composed of fragments that may range from 1/16 to 2 mm in diameter. Most sand is composed of rounded quartz grains, but significant amounts of small rock fragments and feldspar may be present. Sandstone commonly grades into either siltstone and shale or into conglomerate.

(C) Shale is consolidated mud. The fragments that make up a shale are less than 1/256 mm in diameter and are composed mostly of clay minerals. Coarser grains, called silt (1/256 to 1/16 mm in diameter), may make up a significant part of the rock. Shales accumulate in many different environments. The main sediment carried by the great rivers to the sea is mud and fine sand, so it is not surprising that shale is the most abundant sedimentary rock. Black shales are rich in organic matter and accumulate in a variety of quiet-water environments such as lagoons and tidal flats. Red shales are colored with iron oxide and indicate an oxidizing environment.

(D) Limestone is composed principally of calcium carbonate ($CaCO_3$) and is by far the most abundant nonclastic rock. It originates by both chemical and organic processes that produce a variety of rock textures. Some of the major groups are (1) skeletal limestone (composed of fragments of fossil shells), (2) oolitic limestone, and (3) microcrystalline limestone.

Figure 6.4
Examples of sedimentary rocks.

pacted and become recrystallized. This modification produces microcrystalline limestone, a rock with a dense, very fine-grained texture. Its individual crystals can be seen only under high magnification. Microcrystalline limestone also is precipitated from springs and from the dripping water in caves, but the total worldwide amount of dripstone, compared to the amount of marine limestone, is negligible.

Dolostone. Dolostone is a rock composed of the mineral dolomite, a calcium magnesium carbonate, $CaMg(CO_3)_2$. It is similar to limestone ($CaCO_3$) in most textural and structural features and in general appearance. It can develop by direct precipitation from seawater, but most dolostones are apparently formed by the substitution of magnesium for some of the calcium in limestones.

Rock Salt and Gypsum. Rock salt is composed of the mineral halite (NaCl). It forms by evaporation in saline lakes (for example, the Great Salt Lake and the Dead Sea) or in restricted bays along the shore of the ocean. Gypsum, composed of $CaSO_4\ 2H_2O$, also originates from the evaporation of saline water. It collects in layers as calcium sulfate is precipitated. Because evaporites (rocks formed by evaporation) accumulate only in restricted basins that have been subjected to periods of prolonged evaporation, they are important indicators of ancient climatic and geographic conditions.

SEDIMENTARY STRUCTURES

Why are sedimentary structures important in the study of sedimentary rocks?

Statement

Most sediment is transported by currents in streams, along shorelines, and in shallow seas before it is finally deposited. For this reason, sedimentary rocks commonly show layering and other structural features that formed as the material was moved, sorted, and deposited by currents. The following are the most important sedimentary structures:

1. Stratification
2. Cross bedding
3. Graded bedding
4. Ripple marks and mud cracks

Discussion

Stratification

One of the most obvious characteristics of sedimentary rocks is that they occur in distinct layers, which

are expressed by color, texture, and the way the different rock units weather and erode. These layers are termed strata, or simply beds. Stratification occurs on many scales and reflects the changes that occur during the formation of a sedimentary rock. Large-scale stratification is expressed by major changes in rock types, which can be seen in large exposures, such as the Grand Canyon (Figure 6.1), where cliffs of limestone or sandstone alternate with slopes of weaker shale. These major units are called formations. Within each formation, stratification—or bedding—occurs on several smaller scales, expressed by differences in the texture, color, and composition of the rock.

An important aspect of stratification is that the rock layers do not occur in a random fashion but overlie one another in definite sequences and patterns. One of the simpler and more common patterns in a vertical sequence is the cycle of sandstone, shale, and limestone. This pattern is produced by the advance and retreat of shallow seas over the continental platform (Figure 6.5). In diagram (a) in the figure, the sea begins to expand over a lowland drained by a river system. Sand accumulates along the shore, mud is transported in suspension offshore, and lime is precipitated from solution farther offshore, beyond the mud zone. All three types of sediment are deposited simultaneously, each in a different environment. Stream deposits (not shown in the figure) accumulate on the floodplain of the river system.

As the sea expands over the lowland, each environment shifts landward, following the shoreline (Figure 6.5a, b, and c). Beach sands are consequently deposited over the stream sediments, offshore mud is deposited over the previous beaches, and lime is deposited over the mud. As the sea continues to expand, the layers of sand, mud, and lime are deposited farther and farther inland.

As the sea withdraws (Figure 6.5d), the mud is deposited over the lime and the sand near the shore is deposited over the mud. The net result is a long wedge, or layer, of limestone encased in a wedge of shale, which in turn is encased in a wedge of sandstone. Below and above the marine deposits are fluvial (river) sediments deposited by the river system (not shown in the diagram). Subsequent uplift and erosion of the area reveal a definite sequence of rock (Figure 6.5e). Beginning at the base, sandstone is overlain by shale and limestone, which are in turn overlain by shale and sandstone. (See the lower rock layers in Figure 6.2.)

Cross Bedding

Cross bedding is a type of stratification in which the layers within a bed are inclined at an angle to the upper and lower surfaces of the bed. The formation of cross bedding is shown in Figure 6.6. As sand grains

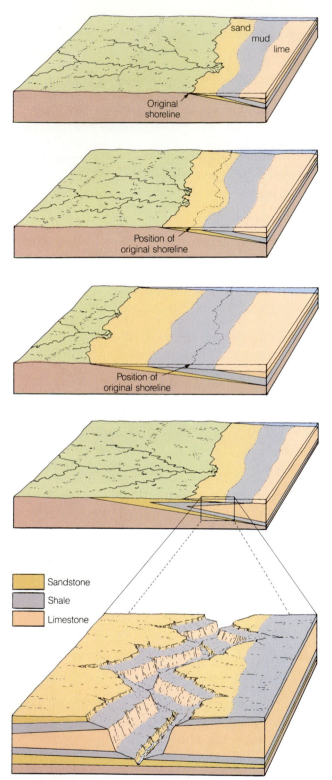

Figure 6.5
A sequence of sediments deposited by the expansion and contraction of a shallow sea is represented in these schematic diagrams. Sand accumulates along the beach, mud is deposited offshore, and lime is precipitated farther offshore, beyond the mud. As the sea expands, these environments move inland, producing a vertical sequence of sand-mud-lime. As the sea recedes, mud is deposited over the lime and sand is deposited over the mud. The net result is a vertical sequence of sedimentary layers: sand-mud-lime-mud-sand.

are moved by currents (either wind or water), they form ripples, or dunes, called sand waves. These range in scale from small ripples less than a centimeter high to giant sand dunes several hundred meters high. Typically they are asymmetrical, with the gentle slope facing in the direction from which the current is coming. As the particles migrate up and over the sand wave, they accumulate on the steep down-current face and form inclined layers. The direction of flow of the ancient currents that formed a given set of cross strata can be determined by measuring the direction in which the strata are inclined. We can determine the patterns of ancient current systems by mapping the direction of cross bedding in sedimentary rocks.

Graded Bedding

A distinctive type of stratification called graded bedding is characterized by a progressive decrease in grain size upward through the bed (Figure 6.7). This type of stratification commonly is produced on the deep-ocean floor by turbidity currents, which transport sediment from the continental slope to the adjacent deep ocean. A turbidity current is generated by turbid (muddy) water, which, being denser than the surrounding clear water, sinks beneath it and moves rapidly down the continental slope (Figure 6.8).

Turbidity current flow can be easily demonstrated in the laboratory by pouring muddy water down the side of a tank filled with clear water. The mass of muddy water moves down the slope of the tank and across the bottom at a relatively high speed, without mixing with the clear water. Turbidity currents can also be observed where streams discharge muddy water into a clear lake or reservoir. The denser, muddy water moves out along the bottom of the basin and can flow for a considerable distance, even along the flat surface of the abyssal floor.

Turbidity currents can also be generated by an earthquake or an underwater landslide, during which mud, sand, and even gravel can be thrown into suspension. In 1929 one of the best-documented large-scale turbidity currents was triggered by an earthquake near the Grand Banks, off Newfoundland. When the earthquake occurred it caused the slumping of a large mass of soft sediment (estimated to be 100 km³) to move as a turbidity flow down the continental slope and onto the abyssal plain; it eventually covered an area of 100,000 km². As the turbidity current moved downslope, it broke a series of transatlantic cables at different times. The speed of the current, determined from the intervals between the times when the cables broke, was from 80 to 95 km per hour. This mass of muddy water formed a graded layer of sediment over a large area of the Atlantic floor.

As a turbidity current moves across the flat floor of

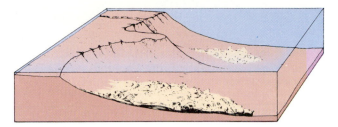

Figure 6.8
The movement of turbidity currents down the slope of the continental shelf can be initiated by a landslide or an earthquake. Sediment is moved largely in suspension. As the current slows down, the coarse grains are deposited first, followed by the deposition of successively finer-grained sediment. A layer of graded bedding is thus produced from a single turbidity current.

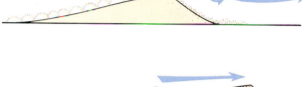

Figure 6.6
Cross bedding is formed by the migration of sand waves (ripple marks or dunes). Particles of sediment, carried by currents, travel up and over the sand wave and are deposited on the steep down-current face to form inclined layers.

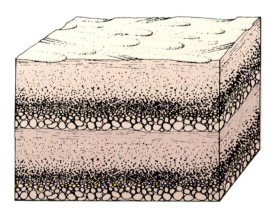

Figure 6.7
Graded bedding is produced by turbidity currents. It occurs in widespread layers, each layer generally less than a meter thick. Deep-marine environments commonly produce a great thickness of graded layers, which can easily be distinguished from sediment deposited in most other environments.

a basin, its velocity at any given point gradually decreases. The coarser sediment is deposited first, followed by successively smaller particles. After the turbid water ceases to move, the sediment remaining suspended in water gradually settles out. One pulse of sedimentation therefore deposits a single layer of sediment, which exhibits continuous gradation from coarse material at the base to fine material at the top. Subsequent turbidity flows can deposit more layers of graded sediment, with sharp contacts between layers. The result is a succession of widespread horizontal layers, each being a graded unit deposited by a single pulse of sedimentation from a turbidity flow.

Ripple Marks, Mud Cracks, and Other Surface Impressions

Ripple marks are commonly seen in modern streambeds, in tidal flats, and along the shores of lakes and the sea. Preserved in rocks, they provide information concerning the environment of deposition, such as the depth of the water, ancient current directions, and the trends of ancient shorelines. Mud cracks are also commonly preserved in sedimentary rock and show that the sedimentary environment occasionally was exposed to the air during the time period the sediment was deposited. Mud cracks in rocks suggest that the original sediment was deposited in shallow lakes, on tidal flats, and on exposed stream banks.

Tracks and trails of animals are typically associated with ripple marks and mud cracks and can provide additional important clues about the environment in which the sediment accumulated.

ORIGIN OF SEDIMENTARY ROCKS

What are the major environments of sedimentation? How can the sedimentary environment be interpreted from an outcrop of rock?

Statement

The genesis of sedimentary rocks involves four major processes:

1. Weathering
2. Transportation (erosion)
3. Deposition
4. Compaction and cementation

Discussion

Weathering

Weathering is the interaction between the elements in the atmosphere and the rocks exposed at the Earth's surface. The atmosphere can physically break down the rock through processes such as frost action, and it can chemically decompose the rock by a variety of reactions. We will study the details of weathering in Chapter 9. For now, note that weathering is the first step in the genesis of sedimentary rock. The atmosphere breaks down and decomposes preexisting solid rock and forms a layer of loose, decayed rock debris, or soil. This unconsolidated material can then be transported easily by various agents, such as streams, wind, groundwater, and glaciers.

Transportation

Running water is the most effective form of sediment transport. All rivers carry large quantities of sediment toward the sea, especially during flood stage. This fact is readily appreciated if you consider the great deltas of the world, each formed from sediment transported by rivers (see Figure 3.3). Indeed, sediment is so abundant in most rivers that a river might best be thought of as a system of water and sediment rather than simply a channel of flowing water.

As sediment is transported by a river, it is sorted and separated according to grain size and composition. Large particles accumulate as gravel, medium-sized grains are concentrated as sand, and finer material settles out as mud. Dissolved minerals are carried in solution and are ultimately precipitated as limestone, salt, or other chemical deposits.

Wind and glaciers also transport sediment, although their activity is somewhat restricted to special climatic zones.

The sorting that occurs during transportation is an important factor in the genesis of sedimentary rock. Indeed, most sedimentary differentiation occurs during transportation, so when the sediment is delivered to the site where it is deposited, it is commonly already sorted and differentiated to some degree according to particle size and composition.

Deposition

Perhaps the most significant factor in the origin of sedimentary rocks is the sedimentary environment—that is, the physical, chemical, and biological conditions that exist at the place where the sediment is deposited. The idealized diagram in Figure 6.9 shows in a general way the regional setting of the major types of sedimentary environments. Continental environments include areas of sedimentation that occur exclusively on the land surface. Most important are major river systems, alluvial fans, desert dunes, lakes, and margins of glaciers. Marine environments include the shallow seas, which cover parts of the continental platform, and the floors of the deep-ocean basins. Between continental areas and marine areas are the transitional, or mixed, environments, which occur along the coasts and are influenced by both marine and nonmarine processes. These include deltas, beaches, tidal flats, reefs, and lagoons (Figure 6.9).

Each of these environments is characterized by certain physical, chemical, and biological conditions. Distinctive types of texture, composition, internal structure, and fossil assemblages are thus developed in each environment. Selected examples of rock sequences deposited in various sedimentary environments are shown in Figure 6.10.

Compaction and Cementation

The final stage in the formation of sedimentary rocks is the transformation of loose, unconsolidated sediment into solid rock. This is accomplished by compaction and cementation. The weight of overlying material, which continually accumulates in a sedimentary environment, compresses and compacts buried sediment into a tight, coherent mass. Cementation occurs as mineral matter, carried by water seeping through the pore spaces of the grains, is precipitated. The most common cementing minerals are calcite, quartz, and limonite. Cementation is an important process that transforms sediment into solid rock.

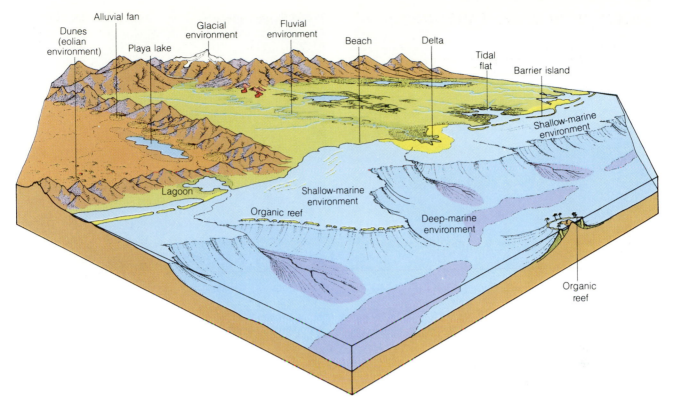

Figure 6.9
The major sedimentary environments are represented in this idealized diagram. Most sediment moves downslope from continental highlands toward the oceans, so the most important environments of sedimentation are found along the shores and in the shallow seas beyond. Sedimentary environments can be categorized in three groups: continental, shoreline (transitional), and marine. Their important characteristics and the types of sediment that accumulate in each are outlined below.

CONTINENTAL ENVIRONMENTS

Alluvial fans are fan-shaped deposits of gravel, sand, and mud that accumulate in dry basins at the bases of mountain ranges.

Eolian (wind) environments include sand seas of deserts, where sand dunes are built and transported by wind, and areas where windblown dust accumulates.

Fluvial (river) environments are the river channels, river bars, and the adjacent floodplain. Lakes are bodies of nonmarine water, including freshwater lakes on continental lowlands and saline lakes in isolated basins.

Glacial environments are the areas where sediment is deposited by glaciers. Most obvious are the margins of the ice where sediment carried by the glacier is dropped as the ice melts. Other subenvironments of a glacier are the lakes and meltwater streams.

SHORELINE ENVIRONMENTS

Deltas are deposits of mud, silt, and sand that form at the mouths of rivers, where they empty into the sea or a lake.

Beaches are shoreline accumulations of sand.

Barrier islands are linear bodies of sand built offshore by the action of ocean waves.

Lagoons are elongated bodies of seawater located between the mainland and barrier islands or reefs. Low wave energy permits the deposition of mud.

Tidal flats are shoreline areas that are covered with water at high tide and uncovered at low tide. Mud is the major type of sediment deposited.

MARINE ENVIRONMENTS

Shallow-marine environments extend from the shore to the edges of the continental shelves. Lime and mud are the principal types of sediment deposited.

Organic reefs are solid structures built from corals, algae, and shells of other marine organisms. Reefs grow in warm, shallow water near islands and continents.

Deep-marine environments characterize the deep oceans beyond the continental slopes and include deep-sea fans and abyssal plains. Turbidites are the major types of sediment deposited.

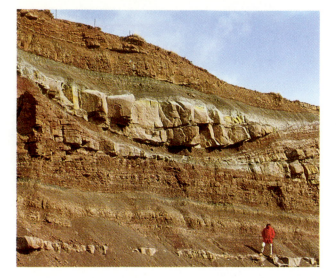

(A) *Ancient stream channel* in Tertiary sediments in central Utah.

(B) *Ancient dune deposits* in Zion National Park, Utah.

(C) *Glacial sediments* deposited during the last Ice Age.

(D) *Lagoonal deposits* of the Cretaceous age in central Utah.

(E) *Ancient shallow-marine sediments* of the Pennsylvanian age in eastern Kansas.

(F) *Ancient deep-marine sediment* originally deposited in a horizontal position near the southern coast of France.

Figure 6.10
Examples of rock sequences deposited in various sedimentary environments

SIGNIFICANCE OF SEDIMENTARY ROCKS

Why are sedimentary rocks important?

Statement

Sedimentary rocks are the product of the hydrologic system and its interaction with the Earth's crust. They, therefore, provide a record of how the surface of Earth has changed throughout geologic time. They are also important as a natural resource for most of the products used in our culture.

Discussion

Sedimentary rocks are the most common rock exposed at the Earth's surface, and in most continents they form a thin but extensive blanket over the igneous and metamorphic rocks below. From a scientific point of view, they are important for the geologic history they record. From the record of sedimentary rocks, geologists are able to interpret such things as ancient mountain building, the erosion of continents, climatic changes, the evolution of life, and the constantly changing patterns of land and sea.

An excellent example of a sequence of sedimentary rocks and the events they record is found in the coal-bearing rocks of western North America. These rocks, exposed throughout the Rocky Mountains and in the Great Plains as far east as Minnesota and Iowa, are composed mainly of sand, mud, and lime. Thick coal beds also abound in the west, and pinch out when traced eastward. From thousands of outcrops and from well-drilling records, geologists have measured the thickness of each rock type and how the rock bodies are related to one another. From these data they have compiled the cross section shown in Figure 6.11.

The relationship of the various rock types indicates a series of former sedimentary environments, which extended in a north-to-south direction. The coarse gravels are interpreted as stream deposits derived from an ancient mountain range, which extends through western Utah and Wyoming, northward into Canada. These grade eastward into coals and sandstone, which were deposited in near-shore environments, such as barrier islands, beaches, and swamps. Eastward in the Great Plains states, mud and calcium carbonate accumulated in a broad, shallow sea to form mudstone or shale and limestone. The ancient geographic conditions under which these rocks accumulated are shown in Figure 6.11.

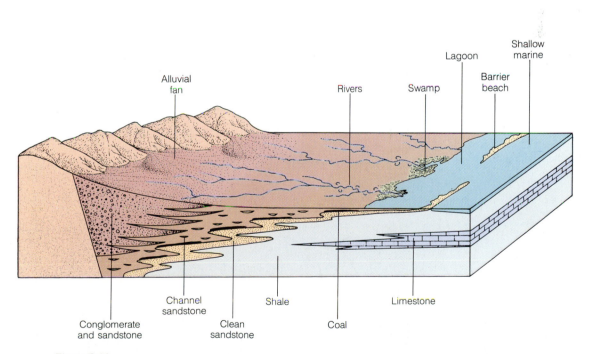

Figure 6.11
Interpretation of ancient sedimentary environments based on sedimentary rocks and their vertical and lateral relationships. The major sedimentary environments are shown on the top of the diagram: alluvial fans, river systems, deltas, lagoons, barrier bars, and shallow marine. The rock sequences—conglomerate, sandstone, coal, shale and limestone—are shown on the sides of the diagram. They were formed in a vertical sequence as each environment shifted back and forth as the sea expanded and contracted with time.

As the seas expanded and contracted, the shoreline environments of beaches and swamps moved back and forth to produce the interfingering relations of the sand and coal with the marine shale. This basic geographic pattern existed for a long time, until finally the sea withdrew completely.

From an economic point of view, sedimentary rocks are extremely important because many of our natural resources are formed by sedimentary processes. Fully 85 to 90 percent of the mineral products produced annually comes from sedimentary rocks. These indicate petroleum, natural gas, coal, oil, shale, and groundwater. Sediments (sand, gravel, and limestone) are the raw materials for cement, the major building material for our culture (buildings, freeways, dams, etc.). Clay is the basis of ceramics. Evaporites provide most of our salt and chemicals. Sandstones are exploited for rare minerals such as gold, diamonds, platinum, and uranium. Glass is made from quartz sand. Most of our iron deposits are found in sedimentary rocks, and sedimentary sequences are the host for many metallic mineral deposits, such as lead and zinc. Bricks are made from clay. The great cathedrals of Europe are made from sedimentary rock, and the statues made by the artists of ancient Greece and Rome and during the Renaissance would have been impossible without limestone. The economic importance of sedimentary rock is so great in fact that it has been a controlling factor in the development of our industry, society, and culture. Without material from sedimentary rocks, humans probably could not have advanced to the neolithic age because flint (a variety of microcrystalline quartz commonly formed in limestone) played such an important role in the development of tools, arrowheads, and axes.

SUMMARY

1. Sedimentary rocks originate from fragments of other rocks, chemically precipitated minerals, and organic matter.
2. Sedimentary rocks typically occur in layers that cover large parts of the continents.
3. Two main types of sedimentary rocks are recognized: (a) clastic rocks, consisting of rock and mineral fragments (clay, silt, sand, and gravel) and (b) chemical and organic rocks.
4. Four major processes are involved in the genesis of sedimentary rocks: (a) weathering, (b) transportation, (c) deposition, and (d) compaction and cementation. During weathering, transportation, and deposition, the sedimentary material is differentiated (sorted and concentrated according to grain size and composition).
5. A sedimentary environment is a place where sediment is deposited; it encompasses physical, chemical, and biological conditions that exist there. The major sedimentary environments include (a) fluvial, (b) alluvial fan, (c) eolian, (d) glacial, (e) delta, (f) shoreline, (g) organic-reef, (h) shallow-marine, and (i) deep-marine environments. Each of these is characterized by certain physical, chemical, and biological conditions and therefore develops distinctive rock types and fossil assemblages.

KEY TERMS

bedding plane (p. 89)
cement (p. 96)
chalk (p. 91)
clastic (p. 91)
conglomerate (p. 91)
cross bedding (p. 93)
deep-marine
   environment (p. 97)
delta (p. 97)
dolostone (p. 93)
dripstone (p. 93)

eolian
   environment (p. 97)
evaporite (p. 93)
fluvial
   environment (p. 97)
fossil (p. 89)
glacial
   environment (p. 97)
graded bedding (p. 94)
gypsum (p. 93)

lagoon (p. 97)
limestone (p. 91)
mud cracks (p. 95)
oolite (p. 91)
reef (p. 97)
ripple marks (p. 95)
rock salt (p. 93)
sandstone (p. 91)
sand waves (p. 94)
sediment (p. 88)

sedimentary
   environment (p. 97)
sedimentary
   structure (p. 93)
shale (p. 91)
siltstone (p. 91)
sorting (p. 96)
stratification (p. 88)
tidal flat (p. 97)
turbidity current (p. 94)

REVIEW QUESTIONS

1. List the characteristics that distinguish sedimentary rocks from igneous and metamorphic rocks.
2. What is the principal mineral in sandstone? Why does this mineral dominate?
3. How does limestone differ from clastic rocks?
4. What is the mineral composition of limestone?
5. Show by a series of sketches the characteristics of stratification, cross bedding, and graded bedding.
6. What major rock types form in the following sedimentary environments: (a) delta, (b) lagoon, (c) alluvial fan, (d) eolian, (e) organic reef, and (f) deep marine?

Multiple-Choice Questions

1. Essentially all sedimentary deposits show evidence of
   a. fossils.
   b. stratification.
   c. cross bedding.
   d. ripple marks.
   e. mud cracks.
2. The major constituent of most sandstone is
   a. feldspar.
   b. olivine.
   c. calcite.
   d. quartz.
   e. garnet.
3. Which of the following is the most abundant nonclastic rock?
   a. gypsum
   b. dripstone
   c. dolostone
   d. limestone
   e. rock salt
4. Cross bedding is commonly found in
   a. rock gypsum.
   b. rock salt.
   c. shale.
   d. sandstone.
   e. coal.
5. Alternating layers of coal, shale and sandstone would have been deposited in which of the following environments?
   a. beach
   b. tidal flat
   c. shallow marine
   d. lagoon
   e. alluvial fan
6. The mineral calcite ($CaCO_3$) is a major constituent of
   a. shale.
   b. limestone.
   c. sandstone.
   d. coal.
   e. dolomite.
7. Transportation and deposition of sediment by turbidity currents is commonly indicated by
   a. cross bedding.
   b. ripple marks.
   c. mud cracks.
   d. graded bedding.
   e. horizontal laminae.
8. A type of current that is so dense with suspended sediments that it flows along the sea floor is called a
   a. longshore current.
   b. tidal current.
   c. turbidity current.
   d. rip current.
   e. wave current.
9. Channels of sandstone and conglomerate in shale probably represent deposits in which of the following environments?
   a. glacial
   b. beach
   c. fluvial
   d. eolian
   e. deep marine
10. A sequence of horizontal beds extending over a large area, with each layer showing graded bedding, would result from deposition in which of the following environments?
    a. deep marine
    b. shallow marine
    c. beach
    d. eolian
    e. glacial

True/False Questions

1. *T F* Clastic rocks are formed from fragments of other rocks.
2. *T F* Conglomerates are accumulating today at the base of many mountain ranges, in stream channels, and on many beaches.
3. *T F* An important aspect of stratification is that the rock layers occur in a random fashion.
4. *T F* The cycle of sandstone-shale-limestone may be produced by the advance and retreat of shallow seas over the continental platform.
5. *T F* Ancient dune deposits are characterized by large-scale graded bedding consisting of well-sorted, well-rounded grains.

Fill in the Missing Terms

1. The single most distinctive feature of sedimentary rocks is _____.
2. Solidified deposits of mud and clay are known as _____.
3. Rock salt is composed of the mineral _____.
4. _____ bedding is a type of stratification commonly produced on the deep-ocean floor by turbidity currents.
5. _____ are deposits of mud, silt, and sand that form at the mouths of rivers, where the river empties into a sea or lake.

7

METAMORPHIC ROCKS

Most of the rocks exposed in the continental shields and in the cores of mountain belts show evidence that their original texture and composition have been changed. Many have been plastically deformed, as indicated by contorted parallel bands of minerals, resembling the swirling layers in a marble cake. Others have recrystallized and developed larger mineral grains, and the constituent minerals of many rocks have a strong fabric (the physical appearance of a rock resulting from the size and orientation of its constituent minerals) with a specific orientation. The mineral assemblages in these rocks are also distinctive. They are characterized by mineral types that form only under high temperature and pressure. Geologists interpret this and other evidence to mean that the original rock has been recrystallized in a solid state. The result is a new rock type with a distinctive texture and fabric and, in some cases, a new mineral composition. These rocks are called metamorphic rocks.

Metamorphic rocks are hosts for many important mineral deposits. They also play an important role in the evolution of continental crust. Consequently, every aspect of metamorphic rock, from the small grain to the regional fabric of a shield, points toward the same theme: Metamorphic rocks dramatically show mobility and change of a dynamic crust.

MAJOR CONCEPTS

1. Metamorphic rocks result from changes in temperature and pressure and from changes in the chemistry of the fluids in their pores. These changes produce new minerals, new textures, and new structures within the rock body.
2. The major types of metamorphic rocks are slate, schist, gneiss, quartzite, marble, amphibolite, metaconglomerate, and hornfels.
3. During metamorphism, new minerals grow in the direction of least stress, producing a planar rock structure called foliation. The three main types of foliation are (a) slaty cleavage, (b) schistosity, and (c) gneissic banding.
4. Rocks with only one mineral (such as limestone and sandstone) do not develop a strong foliation but instead develop a granular texture with large crystals.
5. Regional metamorphism develops in the roots of mountain belts along convergent plate boundaries. Contact metamorphism is a local phenomenon associated with thermal and chemical changes near the contacts of igneous intrusions.

THE NATURE AND DISTRIBUTION OF METAMORPHIC ROCKS

What are the distinguishing features of metamorphic rocks?

Statement

Metamorphic rocks are rocks that have been altered by heat, pressure, and the chemical action of pore fluids (water and early melted mineral matter) to such an extent that new minerals are formed that are stable in an environment of higher temperature and pressure. A new rock texture (or fabric) develops in which the grains commonly have a preferred orientation.

Discussion

Many people know something about various igneous and sedimentary rocks but only vaguely understand the nature of metamorphic rocks. Perhaps the best way to become acquainted with this group of rocks and to appreciate their significance is to carefully study Figures 7.1 and 7.2. The aerial photograph of a portion of the Canadian shield (Figure 7.1) shows that the rocks have been twisted and compressed. Originally, these were sedimentary and volcanic layers deposited in a horizontal position, but they have been deformed so intensely that it is difficult to determine the original base or top of the rock sequence. Light-colored granite batholiths and dikes have been injected into the metamorphic series, and the entire mass has been fractured and displaced by numerous faults.

Figure 7.2a shows a more detailed view of metamorphic rocks. The alteration and deformation of the rock are evident in the contorted layers of dark minerals. The pattern of the distortions shows that the rock was compressed while it was in a plastic or semiplastic state.

The degree of plastic deformation that is possible during metamorphism is best seen by comparing the shape of pebbles in a conglomerate with pebbles in a rock that has been metamorphosed. As is evident from the photograph (Figure 7.2b), the original spherical pebbles in the conglomerate have been stretched into long, ellipsoidal blades (the long axis is as much as thirty times the original diameter), all oriented in the same direction. Even the grains in shale show distortion and deformation of the individual on a microscopic scale.

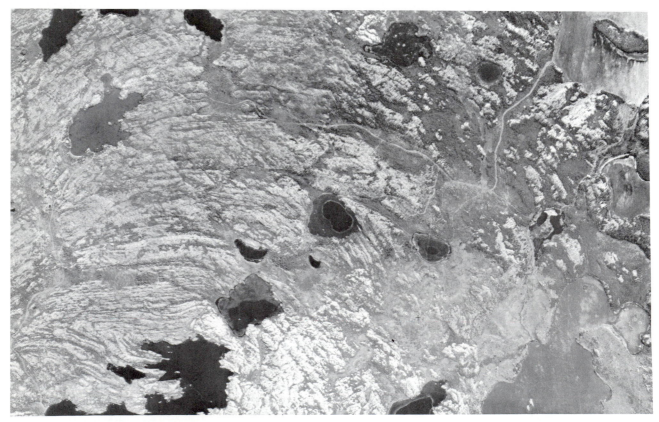

Figure 7.1
The characteristics of metamorphic rocks are shown in this photograph of the Canadian Shield, taken from an altitude of 15 km. These rocks have been compressed and deformed to such an extent that many original features have been obliterated. Metamorphism occurred at great depths, and the area was then deeply eroded, exposing the complex rock sequence. In addition to the tight folding, the rocks have been broken by fractures and intruded by granitic rocks (lighter tones).

(A) Gneiss is a coarse-grained, metamorphic rock with distinct layers or lenses of different minerals. The composition of gneiss is similar to that of granite: the major minerals are quartz, feldspar, amphibole, and mica.

(D) Schist is a coarse-grained, foliated rock in which the foliation results from the parallel arrangement of large grains of platy minerals such as mica, chlorite, talc, or hematite. Schist results from a higher intensity of metamorphism than that which produces slate; hence, it has larger grains than slate.

(B) Metaconglomerate is a metamorphosed conglomerate and clearly illustrates the degree to which a rock can be changed (metamorphosed) while still in a solid state. The original rounded pebbles and cobbles have been stretched, deformed, and fused together into a rock with a distinct, linear fabric.

(E) Marble is formed from the metamorphism of limestones and dolomites. The changes during metamorphism involve recrystallization with the growth of large interlocking crystals.

(C) Slate is a very fine-grained metamorphic rock produced by the metamorphism of shales and other fine-grained rocks. The distinctive feature of this rock is that it tends to split into thin sheets.

(F) Quartzite is a metamorphosed sandstone. The original rounded sand grains have been squeezed and fused into a very tight mass. The mineral grains are commonly flattened, stretched, and elongated so that the rock fabric has a distinct preferred orientation.

Figure 7.2
Examples of metamorphic rocks.

It is important to note that the typical texture of metamorphic rocks does not show a sequence of formation of the individual minerals like that evident in igneous rocks. All grains in metamorphic rocks recrystallize at roughly the same time, and they have to compete for space in an already solid rock body. As a result, the new minerals grow in the direction of least stress. Most metamorphic rocks thus have a layered, or planar, structure resulting from recrystallization.

Metamorphic rocks constitute a large part of the continental crust. Extensive exposures (such as those shown in Figure 7.1) are found in the vast shield areas of the continents, where the cover of sedimentary rocks has been removed. Deep drilling in the stable platform indicates that the bulk of the continental crust immediately beneath the sedimentary cover is also made up of metamorphic rock. These facts permit estimates that metamorphic rocks, together with associated igneous intrusions, make up approximately 85 percent of the continental crust, at least to a depth of 20 km. In addition to the metamorphic rocks beneath the stable platforms of the continents and exposed in the shields, they are also found in the cores of eroded mountain ranges. The nature of Earth's mantle is not known from direct observation, but it is difficult to imagine how the mantle, or at least parts of it, could be other than a type of metamorphic rock.

The widespread distribution of metamorphic rocks in the continental crust, especially among the older rocks, is highly significant evidence that Earth's crust has been subjected to repeated deformation throughout geologic time. The rocks in the continental crust have been compressed almost continually since the crust was formed. This suggests that the tectonic system has operated during most of the Earth's history.

METAMORPHIC PROCESSES

How can the texture and composition of a rock be changed?

Statement

Metamorphism is a series of changes in the texture and composition of a rock that result from environmental conditions different from those under which the rock was originally formed. The changes occur in order to restore equilibrium in rocks that have been subjected to a new environment. The principal agents of metamorphism are changes in

1. temperature.
2. pressure.
3. chemistry of pore fluids.

These agents act in combination, so that distinguishing their individual effects in a given rock body is often difficult.

Discussion

The changes resulting from metamorphism occur in the solid rock. Water is present in most rocks as a pore fluid and is an important component of metamorphism, even if it occurs as only a thin film around the grains. The pore fluid helps atoms and ions to be exchanged and rearranged in a new crystal structure and permits the development of new textural features.

Temperature Changes

Heat is the most important factor in metamorphism. As the rock temperature increases, minerals begin to change from the solid state to the liquid state, and the amount of pore fluid in the rocks increases. Below 200° C, only a small amount of fluid is present, and most minerals will remain unchanged for millions of years at that temperature. As the temperature rises, however, the amount of pore fluid in the rocks increases. Chemical reactions become more vigorous. Crystal lattices are broken down and recreated using different combinations of the same ions but in different atomic structures. As a result, new mineral assemblages begin to appear. At temperatures higher than 700° C, additional components of the rock become fluid and fusible. Considerable evidence suggests that a high-temperature fluid phase, approaching the magma stage, exists in such a rock body. Extreme cases of increasing temperature probably involve layers of solid material mixed with layers of fluid, which gives rise to a rock that is transitional between igneous and metamorphic rocks.

Different minerals are in equilibrium at different temperatures. The mineral composition of a rock provides a key to the temperatures at which it formed. In the field, zones of differing mineral assemblages show how temperature once varied on a regional scale. These zones are particularly obvious around igneous intrusions.

Pressure Changes

High pressure within the Earth's crust causes significant changes in the physical properties of many rocks. Under sufficient pressure, a rock becomes plastic and may be deformed as the constituent grains move and rotate, or the grains may be fractured and sheared. This produces a reorientation of the mineral grains and a new rock texture. Perhaps the most obvious sign of directed pressure is the distinct orientation of the grains of minerals such as mica and chlorite.

Chemically Active Fluids

Metamorphic changes may take place without the addition or removal of chemical constituents from the bulk rock. However, recrystallization is generally accompanied by some change in chemical composition—that is, by a loss or gain of ions and atoms. Especially important is the loss of volatiles, particularly water and carbon dioxide. In the metamorphic process, minerals that crystallize at relatively low temperatures begin to melt, providing a fluid medium for the migration of material. If an atom breaks loose from the crystal structure of a mineral and moves to some other place, that atom is essentially in a fluid state. Although the bulk of the rock remains solid during metamorphism, the rising temperature and pressure free many atoms from their crystal structures. These atoms migrate through fluids in the pore spaces and along the margins of grains. Thus, constant interchange of atoms occurs. Original crystals break down, and new crystal structures, which are stable under the new conditions of temperature and pressure, develop.

KINDS OF METAMORPHIC ROCKS

How are the different types of metamorphic rocks distinguished and classified?

Statement

Because of the great variety of original rock types and the variation in the kinds and degrees of metamorphism, many types of metamorphic rocks have been recognized. A simple classification of metamorphic rocks, based on texture and composition, is usually sufficient for beginning students. This classification distinguishes two major groups.

1. Rocks that possess a definite planar texture, called foliation
2. Rocks that lack foliation and have a granular texture

The foliated rocks are further subdivided on the basis of the type of foliation. The major rock names can then be qualified by adjectives describing their chemical and mineralogical compositions.

Discussion

Foliated Rocks

Foliation is a planar element in metamorphic rocks. It may be expressed by (1) closely spaced fractures (slaty cleavage that is due to the parallel arrangement of microscopic platy minerals such as mica), (2) the parallel arrangement of large platy minerals (schistosity), or (3) the alternating layers of different minerals (gneissic layering). The major types of foliated rocks are slates, schists, and gneiss.

Slate. Slate is a very fine-grained metamorphic rock generally produced by the low-grade metamorphism of shale (metamorphism under conditions of relatively low temperature and low pressure). It is characterized by excellent foliation, called slaty cleavage, in which the planar element of the rock is a series of surfaces along which the rock can easily be split (Figure 7.3). Slaty cleavage is produced by the parallel alignment of minute flakes of platy minerals, such as mica, chlorite, and talc. The mineral grains are too small to be obvious without a microscope, but the parallel arrangement of small grains develops innumerable parallel planes of weakness, so the rock can be split into smooth slabs.

Slaty cleavage should not be confused with the bedding planes of the parent rock. It is completely independent of the original (relict) bedding and commonly cuts across the original planes of sedimentary stratification.

Schist. Schist is a foliated rock ranging in texture from medium-grained to coarse-grained. Foliation results from the parallel arrangement of relatively large grains of platy minerals, such as mica, chlorite, talc, and hematite, and is referred to as schistosity. The mineral grains are large enough to be identified with the unaided eye and produce an obvious planar structure because of their overlapping subparallel arrangement (Figure 7.2d). The foliation of schists differs from that of slate mainly in the size of the crystals. The term *schistosity* comes from the Greek *schistos,* meaning "divided" or "divisible." As the name implies, rocks with this type of foliation break readily along the cleavage planes of the parallel platy minerals.

Schists result from a higher intensity of metamorphism than the type that produces slates. They have a variety of parent rock types, including basalt, granite, sandstone, and tuff, and are one of the most abundant metamorphic rock types.

Gneiss. Gneiss is a coarse-grained, granular metamorphic rock in which foliation results from alternating layers of light and dark minerals, or gneissic layering (Figure 7.2a). The composition of most gneiss is similar to that of granite. The major minerals are quartz, K-feldspar, and ferromagnesian minerals. Feldspar commonly is abundant and, together with quartz, constitutes a light-colored (white or pink) layer of interlocking grains. Mica, amphibole, and other iron-rich minerals form dark layers. Gneissic layering is usually highly contorted, and gneiss can fracture across the planes of foliation, as easily as it

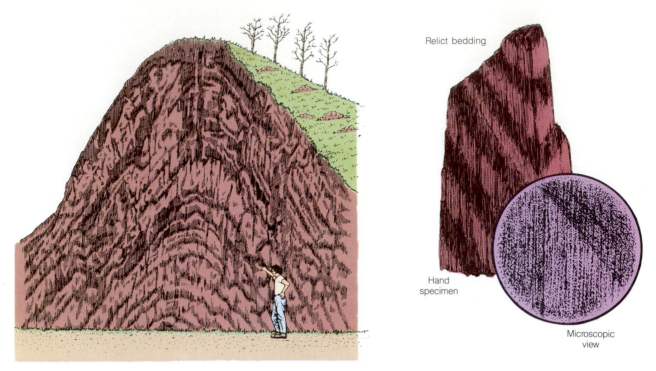

Relict bedding

Hand
specimen

Microscopic
view

Figure 7.3
Slate is a fine-grained metamorphic rock with slaty cleavage. Most of it originates from the metamorphism of shale. Note that slaty cleavage cuts across relict bedding. It is vertical in all views, whereas original bedding surfaces are inclined.

can split along them (see Figure 7.2a). Gneiss forms during regional high-grade metamorphism (metamorphism under conditions of relatively high temperature and high pressure), and in some areas appears to grade into granite.

The Relationship Between Foliation and Larger Structures

The orientation of foliation, therefore, is closely related to the large folds and structural patterns of rocks in the field (Figure 7.3). This relationship commonly extends from the largest folds down to microscopic structures. For instance, the cleavage in slate is generally oriented parallel to the axial planes of the folds, which can be many kilometers apart. A slice of the rock viewed under a microscope shows small wrinkles and folds, having the same orientation as the larger structures mapped in the field.

Nonfoliated Rocks

Rocks such as sandstone and limestone are composed predominantly of one mineral that crystallizes in an equidimensional form. Metamorphism of these rocks does not result in strong foliation, although mica grains scattered through the rock can assume a parallel orientation. The minerals in nonfoliated rock can be flattened, stretched, and elongated and can

show a preferred orientation; but the mass of rock does not develop strong foliation. The resulting texture is best described as granular, or simply nonfoliated.

Quartzite. Quartzite is a metamorphosed, quartz-rich sandstone (Figure 7.2f). It is nonfoliated because quartz grains, the principal constituents, do not form platy crystals. The individual grains commonly are deformed and fused into a tight mass, and thus the rock breaks across the grains as easily as it breaks around them. Pure quartzite is white or light colored, but iron oxide and other minerals often impart various tones of red, brown, green, and other colors.

Marble. Marble is metamorphosed limestone or dolomite. Calcite, the major constituent of the parent rocks, is equidimensional, so the rock is nonfoliated (Figure 7.2e). The grains are commonly large and compactly interlocked, forming a dense rock. Many marbles show bands or streaks resulting from organic matter or other impurities in the original sedimentary rock.

Amphibolite. Amphibolite is a coarse-grained metamorphic rock composed chiefly of amphibole and plagioclase. Mica, quartz, garnet, and epidote also can be present. Amphibolites result from the metamorphism of basalt, gabbro, and other rocks that are rich in iron and magnesium. Some amphibolites develop foliation if mica or other platy minerals are abundant.

Metaconglomerate.   Metaconglomerate is not an abundant metamorphic rock. It is important in some areas, however, and illustrates the degree to which a rock can be deformed in the solid state. Under high pressure, individual pebbles are stretched into a mass that shows distinctive linear fabric (Figure 7.2b).

Hornfels.   A hornfels is a fine-grained, nonfoliated metamorphic rock that is very hard and dense. The grains usually are microscopic and are welded into a regular mosaic. Platy minerals, such as mica, have a random orientation, and grains of high-temperature minerals are present. Hornfels usually are dark colored, and they may resemble basalt, dark chert (flint), or dark, fine-grained limestone. They result from metamorphism around igneous intrusions, which causes partial or complete recrystallization of the surrounding rock. The parent rock usually is shale, although lava, schists, and other rocks can be baked into a hornfels.

Source Material for Metamorphic Rocks

The origin of metamorphic rocks is a complicated process and presents some challenging problems to interpret. A single-source rock can be changed into a variety of metamorphic rocks, depending on the intensity or the degree of metamorphism. For example, shale can be changed to slate, schist, or gneiss (Figure 7.4). Gneiss can also form from many rocks, such as granite or rhyolite. The chart in Figure 7.4, which

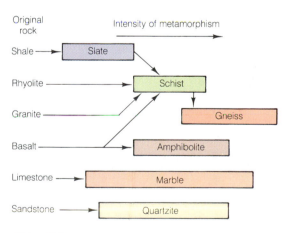

Figure 7.4
The origin of common metamorphic rocks is complex. In some cases, such as quartzite, marble, and metaconglomerate (not shown in diagram), the nature of the original rocks is easily determined. In other cases, such as schist and gneiss, it is difficult and sometimes impossible to determine the type of source rock. This diagram is a simplified flowchart showing the origin of some of the common metamorphic rocks.

relates source rocks and metamorphic conditions to metamorphic rock types, gives a generalized picture of the origin of common metamorphic rocks.

METAMORPHIC ZONES

What features in a rock indicate zones of different degrees of metamorphism?

Statement

Regional metamorphism involves large-scale changes in thick masses of rock in which major recrystallization and structural adjustments occur. These changes are not random but are systematic and generally occur in a well-defined, predictable sequence, controlled by increasing heat and pressure. Regional metamorphic rocks therefore commonly show metamorphic zones that reflect the differences in temperature and pressure.

Discussion

One type of metamorphic zonation can be defined on the basis of the occurrence of certain index minerals (minerals that form over a narrow range of temperature, thus characterizing a particular degree of metamorphism). In the metamorphism of a thick sequence of shale, a typical sequence of index minerals indicating a transition from low-grade to high-grade metamorphism would be chlorite, biotite, almandite, staurolite, kyanite, and sillimanite.

Another type of metamorphic zonation is defined on the basis of a group of associated rocks; each sequence is characterized by a definite set of minerals formed under specific metamorphic conditions. The distinctive group of rocks is called a metamorphic facies and is named after the characteristic rock type or mineral.

By mapping zones of index minerals or the extent of metamorphic facies, geologists can locate the central and marginal parts of ancient mountain belts and interpret something about ancient interactions between tectonic plates. Figure 7.5 is a diagram of the progressive change in associations of metamorphic minerals. This schematic representation shows a series of zones of increasing metamorphic grades produced in the metamorphism of shale. A metamorphic grade reflects the extent or degree of metamorphism. For example, the conversion of shale to slate would be low-grade metamorphism, whereas continued and more intense metamorphism would form a schist, or a higher grade of metamorphism. The higher the degree of metamorphism, the greater the amount of change in the rock.

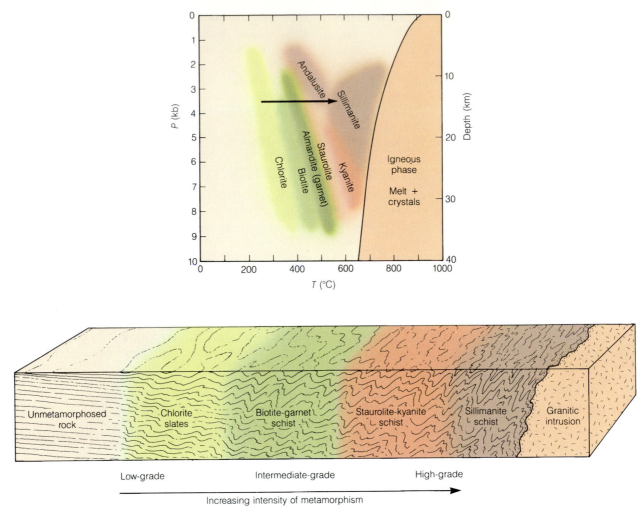

Figure 7.5
The grades of regional metamorphism are related to temperature and pressure. The arrow in diagram (A) shows a typical change from lower to higher grades at a given depth. The sequence of index minerals will commonly be chlorite, biotite, almandite (garnet), staurolite, kyanite, and sillimanite. Diagram (B) is a cross section showing the zones of different grades of metamorphism surrounding a granitic intrusion.

METAMORPHIC ROCKS
AND PLATE TECTONICS

What is the tectonic setting where metamorphic rocks are formed?
What does metamorphism tell us about the mobility of the Earth's crust?

Statement

Regional metamorphism is the most significant type of metamorphism. It results from stresses at converging plate margins and is associated with mountain building, and occurs over large regions. Contact met-amorphism, which is of local extent, occurs around igneous intrusions.

Discussion

Figure 7.6 is a graphic model summarizing some of the major ideas concerning the origin of regional metamorphic rocks. According to the theory of plate tectonics, pressure results from the converging plates, heat is generated in the subduction zone, and shearing occurs where the plates slide past each other along fracture systems and in the subduction zone. Metamorphism is thus best developed in the deep roots of folded mountain belts, which form at convergent plate boundaries. The original material, prior to met-amorphism, may be sediment derived from erosion of a continent, sediment and volcanic material derived from a volcanic arc, or deep-marine sediments and basalt from the oceanic crust. This material is squeezed at convergent plate boundaries as though it were in a vise. Recrystallization tends to produce high-angle or vertical foliation in a linear belt parallel to the margins of the converging plates. Metamorphism theoretically is most intense in the deep mountain roots, where partial melting contributes material to the rising magma generated in the subduction zone. Batholiths and dikes are thus intimately associated with zones of intense metamorphism. Different groups of metamorphic rocks are generated from different source materials—sand, shale, and limestones along continental margins, volcanic sediments and flows along island arcs, and a mixture of deep-marine sediments and oceanic basalt from the oceanic crust.

Close to the subduction zone, high pressure from the converging plates dominates the metamorphic processes. Sediments that have accumulated on the seafloor, together with fragments of oceanic crust,

may be scraped off the descending plate and crushed in a chaotic mass of deep-sea sediment, oceanic basalt, and other rock types; and some material may be derived from the upper plate. This jumbled association of rock is called melange (a heterogeneous mixture). Farther away from the subduction zone, in the mountain root, high-temperature and high-pressure metamorphism occur (Figure 7.6). Metamorphism can also be produced by shearing along fracture zones.

After the stresses from the converging plates are spent, erosion of the mountain belt occurs, and the mountain roots rise because of isostasy. Ultimately, the deep roots and their complex of metamorphic rocks are exposed at the surface, forming a new segment of continental crust. The entire process takes several hundred million years. Repetition of this process causes the continents to grow larger with each mountain-building event. The belts of metamorphic rocks in the shields are thus considered to be the record of ancient continental collisions (see Figure 7.1).

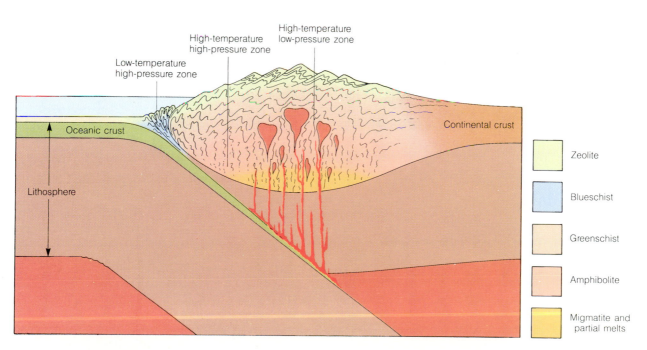

Figure 7.6
The origin of metamorphic rocks at convergent plate margins involves changes in pressure and temperature as a result of the collision of plates plus the chemical action of fluids generated in the subduction zone. Relatively low temperatures and high pressures act on marine sediments and oceanic crust near the continental margin. These conditions produce blue schists. In deep mountain roots, high temperatures and high pressures occur and develop schists and gneiss, which are typical of the continental shields. Contact metamorphism develops around the margins of igneous intrusions.

SUMMARY

1. Metamorphic rocks result mainly from changes in temperature and pressure and changes in the chemistry of pore fluids within a rock body. These changes develop new minerals, new textures, and new structures. The diagnostic features of the original sedimentary and igneous rocks are greatly modified or obliterated.

2. Metamorphic rocks are especially significant because they constitute a large part of the continental crust and indicate that the continents have been mobile and dynamic throughout most of geologic time. They constitute an important record of ancient plate movements.

3. During metamorphism, new minerals grow in the direction of least stress, so rocks are produced with a planar element called foliation.

4. Three major types of foliation are recognized: (a) slaty cleavage, (b) schistosity, and (c) gneissic layering. Rocks originally composed of one mineral (such as limestone and sandstone) do not develop strong foliation. Instead, they develop a granular texture with large mineral grains.

5. Metamorphic rocks are classified on the basis of textures and composition. Two major groups are recognized: (a) foliated and (b) nonfoliated.

6. The major types of foliated rocks are slate, schist, and gneiss. The major types of nonfoliated rocks are quartzite, marble, amphibolite, metaconglomerate, and hornfels.

7. Heat is one of the most important factors in metamorphism. As temperatures increase, the original minerals begin to change from solid to liquid. The pore fluid then recrystallizes to form new minerals.

8. Directed pressures reduce pore spaces and can produce new minerals whose atomic particles are packed closer together.

9. Changes in chemistry can result from the partial melting of low-temperature minerals and the migration of the fluid, which subsequently recrystallizes.

10. An increase in heat and pressure can result from deep burial or from igneous intrusions.

11. Most metamorphic rocks of regional extent, however, develop deep in folded mountain belts as a result of plate collision. The metamorphic mountain roots are welded to the continent and become part of the continental crust.

12. Contact metamorphism occurs along the contacts of igneous intrusion. It results mostly from heat and is of local extent.

KEY TERMS

amphibolite (p. 108)
foliation (p. 107)
gneiss (p. 107)
gneissic layering (p. 107)
high-grade
   metamorphism (p. 108)

hornfels (p. 109)
low-grade
   metamorphism (p. 107)
marble (p. 108)
metaconglomerate (p. 109)
metamorphic rock (p. 104)

metamorphism (p. 106)
plastic
   deformation (p. 104)
pore fluid (p. 106)
quartzite (p. 108)

schist (p. 107)
schistosity (p. 107)
slate (p. 107)
slaty cleavage (p. 107)

REVIEW QUESTIONS

1. Compare and contrast the characteristics of metamorphic rocks with those of igneous and sedimentary rocks.
2. Make a series of sketches showing the changes in texture that occur with a metamorphism of (a) slate, (b) sandstone, and (c) conglomerate.
3. Define foliation and explain the characteristics of (a) slaty cleavage, (b) schistosity, and (c) gneissic layering.
4. Describe the major types of metamorphic rocks.
5. Make a generalized flowchart showing the origin of the common metamorphic rocks.
6. What are the agents of metamorphism?
7. Draw an idealized diagram of converging plates to illustrate the origin of regional metamorphic rocks.
8. What type of metamorphic rock would result if zeolite rocks were subjected to temperatures of about 800° C at a depth of 15 km?

Multiple-Choice Questions

1. The texture of metamorphic rocks shows
   a. that minerals crystallized in a sequence develop interlocking grains.
   b. rounded and abraded grains.
   c. grains that have been recrystallized and commonly have a preferred orientation.
   d. a porphyritic texture.
   e. a clastic texture.
2. The texture exhibited by slate, schist, and gneiss is called
   a. alignment.
   b. cleavage.
   c. foliation.
   d. fracture.
   e. stratification.
3. Which of the following metamorphic rocks represents the highest intensity of metamorphism?
   a. slate
   b. schist
   c. gneiss
   d. marble
   e. quartzite
4. Which of the following forms by the metamorphism of sandstone?
   a. amphibolite
   b. marble
   c. gneiss
   d. quartzite
   e. slate
5. Exposures of metamorphic rocks are most widespread in
   a. young mountains.
   b. stable platforms.
   c. shields.
   d. oceanic islands.
   e. plateaus.
6. The process of metamorphism may involve
   a. the growth of new minerals.
   b. the development of a new rock texture.
   c. recrystallization.
   d. all of the above.
   e. none of the above.

7. Slaty cleavage results from
   a. parallel alignment of minute plates of mica, chlorite, or talc.
   b. parallel arrangement of large plates of mica.
   c. alternating layers of light and dark minerals.
   d. growth of large equidimensional crystals.
   e. deformation of mineral grains.
8. Marble commonly forms from the metamorphism of
   a. sandstone.
   b. basalt.
   c. limestone.
   d. shale.
   e. all of the above.
9. A foliated metamorphic rock with alternating layers of light and dark minerals is called a
   a. schist.
   b. slate.
   c. marble.
   d. gneiss.
   e. quartzite.
10. The bulk of metamorphic rock is believed to have formed in the
    a. vicinity of volcanoes.
    b. oceanic crust.
    c. zones surrounding laccoliths.
    d. roots of folded mountain belts.
    e. sedimentary rock covering stable platforms.

True/False Questions

1. *T F* Metamorphism obscures or obliterates most structural and textural features of the original rock, such as stratification, vesicles, and porphyritic texture.
2. *T F* Metamorphic rocks can be formed from igneous or sedimentary rocks, but not from previously metamorphosed rocks.
3. *T F* Quartzite is a foliated, metamorphosed, quartz-rich sandstone.
4. *T F* The mineral composition of a metamorphic rock provides a key to the temperatures at which the rock formed.
5. *T F* Rocks comprised of one mineral (such as limestone or sandstone) commonly develop marked foliation.

Fill in the Missing Terms

1. _____ metamorphism is of local extent, and is restricted to zones around igneous intrusions.
2. A fine-grained metamorphic rock produced by the low-grade metamorphism of shale is called _____.
3. _____ is a medium-to-coarse-grained metamorphic rock in which foliation results from the parallel arrangement of relatively large platy minerals such as mica, chlorite, talc, and hematite.
4. The parallel alignment of minute flakes of platy minerals produces _____ cleavage.
5. _____ is a coarse-grained, granular metamorphic rock in which foliation results from alternating layers of light and dark minerals.

8

GEOLOGIC TIME

Some sciences deal with incredibly large numbers, others with great distances, still others with infinitesimally small particles. In every field of science, students must expand their conceptions of reality—a sometimes difficult, but very rewarding, adjustment to make.

Geology students must expand their conceptions of the duration of time. Because our lifetimes are short, we tend to think twenty or fifty years is a long time. A hundred years in most frames of reference is a very long time, yet in studying Earth and the processes that operate on it, you must attempt to comprehend time spans of one million years, 100 million years, and even several billion years. How do scientists measure such long periods of time? Nature contains many types of time-measuring devices. Earth itself acts like a clock, rotating on its axis once every 24 hours.

Furthermore, rocks are records of time, and from their interrelationships, the events of Earth's history can be arranged in proper chronological order. Fossils within rocks constitute a separate organic clock by which geologists can "tell time" and identify synchronous events in Earth's history. Rocks also contain radioactive clocks, which permit us to measure with remarkable accuracy the numbers of years that have passed since the minerals forming the rocks crystallized.

In this chapter we will consider the fascinating subject of geologic time, believed by many to be the most significant contribution geology has made to modern thought.

MAJOR CONCEPTS

1. The interpretation of past events in the Earth's history is based on the principle of uniformitarianism (the laws of nature do not change with time).
2. Relative dating (determining the chronologic order of a sequence of events) is achieved by applying the principles of (a) superposition, (b) faunal succession, (c) crosscutting relations, and (d) inclusion.
3. The standard geologic column was established from studies of the rock sequence in Europe. This system is now used worldwide. Rocks are correlated from different parts of the world on the basis of the fossils they contain.
4. Absolute time designates a specific duration of time in units of hours, days, or years. In geology, long periods of absolute time can be measured by radiometric dating.

THE DISCOVERY OF TIME

Why are rocks considered to be records of time?

Statement

Time was discovered in Edinburgh in the 1770s by a small group of scholars led by James Hutton. These men challenged the conventional thinking of their day, in which the largest unit of time was the human life span (the lives of the patriarchs), and the age of Earth was accepted to be 6,000 years, as established by Bishop Ussher's summation of biblical chronology. Hutton and his friends studied the rocks along the Scottish coast and observed that every formation, no matter how old, was the product of erosion from other rocks, older still. Their discovery showed that the roots of time were far deeper than anyone had supposed. Hutton's discovery of time was based on the interpretation of rocks as products of events in Earth's history. It was perhaps the most significant discovery of the eighteenth century because it changed forever the way we look at Earth, the planets, the stars, and, as a consequence, the way we look at ourselves.

The interpretation of rocks as products and records of events in Earth's history is based on the principle of uniformitarianism, which states that the laws of nature do not change with time. It is one of the fundamental assumptions of scientific inquiry.

Discussion

Hutton's principle of uniformitarianism was radical for the time and was slow to be accepted. In the late eighteenth century, before modern geology had developed, the Western world's prevailing view of Earth's origin and history was derived from the biblical account of creation. Earth was believed to have been created in six days and to be approximately 6,000 years old. Creation in so short a time was thought to have involved forces of tremendous violence, surpassing anything experienced in nature. This type of creation theory was called catastrophism. Foremost among its proponents in the eighteenth century was Baron Georges Cuvier (1769–1832), a noted French naturalist. Cuvier, an able student of fossils, concluded that each fossil species was unique to a given sequence of rocks. He cited this discovery in support of the theory that each fossil species resulted from a special creation and was subsequently destroyed by a catastrophic event.

This theory was generally supported by scholars until 1785, when it was challenged by James Hutton (1762–1797). He saw evidence that Earth had evolved by uniform, gradual processes over an immense span of time, and he developed a concept that became known as the principle of uniformitarianism. According to this principle, past geologic events can be explained by natural processes we observe operating today, such as erosion by running water, volcanism, and the gradual uplift of the Earth's crust. Hutton assumed that these processes occurred in the distant past just as they occur now. He saw that in the vast abyss of time enormous work could be achieved by what appeared to be small and insignificant processes. Rivers could completely erase a mountain range. Volcanism and earth movements could form new ones. On the basis of his observations of the rocks of Great Britain, he visualized "no vestige of a beginning—no prospect of an end." In a way, what Copernicus did for space, Hutton did for time. The universe does not revolve around Earth, and time is not measured by the human life span. Before Hutton, human history was all of history. Since Hutton, we know that we are but a tiny pinpoint on an extraordinarily long time line.

Modern Views of Uniformitarianism (Naturalism)

With the help of modern scientific instruments, geologists have studied much more of the geologic record than did Hutton, and they have observed and measured many subtle details that earlier scientists could not measure. Modern science is making significant advances in understanding Earth, its long history, and how it was formed. By applying principles of thermodynamics, electromagnetism, chemistry, and related scientific disciplines, geologists are discovering more clues about Earth's genesis and evolution.

The assumptions of constancy in natural law are not unique to the interpretations of geologic history; they constitute the logical essentials in deciphering recorded history, as well. We observe only the present and interpret past events on inferences based on present observations. We thus conclude that books or other records of history—such as fragments of pottery, cuneiform tablets, flint tools, temples, and pyramids—which were in existence prior to our arrival, have all been the works of human beings, despite the fact that postulated past activities have been outside the domain of any possible present-day observations. Having excluded supernaturalism, we draw these conclusions because humankind is the only known agent capable of producing the effects observed. Similarly, in geology we conclude that ripple marks in a sandstone formation in the folded Appalachian Mountains were in fact formed by currents or wave action, or that coral shells found in limestones exposed in the high Rocky Mountains are indeed the skeletons of corals that lived in a now nonexistent sea.

Many features of rocks serve as records or documents of past events in Earth's history. Igneous rocks are records of thermal events; the texture and composition of an igneous rock indicate whether volcanic eruptions occurred or the magma cooled beneath the surface (see page 69). Sedimentary rocks record changing environments on Earth's surface—the rise and fall of sea level, changes in climate, and changes in life forms. A layer of coal is a record of the lush growth of vegetation, commonly in a swamp. Limestone composed of fossil-shell debris indicates the former existence of a shallow sea. Salt is precipitated from seawater or from salt lakes only in a dry climate, so a layer of salt carries specific information about past climatic conditions. The list of examples could go on and on. For more than two centuries, geologists have extracted from the rocks a remarkably consistent record of events in Earth's history: a record of time.

UNCONFORMITIES

What are conformities?
Why are they significant in the study of geologic history?

Statement

Geologic time is continuous; it has no gaps. But the information on which it is based comes from a rock record that is discontinuous. In any sequence of rocks, there are many major discontinuities that indicate significant interruptions in the rock-forming processes. If sedimentation stops in an area that is subsequently subjected to erosion, an unconformity (a physical discontinuity in the succession of strata) is produced. This is of paramount importance in the interpretation of geologic events.

Discussion

James Hutton was a very perceptive observer who clearly recognized the historical implications of the relationships between rock bodies. He not only recognized the vastness of time recorded in the rocks of Earth's crust, but he also recognized breaks, or gaps, in the record. In 1788 Hutton, together with Sir James Hall and John Playfair, visited Siccar Point in Berwickshire, Scotland, and saw for the first time the Old Red Sandstone resting upon the upturned edges of the older strata (Figure 8.1). This exposure proved that the older rocks, called Primary Strata, had been uplifted, deformed, and partly eroded before the deposition of the "secondary strata." They soon

Figure 8.1
Angular unconformity at Siccar Point, southeastern Scotland. It was here that the historical significance of an unconformity was first realized by James Hutton in 1788.

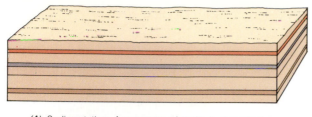

(A) Sedimentation: A sequence of rocks is deposited over time.

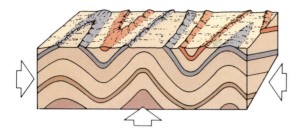

(B) Deformation: The sequence of rocks is deformed by mountain-building processes or by broad upwarps in the Earth's crust.

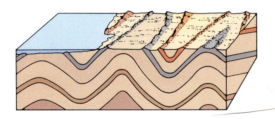

(C) Erosion: The deformed rocks are eroded and part of the sequence is removed.

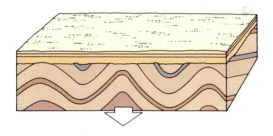

(D) Renewed sedimentation: A new sequence of rocks is deposited on the eroded surface of the older deformed rocks.

Figure 8.2
The geologic events implied from an angular unconformity represent a sequence of major events in the geologic processes operating within the area.

discovered comparable relationships in other parts of Great Britain. This relationship between rock bodies became known as an angular unconformity.

To appreciate the significance of an angular unconformity, consider what the angular discordance implies by studying the sequence of diagrams in Figure 8.2. At least four major events are involved in the development of an angular unconformity: (1) an ini-

tial period of sedimentation during which the older strata are deposited in a near-horizontal position, (2) a subsequent period of deformation during which the first sedimentary sequence is folded, (3) development of an erosional surface on the folded sequence of rock, and (4) a period of renewed sedimentation and the development of a younger sequence of sedimentary rocks on the old erosional surface.

RELATIVE DATING

How is geologic time measured?

Statement

Two different concepts of time, and hence two different, but complementary, methods of dating, are used in geology: relative time and absolute time.

Relative dating is simply determining the chronologic order of a sequence of events. No quantitative or absolute number of days is deduced, only that one event occurred earlier or later than another. Absolute time, in contrast, designates a specific duration in units of hours, days, years, etc.

In studying Earth, relative dating is important because many physical events such as volcanism, canyon cutting, the deposition of sediment, or the upwarping of Earth's crust can be identified. To establish the relative age of these events is to determine their proper chronologic order. This can be done by applying several principles of remarkable simplicity and universality. The most significant of these are:

1. The principle of superposition
2. The principle of faunal succession
3. The principle of crosscutting relations
4. The principle of inclusion

Discussion

The Principle of Superposition

The principle of superposition is the most basic guide in the relative dating of rock bodies. It states that in a sequence of undeformed sedimentary rock, the oldest beds are on the bottom and the higher layers are successively younger. The relative ages of rocks in a sequence of sedimentary beds can thus be determined from the order in which they were deposited.

In applying the principle of superposition, we make two assumptions: (1) layers were essentially horizontal when they were deposited, and (2) the rocks have not been so severely deformed that the beds are overturned. (Rock sequences that have been overturned are generally easy to recognize by their sedimentary structures, such as cross bedding, ripple marks, and mud cracks.)

The Principle of Faunal Succession

In addition to superposition, the sequence of sedimentary rocks in Earth's crust contains another independent element that can be used to establish the chronologic order of events: the changes in groups of fossils contained in the rocks. Fossils are the actual remains of ancient organisms, such as bones and shells, or the evidence of their presence, such as trails and tracks. Their abundance and diversity are truly amazing. Some rocks (such as coal, chalk, and certain limestones) are composed almost entirely of fossils, and others contain literally millions of specimens. Invertebrate marine forms are most common, but even large vertebrate fossils of mammals and reptiles are plentiful in many formations. For example, it is estimated that more than 50,000 fossil mammoths have been discovered in Siberia, and many more remain buried.

Even before Darwin developed the theory of natural selection, the principle of faunal succession was recognized by William Smith (1769–1839), a British surveyor. Smith worked throughout much of southern England and carefully studied the fresh exposures of rocks in quarries, road cuts, and excavations. In a succession of interbedded sandstone and shale formations, he noted that although the several shales were very much alike, the fossils they contained were not. Each shale had its own particular groups of fossils. By correlating types of fossils with rock sequences, Smith developed a practical tool that enabled him to predict the location and properties of rocks beneath the surface.

Fossils provide geologists with a means of establishing relative dates, in much the same way that archaeologists use artifacts. Both show evolution and change with time. For example, in a city dump where refuse is buried in succession, we could recognize a period of time prior to the automobile by the remains of wagon wheels, saddles, and similar equipment. A layer containing abundant scraps of Model-T Fords would be recognized as being older than a layer containing remains of Ford Mustang, and layers containing new models such as the Porsche would be recognized as being younger, even though they might not rest on layers containing any of the older materials.

Today, the principle of faunal succession has been confirmed beyond doubt. It has been used extensively to locate valuable natural resources, such as petroleum and mineral deposits. It is also the foundation for the standard geologic column, which divides geologic time into eras, periods, epochs, and ages (see page 122).

The Principle of Crosscutting Relations

The principle of crosscutting relations states that faults or igneous intrusions or other rock bodies are younger than the rocks they cut across or intrude into (Figure 8.3). Faults (surfaces along which a rock body has been fractured and displaced) and igneous intrusions are obviously younger than the rocks they cut. Dikes, sills, stocks, and batholiths are all younger than the rocks in which they have been emplaced. Lava flows are younger than the rocks they cover. If the flows are cut by a dike, then the dike is younger than the flow. Crosscutting relations can be complex, however, and careful observation may be required to establish the correct sequence of events. The scale of crosscutting features is highly variable, ranging from large faults with displacements of hundreds of kilometers to small fractures less than a millimeter long (Figure 8.3).

The Principle of Inclusion

The relative age of intrusive igneous rocks (with respect to the surrounding rock) is commonly apparent from inclusions, or fragments of older rocks, in the younger rocks (Figure 8.4). As a magma moves upward through the crust, it dislodges and engulfs large fragments of the surrounding material, which remain as unmelted foreign inclusions.

The principle of inclusion states that a rock body is younger than the fragments included or incorporated within it. The principle can also be applied to conglomerates, in which relatively large pebbles and boulders that have been eroded from preexisting rocks have been transported and deposited in a new formation. The conglomerate is obviously younger than the formations from which the pebbles and cobbles were derived. In areas where superposition or other methods do not indicate relative ages, a limit to the age of a conglomerate can be determined from the rock formation represented in its pebbles and cobbles.

Succession in Landscape Development

Surface features of Earth's crust are continually being modified by erosion and commonly show the effects of successive events through time. Many landforms evolve through a definite series of stages, so that the relative age of a feature can be determined from the degree of erosion. This is especially obvious in volcanic features such as cinder cones and lava flows. These features are created during a period of volcanic activity and then subjected to the forces of erosion until they are completely destroyed or buried by erosional debris.

Figure 8.3
Crosscutting relationships clearly indicate the relative age of rock bodies and geologic structure. In this photograph several generations of dikes cut across the green metamorphic rock. The thick dike is the youngest because it cuts across all other rock bodies.

Figure 8.4
Inclusions of one rock in another provide a means of determining relative age. In this example, fragments of granite are included in the basalt, clearly indicating that the granite is the oldest.

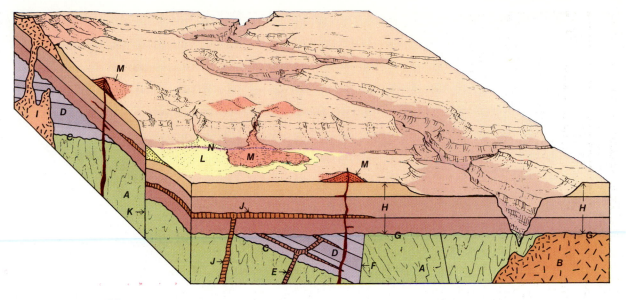

Figure 8.5
The sequence of geologic events can be determined using the principles of superposition and crosscutting relations.

Example

The composite diagram in Figure 8.5 shows several kinds of crosscutting relationships, as well as several unconformities and the superposition of various rock bodies. While this is a hypothetical situation, these types of relationships are found in the Grand Canyon area of the Colorado Plateau. Although this diagram covers a large area, the relationships between rock bodies are in canyons, valley walls, and on the plateau surface. The major rock bodies, faults, and unconformities are labeled by letters arranged in alphabetical order from oldest to youngest.

A Metamorphic rocks
B Granite
C Unconformity
D Tilted shale
E Injection into tilted strata
F Fault displacing tilted strata
G Unconformity
H Horizontal strata
I Laccolithic intrusion
J Basaltic dikes and sills
K Major faulting
L Fans
M Volcanism
N Recurrent movement

The oldest rocks in the diagram are the metamorphic rocks, A; the granite, B, intrudes into these rocks and is younger; but the granite is not in contact with the tilted strata, D, so their age relationship is not certain. An erosional surface, C, developed on the metamor-

phic terrain, and then a sequence of sedimentary rocks, D, was deposited. These rocks were then intruded by dikes and sills, E. In order to establish the relative age of the granite, B, more accurately, we would need some absolute dates based on radiometric age determination for units B and E. Faults, F, displaced the sequence, D. Widespread erosion then occurred, developing the unconformity, G, which cuts across all of units A–F. The sequence of horizontal rocks, H, was then deposited. Two igneous intrusions, I and J, occurred. Intrusion I formed a laccolith, whereas J formed a dike and sill. Because we cannot tell the relative age relationship of these intrusions, we would have to obtain radiometric dates to place these events more accurately in the sequence. We do know, from crosscutting relationships, that intrusion I is older than the fault, K, and the volcanics, M.

The next group of events—erosion, volcanic eruptions, and sedimentation—has surface expressions. Judging from the deep canyon, erosion appears to have been initiated relatively early. The surface upon which alluvial fans, volcanic cones, and lava flows were formed is related to the erosion by the major river. Lava flow, M, is younger than the alluvial fan, L. Both are cut by recurrent movement on fault K. Note the amount of displacement along the fault of the sedimentary rocks, H, and the small amount of displacement of the fan, L, and lava flow, M. Judging from the lack of erosion on the volcanoes, it would appear that the cones are very young features.

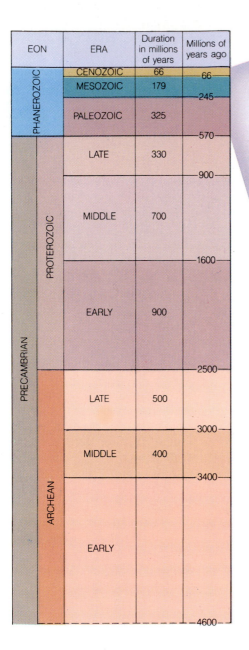

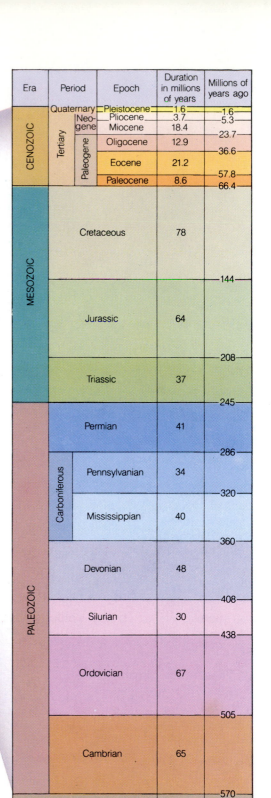

Figure 8.6
The standard geologic column was developed in Europe during the mid-nineteenth century on the basis of the principles of superposition and faunal succession. Later, radiometric dates provided a scale of absolute time for the standard geologic periods.

The Standard Geologic Column

Using the principles of superposition and faunal succession, geologists have determined the chronologic sequence of rocks throughout broad regions and

have constructed a standard geologic time scale (Figure 8.6). Most of the original geologic time scale was pieced together from sequences of strata studied in Europe during the mid-nineteenth century. Major units of rock (such as the Cambrian, Ordovician, and

Silurian) generally were named after a geographic area in Europe where they are well exposed. The rock units are distinguished from each other by major changes in rock type, unconformities, or abrupt vertical changes in the fossil groups they contain. The original subdivision of the geologic column was based simply on the sequence of rock formations in their superposed order as they are found in Europe. Rocks in other areas of the world that contain the same groups of fossils as a given part of the European succession are considered to be of the same age and commonly are referred to by the same names.

The geologic column by itself indicates only the relative ages of the major periods in Earth's history. It tells us nothing about the specific duration of time represented by a period. With the discovery of the radioactive decay of uranium and other elements, a new tool for measuring geologic time became available. It greatly enhanced our understanding of time and the history of Earth and provided benchmarks of absolute time for the standard geologic column.

RADIOMETRIC MEASUREMENTS OF ABSOLUTE TIME

How do we measure the magnitude of geologic time?

Statement

Unlike relative time, which specifies only the chronologic relationships among events, absolute time, or finite time, geologic time designates specific durations measured in units of hours, days, or years. One very useful natural geologic clock measures time by the process of radioactive decay. In this process, atoms of certain elements lose particles from their nuclei and thus change into atoms of other elements. Because the rate at which these elements decay is unaffected by conditions such as pressure, temperature, and chemical binding forces, it can be used as a very precise and accurate measure of geologic time. The time that has elapsed since a radioactive element was locked into the crystal of a mineral can be determined if the rate of decay is known, and the proportions of the original element and the decayed product can then be measured. The most important clocks for geologic studies are the radioactive isotopes of uranium, thorium, rubidium, potassium, and carbon.

Discussion

When Henri Becquerel (1852–1908), a French physicist, discovered natural radioactivity in 1896, he opened new vistas in every field of science. Among the first to experiment with radioactive substances was the distinguished British physicist, Lord Rutherford (1871–1937). After defining the structure of the atom, Rutherford made the first clear suggestion that radioactive decay could be used to date geologic events in absolute time.

Radioactive isotopes are unstable: Their nuclei spontaneously disintegrate, transforming them into completely different atoms. In the process, radiation is given off and heat is liberated. Initially, scientists assumed that each radioactive substance disintegrates at its own rate, and that for many substances the rate is extremely slow. This assumption has been proven by experiment.

The rate of radioactive decay is defined in terms of half-life, the time it takes for half of the nuclei in the sample to decay. In one half-life, half of the original atoms decay. In a second half-life, half of the remainder (or a quarter of the original atoms) decay. In a third half-life, half of the remaining quarter decay, and so on (see Figure 8.7). The time elapsed since the formation of a crystal containing a radioactive ele-

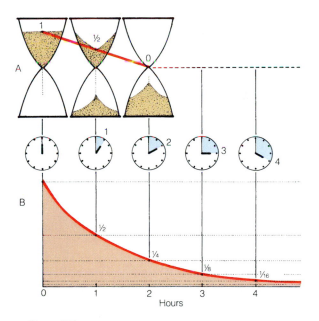

Figure 8.7
Rates of depletion can be linear or exponential. (A) Most processes are characterized by uniform, straight-line depletion, like sand moving through an hourglass. If half of the sand is gone in one hour, all of it will be gone in two hours. (B) Radioactive decay, in contrast, is exponential. If half is depleted in one hour, half of the remainder, or one-fourth, will be depleted in two hours, leaving one-fourth. Rates of radioactive decay are expressed in half-lives, the time required for half of the remaining amount to be depleted. In this case, the half-life is one hour.

Table 8.1 Radioactive Isotopes Used in Determining Geologic Time

Parent Isotope	Daughter Isotope	Half-Life
Uranium-238	Lead-206	4.5 billion years
Uranium-235	Lead-207	713.0 million years
Thorium-232	Lead-208	13.9 billion years
Rubidium-87	Strontium-87	50.0 billion years
Potassium-40	Argon-40	1.5 billion years

ment can be calculated from the rate at which that particular element decays. The amount of the radioactive element remaining in the crystal (parent isotope) is simply compared with the amount of the disintegration product (daughter isotope).

There are numerous radioactive isotopes. Most decay rapidly; that is, they have short half-lives and lose their radioactivity within a few days or years. Some decay very slowly, however, with half-lives of hundreds of millions of years. These can be used as atomic clocks for measuring long periods of time. The parent isotopes and their daughter products that are most useful for geologic dating are listed in Table 8.1.

The theory of radiometric dating is simple enough, but the laboratory procedures are complex. The principal difficulty lies in the precise measurement of minute amounts of isotopes. The accuracy of the method also depends on the accuracy with which the half-life is known. (Measurements of the decay of uranium-235 to lead-207 are considered accurate within 2 percent.)

At present, the potassium-argon method of radiometric dating is of great importance. It can be used on micas and amphiboles, which are widely distributed in igneous rocks, and it can be used both on rocks as young as a few thousand years or on older rocks.

Another important radioactive clock uses the decay of carbon-14 (^{14}C), or radiocarbon, which has a half-life of 5,730 years. Carbon-14 is produced continually in the Earth's atmosphere as a result of the bombardment of nitrogen-14 (^{14}N) by cosmic rays. The newly formed radioactive carbon becomes mixed with ordinary carbon atoms in carbon dioxide gas (CO_2). The formation of radiocarbon is in balance with its decay, so the proportion of ^{14}C in atmospheric CO_2 is essentially stable.

Plants use carbon dioxide in photosynthesis, and animals eat plants. Both thus maintain a fixed proportion of ^{14}C while they are alive. After death, however, no additional ^{14}C can replenish what is lost by radioactive decay. The ^{14}C steadily reverts to ^{14}N. The time elapsed since an organism died can therefore be determined by measuring the remaining proportion of ^{14}C. The longer the time elapsed since death, the less ^{14}C remains. Since the isotope's half-life is 5,730 years, the amount of ^{14}C remaining in

organic matter older than 50,000 years is too small to be measured accurately. The radiocarbon method is therefore useful for dating very young geologic events involving organic matter and for dating more recent archaeologic material.

To safeguard against errors, radiometric dating is subjected to constant checks. A number of radioactive isotopes are suitable for absolute dating, so one obvious test is to date a mineral or rock by more than one method. If the results agree, the probability is high that the date is reliable. If the results differ significantly, additional methods must be used to determine which date, if either, is correct. Another independent check can be made by comparing the absolute date, determined radiometrically, with the relative age of the rocks, determined from such evidence as superposition and fossils. Through this system of tests and cross-checks, many very reliable radiometric dates have been determined.

The Radiometric Time Scale

Absolute dates of numerous geologic events have been determined. These, in combination with the standard geologic column, provide a radiometric time scale from which the absolute age of other geologic events can be estimated. (Note the radiometric time scale on the two right-hand columns in Figure 8.6.)

The presently accepted geologic time scale is based on the standard geologic column, established by faunal succession and superposition, plus the finite radiometric dates of rocks that can be placed precisely in the column. Each dating system provides a crosscheck on the other because one is based on relative time and the other on absolute time. Agreement between the two systems is remarkable, and discrepancies are few. In a sense, the radiometric dates act as the scale on a ruler, providing reference markers between which interpolation can be made. Enough dates have been established so that the time span of each geologic period can be estimated with considerable confidence. The age of a rock can be determined by finding its location in the geologic column and interpolating between the nearest radiometric time marks.

From this radiometric time scale, we can make several general conclusions about the history of Earth and geologic time.

1. Present evidence indicates that the age of the Earth is about 4.5 to 4.6 billion years.
2. The Precambrian constitutes more than 80 percent of geologic time and ended about 570 million years ago.
3. Phanerozoic time (the Paleozoic and later) began about 570 million years ago. Rocks deposited since Precambrian time can be correlated worldwide by means of fossils, and the dates of many

important events during their formation can be determined from radiometric dating.

4. Some major events in the Earth's history are difficult to place in their relative positions on the geologic column but can be dated by radiometric methods.

MAGNITUDE OF GEOLOGIC TIME

How can we conceive of vast periods of time?

Statement

Great time spans are difficult for most people to comprehend. The norms established through sensory experience are short intervals, such as the day, the week, and the changing seasons. Students of geology must continually attempt to enlarge their temporal norms to encompass the magnitude of geologic time. Without an expanded conception of time, extremely slow geologic processes, considered only in terms of human experience, have little meaning.

Discussion

In studying the magnitude of geologic time it might be well to remember the old Chinese parable: we are like insects who live but a day in the forest, so we cannot watch the trees grow. Yet, if we are observant we can find seeds, saplings, mature trees, and fallen logs, and through logic we can reconstruct the life pattern of a tree. The same is true for the geologic history of Earth. Although our planet has existed through billions of human lifetimes, by examining the processes of how rocks are being formed today and how rock bodies occur in the Earth's crust, we can understand how our planet evolved. Creation is a continuing process we can observe.

Still, many of us are reluctant to acknowledge the immense span of time that passed before our existence. Western religions in particular have fostered the belief that Earth is but a few thousand years old. This is indeed unfortunate because when we reduce the scope of the creation, we limit our appreciation for its grandeur. Perhaps the greatest contribution of modern geologic thought is that it has not only revealed a planet of pulsing energy, spectacular beauty, and great mystery, but it has also given us a perspective on our ultimate origins.

In an effort to appreciate the magnitude of geologic time, let us abandon for a moment the large, intimidating numbers and refer instead to something tangible and familiar. In Figure 8.8, the length of a football field represents the lapse of time from the beginning of the Earth's history to the present. An absolute time scale and the standard geologic periods are shown on the left. Precambrian time constitutes the greatest portion of the Earth's history (87 yards). The Paleozoic and later periods are equivalent to only the last 13 yards. To show events with which most people are familiar, the upper end of the scale must be enlarged. The first abundant fossils occur at the 13-yard line. The great coal swamps are at about the 6-yard line. The dinosaurs became extinct about a yard from the goal line, and the Ice Age occurred an inch from the goal line. Recorded history corresponds to less than the width of a blade of grass.

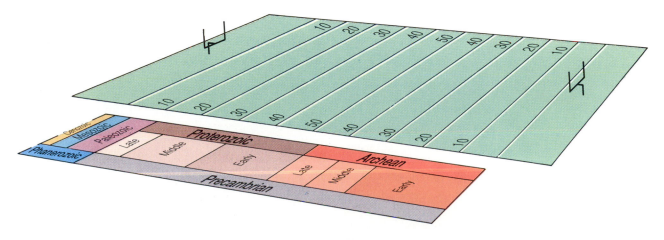

Figure 8.8
If the length of geologic time is compared to a football field, Precambrian time represents the first 87 yards, and all events since the beginning of the Paleozoic are compressed into the last 13 yards. Dinosaurs first appeared 5 yards from the goal line. The glacial epoch occurred in the last inch, and historic time is so short that it cannot be represented, even on the enlarged part of the figure.

SUMMARY

1. The basic assumption endorsed by essentially all geologists today in studying and interpreting the history of Earth is the principle of uniformitarianism, which states that natural laws do not change with time. This principle is not unique to geology. It is a fundamental law in all fields of science.
2. Relative dating determines the chronologic order of a sequence of events. The most important methods of relative dating are (a) superposition, (b) faunal succession, (c) crosscutting relations, and (d) inclusions.
3. The standard geologic column was established during the early and middle 1800s by means of the principles of relative dating.
4. Radiometric dating provides a method for measuring geologic time directly in terms of a specific number of years (absolute dating). It has been used extensively during the last fifty years to provide an absolute time scale for the events in the Earth's history.

KEY TERMS

absolute time (p. 123)

angular
    unconformity (p. 117)

carbon-14 (p. 124)

catastrophism (p. 116)

Cenozoic (p. 122)

crosscutting
    relations (p. 119)

daughter isotope (p. 123)

faunal succession (p. 119)

geologic column (p. 122)

geologic time scale (p. 122)

half-life (p. 223)

inclusion (p. 119)

Mesozoic (p. 122)

Paleozoic (p. 122)

parent isotope (p. 123)

Precambrian (p. 122)

radioactivity (p. 123)

radiocarbon (p. 124)

radiometric dating (p. 123)

relative age (p. 118)

relative dating (p. 118)

superposition (p. 118)

uniformitarianism (p. 116)

REVIEW QUESTIONS

1. Explain the modern concept of uniformitarianism.
2. Explain the concept of relative dating.
3. Explain how the following principles are used in determining the relative age of rock bodies: (a) superposition, (b) faunal succession, (c) crosscutting relations, (d) inclusions.
4. Discuss the sequence of events illustrated in Figure 8.2.
5. What is the standard geologic column? How did it originate?
6. Explain the meaning of half-life in radioactive decay.
7. How is the absolute age of a rock determined?
8. How is the half-life of a radioactive isotope used to determine the radiometric age of a rock?

Multiple-Choice Questions

1. The most accurate expression of the modern view of uniformitarianism is that
   a. the present is the key to the past.
   b. types of geologic processes are uniform, but rates may change.
   c. rates of geologic processes are uniform, but types may change.
   d. natural laws do not change.
   e. past geologic events can be explained only by processes we observe at the present.
2. The concept of uniformitarianism was proposed by
   a. James Hutton.
   b. Charles Darwin.
   c. Charles Lyell.
   d. Georges Cuvier.
   e. Alfred Wegener.
3. Relative dating is determining the
   a. age of a rock in terms of numbers of years.
   b. length of time since a rock formed.
   c. chronological order of a sequence of events.
   d. time span between geologic events.
   e. approximate age of a rock.
4. Which is the most useful in absolute dating?
   a. crosscutting relations
   b. radioactive age dating
   c. inclusions
   d. fossils
   e. superposition
5. Determining the relative age of a rock body is based largely on
   a. half-life measurements.
   b. Carbon-14.
   c. inclusions.
   d. radioactive dates.
   e. superposition.
6. Which of the following applies to the standard geologic time scale?
   a. It was developed through radioactive dating.
   b. It is based on superposition and faunal succession.
   c. It is divided into periods of equal length.
   d. It was developed in North America.
   e. It is divided into periods named after rock types.

7. The most useful absolute dating technique is
   a. varves.
   b. radioactive age dating.
   c. dendrochronology.
   d. fossils.
   e. superposition.
8. Half-life in a radioactive element is
   a. the amount of radioactive material left after a given period of time.
   b. the amount of radioactive material left after one million years.
   c. the amount of C-14 left after half of the uranium had decayed.
   d. the time required for half of the nuclei in the radioactive sample to decay.
   e. the time required for half of the sample to become radioactive.
9. The best geologic evidence indicates that Earth was formed about
   a. 4 million years ago.
   b. 3.5 billion years ago.
   c. 4.5 billion years ago.
   d. 10 billion years ago.
   e. 10 million years ago.
10. The geologic time scale is divided into four main units called eras. The most recent era is the
    a. Precambrian.
    b. Mesozoic.
    c. Cenozoic.
    d. Paleozoic.
    e. Quaternary.

True/False Questions

1. *T F* The principle of uniformitarianism states that the laws of nature do not change with time.
2. *T F* Time is measured by change.
3. *T F* Relative dating implies a quantitative or absolute length of time.
4. *T F* Absolute time, or finite time, designates specific durations measured in units, such as hours, days, or years.
5. *T F* Cambrian refers to the oldest known rocks.

Fill in the Missing Terms

1. The principle of _____ states that the laws of nature do not change with time.
2. The principle of _____ _____ states that groups of fossil plants and animals occur in the geologic record in a definite and determinable order, and that a period of geologic time can be recognized by its respective fossils.
3. The principle of _____ _____ states that igneous intrusions and fractures, such as faults, are younger than the rocks they cut.
4. Rocks formed in _____ time contain only a few fossils of the more primitive life forms.
5. Present evidence indicates that the age of Earth is approximately _____ billion years.

9

WEATHERING

A new building gradually deteriorates. The paint chips and peels, its wood dries and splits, and even the bricks, building stone, and cement eventually decay and crumble. Left alone, most buildings decompose into a pile of rubble within a few hundred years. This process of natural decay is called weathering. Solid bedrock is also subject to weathering and eventually decomposes into piles of rubble. In fact, weathering affects some rocks more than it does the paint on a house.

Weathering is a general term describing all the changes that result from the exposure of rock materials to the atmosphere. Its effects can be seen wherever rocks are exposed. It is seen in the jagged outcrops of rocks, broken by ice wedging. It is seen in piles of loose rock fragments at the base of a cliff and in soil, which covers much of the landscape. Weathering occurs because most rocks are in equilibrium with higher temperatures and pressures deep within Earth. If they are exposed to the much lower temperatures and pressures at the surface, to the gases in the atmosphere, and to the elements in water, they become unstable and undergo various chemical changes and mechanical stresses. As a result, the solid bedrock breaks down into loose, decomposed products.

From a geological point of view, the importance of weathering is that it transforms the solid bedrock into small, decomposed fragments and prepares it for removal by the agents of erosion. It is this weathered rock material that is washed away by river systems in the process of eroding the surface of the land. Without weathering, there would be no erosion as it is known today, and the landscape would be strikingly different. In addition, the products of weathering form a blanket of soil over the solid bedrock, and soil is the basis for most terrestrial life. Therefore, weathering should be considered a part of the geologic system with tremendous ecological significance.

MAJOR CONCEPTS

1. The major types of weathering are mechanical disintegration and chemical decomposition.
2. Ice wedging is the most important form of mechanical weathering.
3. The major types of chemical weathering are oxidation, dissolution, and hydrolysis.
4. Joints facilitate weathering because they permit water and gases in the atmosphere to attack a rock body at considerable depth. They also greatly increase the surface area on which chemical reactions can occur.
5. The major products of weathering are a blanket of soil (regolith) and spheroidal rock forms.
6. Climate greatly influences the type and rate of weathering. The major controlling climatic factors are precipitation and temperature.

MECHANICAL WEATHERING

How does mechanical weathering break down a solid mass of rock into small fragments?

Statement

Mechanical weathering is strictly a physical process, involving no change in the chemical composition of the rock. No chemical elements are added to, or subtracted from, the rock. The rock is simply broken down into small fragments by various physical stresses. The most important types of mechanical weathering are

1. ice wedging—in which freezing water expands in cracks or bedding planes and wedges the rock apart.
2. sheeting, or unloading—in which a series of fractures is produced by expansion of the rock body itself, as a result of the removal of overlying material by erosion.

Discussion

Ice Wedging

Figure 9.1 is a simple diagram showing how ice wedging breaks a rock mass into small fragments. Water from rain or melting snow easily penetrates cracks, bedding planes, and other openings in the rock. As it freezes, it expands about 9 percent, exerting great pressure on the rock walls, similar to the pressure produced by driving a wedge into a crack. Eventually the fractured blocks and bedding planes are pried free from the parent material. The stress generated each time it freezes is approximately 110 kg/cm^2, roughly equivalent to that produced by dropping an 8-kg ball of iron about the size of a large sledgehammer from a height of 3 m. Stress is exerted with each freeze so that, over a period of time, the rock is literally hammered apart.

Ice wedging occurs under the following conditions: (1) when there is an adequate supply of moisture, (2) where preexisting fractures, cracks, or other voids occur within the rock, into which water can enter; and (3) where temperatures frequently rise and fall across the freezing point. Temperature fluctuation above and below the freezing point is especially important because pressure is applied with each freeze. In areas where freezing and thawing occur many times a year, the ice wedging is far more effective than in exceptionally cold areas, where water is permanently frozen. Ice wedging thus occurs most frequently above the timberline. It is especially active on the steep slopes above valley glaciers, where meltwater produced during the warm summer days seeps into cracks and joints and freezes during the night.

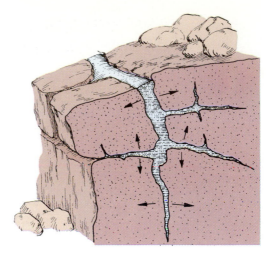

Figure 9.1
Ice wedging occurs when water seeps into fractures and expands as it freezes. The expanding wedge forces the rock apart and produces loose, angular fragments that move downslope by gravity and accumulate at the base of the cliff as talus cones.

Sheeting

Rocks formed deep within Earth's crust are under great confining pressure from the weight of thousands of meters of overlying rocks. As this overlying cover is removed by erosion, the confining pressure is released, and the buried rock body tends to expand. The internal stresses set up by expansion can cause large fractures, or expansion joints, parallel to Earth's surface (Figure 9.2). This result is called sheeting. It can be observed directly in quarries, where the removal of large blocks is sometimes followed by the rapid, almost explosive expansion of the quarry floor. A sheet of rock several centimeters thick may burst up, and at the same time, numerous new parallel fractures will appear deeper in the rock body. The same process occasionally causes rock bursts in mines and tunnels, when the confining pressure is released during the tunneling operation. It can also be seen in many valley walls and in excavations for roads, where rock slumping due to sheeting can cause serious highway problems.

Organic Activity

Animals and plants play a variety of relatively minor roles in mechanical weathering. Burrowing animals, such as rodents, mechanically mix the soil and loose rock particles, a process that facilitates further breakdown by chemical means. Pressure from growing roots widens cracks and contributes to the rock breakdown. Lichens can live on the surface of bare rock and extract nutrients from its minerals by ion exchange. This results in both mechanical and chemical alteration of the minerals. These processes

Figure 9.2
Sheeting in granite of the Sierra Nevada occurs as erosion removes the overlying rock cover and reduces the confining pressure. The bedrock expands and large fractures develop parallel to the surface. The fractures may subsequently be enlarged by frost action.

may seem trivial, but the work of innumerable plants and animals over a long period of time adds significantly to the disintegration of the rock.

Example

The products of mechanical weathering are best seen in the high mountain country, where ice wedging dominates and produces a large volume of angular rock fragments. This material commonly accumulates in a pile at the base of the cliffs from which it was derived. Because most cliffs are notched by steep val-leys and narrow ravines, the fragments dislodged from the high valley walls are funneled through the ravines to the base of the cliff, where they accumulate in cone-shaped deposits called talus cones (Figure 9.3).

The talus cones are built up by isolated blocks loosened by ice wedging. The blocks commonly fall separately, as almost any mountain climber can testify, but large masses of the material on steep slopes may be moved by an avalanche. Earthquakes may also suddenly activate a magnitude of blocks loosened by many seasons of ice wedging.

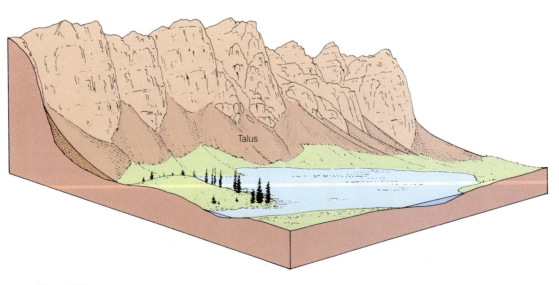

Talus

Figure 9.3
Talus cones are piles of rock debris that accumulate at the base of a cliff as the result of rockfalls. Most rock fragments in talus cones are produced by frost action.

CHEMICAL WEATHERING

What are the major chemical reactions in weathering? What do they produce?

Statement

Chemical weathering (chemical decomposition) is the breakdown of rocks by the chemical alteration of the constituent minerals. It involves several important reactions between the elements in the atmosphere and those in the minerals of Earth's crust. In these processes the internal structures of the original minerals are destroyed, and new minerals are formed, with new crystal structures that are stable under conditions at Earth's surface.

The chemical reactions involved in the decomposition of rock are complex, but three main groups are recognized:

1. Hydrolysis
2. Dissolution
3. Oxidation

Discussion

What processes are involved in the weathering of feldspar?
What is the origin of clay?

Hydrolysis

The chemical union of water and a mineral is known as hydrolysis. The process involves not merely absorption of water, as in a sponge, but a specific chemical change in which a new mineral is produced. In hydrolysis, ions derived from one mineral react with the H+ or OH− ions of the water to produce a different mineral.

A good example of hydrolysis is the chemical weathering of feldspar. As you recall from previous chapters, feldspar is an abundant mineral in a great many igneous, metamorphic, and sedimentary rocks, so it is important to understand how feldspars weather and decompose into clay minerals, which form the most abundant sedimentary rock, shale. Two substances are essential in the weathering of feldspars: carbon dioxide and water. The atmosphere and the soil contain carbon dioxide, which unites with rainwater to form carbonic acid. If K-feldspar comes in contact with carbonic acid, the following chemical reaction occurs.

$$2 \text{ Kalsl}_3\text{O}_8 \ + \ \text{H}_2\text{CO}_3 \ + \ \text{H}_2\text{O} \ \longrightarrow$$
(K-feldspar) (carbonic acid) (water)

$$\text{K}_2\text{CO}_3 \ + \ \text{Al}_2\text{Si}_2\text{O}_5(\text{OH})_4 \ + \ 4 \text{ SiO}_2$$
(potassium   (a clay mineral)   (soluble
carbonate)   hydrated silica)

The hydrogen ion of the H_2CO_3 displaces the potassium ion of the feldspar and thus disrupts the crystal structure. It then combines with the aluminum silicate of the feldspar to form a clay mineral. The potassium combines with the carbonate ion to form potassium carbonate, a soluble salt. Silica is also released but may remain in solution. The new clay mineral does not contain potassium, which was present in the original feldspar. The new mineral also has a new crystal structure, consisting of sheets of silica tetrahedra that form submicroscopic crystals.

Dissolution

How are limestones weathered?

Dissolution is a process whereby rock material passes directly into solution, like salt in water. Quantitatively, the most important minerals involved in dissolution are the carbonate minerals, calcite and dolomite. These minerals make up the limestones of the world. Dissolution occurs because water is one of the most effective and universal solvents known.

Some rock types can be completely dissolved and flushed away by water. Rock salt is perhaps the best known example. It is extremely soluble, surviving at Earth's surface only in the most arid regions. Gypsum is less soluble than rock salt but is also easily dissolved by surface water. Few, if any, large outcrops of these rocks occur in humid regions. Limestone is also soluble in water, especially if the water contains carbon dioxide. Where water is abundant, limestone commonly weathers into valleys, but in arid regions it forms cliffs.

The chemical analysis of river water illustrates the effectiveness of dissolution in the weathering of rocks. Fresh rainwater contains relatively little dissolved mineral matter, but running water soon dissolves the more soluble minerals in the rock and transports them in solution. Each year the rivers of the world carry about 3.9 million metric tons of dissolved minerals to the oceans. It is not surprising, then, that seawater contains 3.5 percent (by weight) dissolved salts, all of which were dissolved from the continents by rainwater.

Oxidation

Oxidation is the combination of atmospheric oxygen with a mineral to produce an oxide. The process is especially important in the weathering of minerals that have a high iron content, such as olivine, pyroxene, and amphibole.

Like many chemical reactions, the rate of chemical weathering increases as temperature rises. Chemical decomposition, therefore, is most intense in warm, wet, tropical climates.

Plants and bacteria are also important agents in chemical weathering because some acids and organic compounds in soils are produced by the bacterial decay of plant and animal remains. Water seeping through organic remains in soils commonly becomes more acidic, which increases its effectiveness as a weathering agent.

Inasmuch as feldspars and other silicate minerals that weather into clay compose a large percentage of igneous and metamorphic rocks, an enormous amount of clay has been produced by the weathering of these minerals throughout geologic time. It has been calculated that sediment and sedimentary rocks have average thicknesses of 3 km throughout the ocean basins, 5 km on the continental shelves, and 1.5 km on the continents. Because clay makes up about one-third of all sedimentary rocks, the total amount of clay would form a layer almost 2 km thick over the entire surface of Earth.

A Concluding Note

Water is of prime importance in chemical weathering. It takes part directly in chemical reactions. It acts as a medium to transport elements of the atmosphere to the minerals of the rocks, where reactions can occur; and it removes the products of weathering to expose fresh rock. The rate and degree of chemical weathering, therefore, are greatly influenced by the amount of precipitation.

No area of Earth's surface is continually dry. Even in the most arid deserts, some rain falls. Chemical weathering is therefore essentially a global process, but it is least effective in deserts and in climates where water is frozen the entire year.

We should remember that although we have considered mechanical and chemical weathering as separate, individual processes, in nature these processes cannot be separated because many types of weathering processes are usually involved in the weathering of any outcrop. Mechanical fracturing of a rock increases the surface area, where chemical actions take place and permit deeper penetration for chemical decomposition. Chemical decay in turn facilitates mechanical disintegration. One process may dominate in a given area, depending on the climate and rock composition, but mechanical and chemical weathering processes generally attack the rock at the same time.

GEOMETRY OF ROCK DISINTEGRATION

How do joints affect weathering?

Statement

The breakup of a solid mass of rock into smaller particles may at first seem to be a random process in which an infinite variety of shapes may be reproduced. Careful study, however, shows that there is system and order in the process. The mechanical breakdown of rocks and the shape of most rock fragments are inherited from patterns of joints, bedding, cleavage, and other planes of structural weakness in the parent rock material.

Discussion

The Importance of Joints in Weathering

Almost all rocks are broken into a system of fractures called joints. Joints result from strain that occurs when the rocks are uplifted, tilted, folded, or fractured by tectonic forces; from the release of confining pressure when material is removed by erosion; and from contraction produced by the cooling of lava. Joints greatly influence the weathering of rock bodies in two ways.

1. They effectively cut large blocks of rock into smaller ones and thereby greatly increase the surface area where chemical reactions take place.
2. They act as channelways through which water can penetrate to break down the rock by ice wedging.

The importance of joints in weathering processes can be appreciated by considering the amount of new surface area produced by jointing. Consider, for example, a cube of rock that measures 10 m on each side (as shown in Figure 9.4). If only the upper surface of the cube were exposed and the rock were not jointed, weathering could attack only the exposed top surface of 100 m^2. If the block were bounded by intersecting joints 10 m apart, however, the surface area exposed to weathering processes would be 600 m^2. If three additional joints cut the cube into eight smaller cubes, the surface exposed to weathering would be 1200 m^2. If joints 1 m apart cut the rock, 6000 m^2 of rock surface would be exposed. Obviously, a highly jointed rock body weathers much more rapidly than a solid one. The breakdown of a rock along a system of jointing planes is called joint-block separation (Figure 9.5A).

Other Planes of Weakness

Bedding planes in many sedimentary rocks form planes of weakness, which cause the rock to break into slabs or plates (Figure 9.5C). Foliation in metamorphic rock is similar. Schists tend to break into small, splintery pieces with flat sides parallel to planes of foliation. Slate is an even better example of how foliation influences the way in which a rock breaks into smaller pieces.

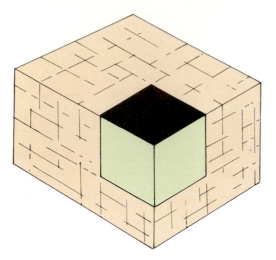

(A) A surface of bedrock 10 m long and 10 m wide, with no joints, exposes a total area of 100 m² to weathering processes. If a set of joints divides the rock into a 10-m cube, the surface area exposed to weathering is increased to 600 m².

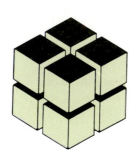

(B) Three additional joints, dividing the block into eight cubes, would increase the surface area to 1200 m².

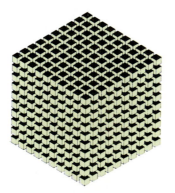

(C) If joints 1 m apart cut the rock, the surface area exposed to weathering would be increased to 6000 m².

Figure 9.4
A system of joints cutting a rock body greatly increases the surface area exposed to weathering.

(A) Joint block separation results when prominent fractures divide the rock into small blocks. The Devil's Post Pile in California is an excellent example of columnar joints controlling the geometric patterns of rock breakup.

(B) Granular Disintegration is also produced when cementing material in conglomerate and sandstone dissolves so the grains fall apart.

Granular Disintegration

In many sandstones the quartz grains are weakly cemented with calcite, which is readily dissolved and removed by water. Once the cementing material is removed or weakened, there is nothing to hold the grains together, and they fall apart, or desegregate. This process of grain-by-grain breakdown is known as granular disintegration (Figure 9.5). Individual mineral grains simply separate from one another along their natural contact and produce sand or gravel, in which each particle has the same shape and size it had in the original rocks. Similarly, the quartz grains in granite are desegregated as the feldspars are decomposed into clay. The soft clay loosens the interlocking grains, and the once solid piece of granite starts to crumble into individual grains or groups of grains. The result is that the small, irregular quartz grains that once filled the spaces between feldspar are liberated and become the source of most sand.

(C) Bedding planes frequently form a prominent zone of weakness in sedimentary rocks and cause the rock to break up into slabs.

(D) Shattering occurs when a rock is subjected to severe stress which causes the rock to rupture into sharp, irregular, angular blocks. Ice wedging produces shattering in nature of rock outcrops; blasting bedrock with explosives produce shattering artificially.

(E) Granular disintegration is the grain-by-grain breakdown of rock masses. It is common in coarse-grained igneous rocks, sandstones, and conglomerate and results when different minerals in a rock weather at a different rate. Feldspars weather rapidly in a granite, permitting the quartz grains simply to separate from one another along their contacts.

(F) Granular Disintegration in sandstone commonly occurs and produces spheroidal boulders. Each grain in the disintegrated material has the same size and shape as it did in the original rock.

(G) Exfoliation occurs when the solid rock mass comes apart in a series of shells or plates that conform roughly to the configuration of the outer surface of the rock mass. Exfoliation can occur on a very large scale, such as Half Dome in Yosemite National Park, or on a very small scale, with the individual plates being only a millimeter or less thick. Exfoliation affects many rock types and is caused by both physical and chemical processes.

(H) Jointing is commonly the major type of structural weakness in granites and related rocks, and causes the rock to break up into large blocks.

Figure 9.5
The geometric patterns of rock disintegration depend upon the composition, texture, and structure of the rock body.

Differential Weathering

Different rock masses and even different sections of the same rock weather at different rates. This is known as differential weathering. It occurs on a broad scale, from the great cliffs and slopes of the Grand Canyon to thin layers of sedimentary rock. The more resistant zones stand out as ridges, and the weaker zones form depressions. Differential weathering can lead to the formation of unusual shapes and forms, such as the spindles and pinnacles in Bryce Canyon (Figure 9.6) to pits and caverns on a rock face. Differential erosion on dikes of igneous rocks can form trenches or walls, depending on whether the dike is more or less resistant than the surrounding rock.

Differential weathering can be seen everywhere a rock is exposed. Study the photo in Figure 9.6, and you will notice that each layer has its own weathering characteristics. The white layers erode most rapidly and tend to form slender columns. The thicker beds of sand are more resistant, whereas the interbeds of siltstone and shale weather rapidly. Thus, the horizontal layers are etched out into ridges and furrows, which are responsible for much of the beauty in this scene.

PRODUCTS OF WEATHERING

What are the major products of weathering?

Statement

The results of weathering can be seen over the entire surface of Earth, from the driest deserts and the frozen wastelands to the warm, humid tropics. The major products include

1. a blanket of loose, decayed rock debris, known as regolith, which forms a discontinuous cover over the solid bedrock.
2. rock bodies modified into spherical shapes.
3. soluble products carried away by streams and groundwater.

Discussion

Regolith and Soil

The term *regolith* comes from the Greek word *rego*, meaning "blanket" (blanket rock). It is a layer of soft, desegregated rock material formed in place by the

Figure 9.6
Differential weathering has produced the spectacular landforms in Bryce Canyon, Utah. Two zones of weakness occur within the rock body: (1) horizontal layers of different material, and (2) a system of intersecting joints that divide the rocks into a series of rectangular columns. Rapid weathering along the joints produces a series of columns, and differential weathering of the sedimentary layers produces an irregular form for the column.

decomposition and disintegration of the bedrock that lies beneath it. Within the regolith individual grains and small rock particles are easily separated from one another.

Gravel, sand, silt, and mud deposited by streams, wind, and glaciers are sometimes referred to as transported regolith, in order to distinguish them from the residual regolith produced by weathering. Many types of transported regolith, or surficial deposits, have been identified. We will learn more about them in later chapters dealing with rivers, glaciers, and wind.

The uppermost layer of the regolith is the soil. It is composed chiefly of small particles of rocks and minerals, plus varying amounts of decomposed organic matter. Soil is so widely distributed and so economically important it has acquired a variety of definitions. You should be aware that the term, as used by engineers, geologists, farmers, and soil scientists, has somewhat different meanings.

The transition from the upper surface of the soil down to fresh bedrock shows a rather constant sequence of layers, or zones, that are distinguished by composition, color, and texture. These are shown in Figure 9.7. The A horizon is the topsoil layer, which often is visibly divided into three layers: A 0 is a thin surface layer of leaf mold, especially obvious on forest floors; A 1 is a humus-rich, dark layer; and A 2 is a light, bleached layer. The B horizon is the subsoil, containing fine clays and colloids washed down from the topsoil. It is largely a zone of accumulation and commonly is reddish in color. The C horizon is a zone of partly disintegrated and decomposed bedrock. The individual rock fragments are often weathered, spheroidal boulders, which may be completely decomposed. The C horizon grades downward into fresh, unaltered bedrock.

The type and thickness of soil depend on a number of factors, the most important of which are climate, parent rock material, and topography. Climate is of major importance because rainfall, temperature, and seasonal changes all directly affect the development of soil. For example, in deserts, arctic regions, and high mountainous regions, mechanical weathering dominates as the means of soil production, and organic matter is minimal. The resulting soil is thin and consists largely of broken fragments of bedrock. In equatorial regions, where rainfall is heavy and temperatures are high, chemical processes dominate and thick soils develop rapidly. As a consequence, soil profiles 60 m thick are common in the tropics and subtropics. In some areas (such as central Brazil) the zone of decayed rock is more than 150 m thick.

Spheroidal Weathering

Why are spheroidal forms the universal result of weathering processes?

In the weathering process there is an almost universal tendency for rounded (or spherical) surfaces to form on a decaying rock body. A rounded shape is produced because weathering attacks an exposed rock from all sides at once, and therefore decomposition is more rapid along the corners and edges of the rock (Figure 9.8). As the decomposed material falls off, the corners become rounded, and the block eventually is reduced to an ellipsoid or a sphere. The sphere is the geometric form that has the least amount of surface area per unit of volume. Once the block attains this shape, it simply becomes smaller with further weathering. This process is known as spheroidal weathering. Examples of spheroidal weathering can be seen in almost any exposure of rock (Figure 9.9). It can also be seen in the rounded blocks of ancient buildings and monuments (see also Figure 9.12). Exfoliation is a special type of spheroidal weathering in which the rock breaks apart by separation along a series of concentric shells, or layers, that look like cabbage leaves (Figure 9.10). The layers, essentially parallel to each other and to the surface, develop both

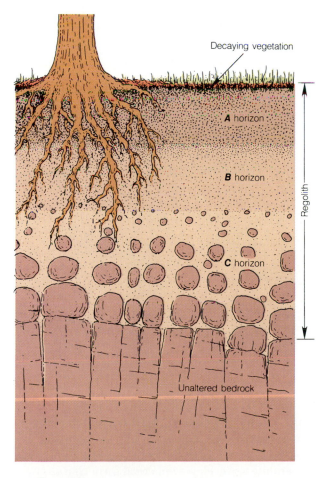

Figure 9.7
A soil profile shows the transition from bedrock to regolith through a sequence of layers, or horizons, consisting of successively smaller fragments of rock.

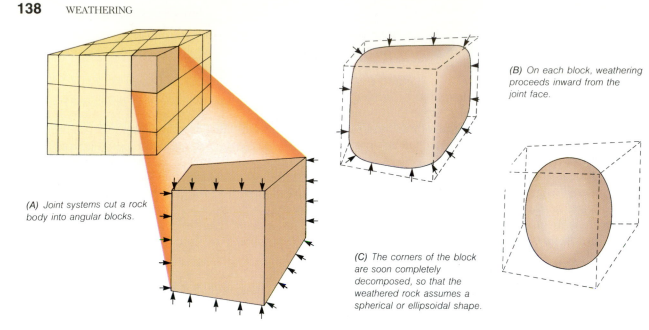

(A) Joint systems cut a rock body into angular blocks.

(B) On each block, weathering proceeds inward from the joint face.

(C) The corners of the block are soon completely decomposed, so that the weathered rock assumes a spherical or ellipsoidal shape.

Figure 9.8
Spheroidal weathering occurs because the edges and corners of a joint block are easily decomposed.

Figure 9.9
Examples of spheroidal weathering can be seen in almost every outcrop. In this exposure of granite in the Sierra Nevadas, joints have cut the rock into rectangular blocks and the corners have been rounded by mechanical and chemical weathering.

by chemical and mechanical means. Sheeting can play an important part because rocks like granite, if they are deeply buried, have a tendency to expand upward and outward as the overlying rock is removed. In cold climates frost wedging along the sheeting joints helps to remove successive layers grad- ually. The increase in volume of mineral grains asso- ciated with the decomposition of feldspar might also promote exfoliation. Exfoliation causes massive rocks like granite to develop a spherical form characterized by a series of concentric layers ranging from boulders to a mountain.

Figure 9.10
Exfoliation domes are well developed in the Sierra Nevada, in California. These mountains are composed of massive granite cut by joints that separate the rock into large blocks. Exfoliation domes developed as huge slabs of rock fell off. The joints separating the rock slabs probably resulted from expansion of the rock as erosion removed the overlying rock.

CLIMATE AND WEATHERING

How does climate influence weathering?

Statement

The type, rate, and extent of weathering are greatly influenced by climate, a fact clearly shown in the striking contrast in the soil and regolith developed in the tropics, deserts, and arctic regions. The most important climatic factors influencing weathering are

1. precipitation.
2. temperature.

Discussion

Why are soils thick in the tropics but thin in deserts and polar regions?

Water is the most important single factor in almost all forms of weathering. The changes in volume caused by freezing and thawing plus the simple addition or removal of water may cause a rock to split apart or crumble. Most chemical reactions—such as hydrolysis, dissolution, and oxidation—require the presence

of water, so that the total amount of precipitation in an area is clearly a major factor. The extent and style of weathering are not controlled entirely by total water supply, and weathering may be greatly affected by other conditions. Many reactions are controlled by the hydrogen ion concentration, which is expressed as the pH value, ranging from 1 (acid) to 14 (alkaline). Iron, for example, becomes 100,000 times more soluble at pH 6 than at pH 8.5.

Temperature is also important in all aspects of weathering. In mechanical weathering perhaps the single most important temperature changes are the ones that produce a continual series of freeze-thaw changes. This results in repeated freezing and expansion of water in rock and soil, which ultimately breaks the rock into smaller fragments.

Temperature is important because the rate of chemical reactions (and biological activity) tends to increase as temperature increases. Commonly, a 10° C increase in temperature doubles reaction rates.

Perhaps the best way to appreciate the influence of climate on weathering is to consider variations in the types and thicknesses of soils from the equator to the poles, as shown in Figure 9.11. This diagram summarizes the relationships between the amount of chemical weathering and variations in precipitation and temperature.

In humid, tropical climates, extreme chemical weathering rapidly develops thick soils to depths greater than 70 m. Under such conditions the feldspars in granites and related rocks are completely altered to clays, and all soluble minerals are leached out. Only the most insoluble materials (such as silica, aluminum, and iron) remain in the thick, deep soil, with the result that the soil commonly is infertile. The high temperatures in tropical zones speed chemical reactions, so chemical decomposition is very rapid. Frost action, of course, is essentially nonexistent in the tropics, except on the tops of high mountains.

In the low-latitude deserts north and south of the tropical rain forests, chemical weathering is minimal because of the lack of precipitation. The soil is thin, and exposures of fresh, unaltered bedrock are abundant. Mechanical weathering is evident, however, in the fresh, angular rock debris that litters most slopes.

In the temperate regions of the higher latitudes, the climate ranges from subhumid to subarid, and temperatures range from cool to warm. Both chemical and mechanical processes operate, and the soil and regolith develop to depths of several meters.

In the polar climates, weathering is largely mechanical. Temperatures are too low for much chemical weathering, so the soil typically is thin and composed mostly of angular, unaltered rock fragments. In permafrost zones (areas where water in the pore spaces of the soil and rock is permanently frozen) the surface layer melts during the summer, but freezes again in

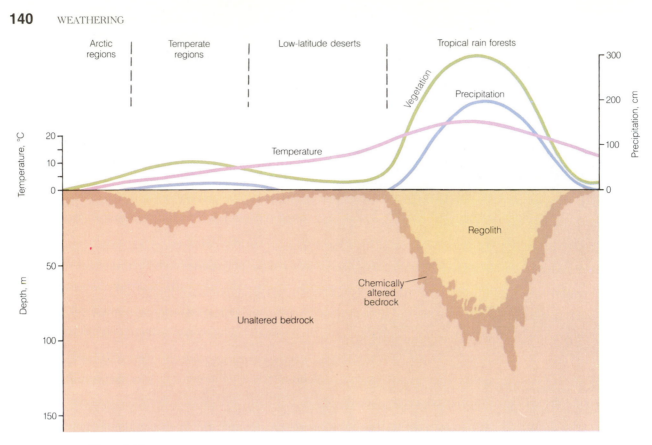

Figure 9.11
The type and extent of weathering vary with climate because of the combined effects of precipitation, temperature, and vegetation. Weathering is most pronounced in the tropics, where precipitation, temperature, and vegetation reach a maximum. Conversely, a minimum of weathering is found in deserts and polar regions where these factors are minimal.

the winter. This unique condition produces polygonal ground patterns, which result from thermal contractions and the differential thawing and freezing. We will have more to say about permafrost areas in Chapter 12.

RATE OF WEATHERING

Why do some rocks weather more rapidly than others?

Statement

The rate at which weathering processes decompose and break down a solid rock body varies greatly. It depends on three main factors:

1. Susceptibility of the constituent minerals to weathering
2. Climate
3. Amount of surface exposed to the atmosphere

Discussion

Minerals vary widely in their ability to resist weathering. Although many details are not completely understood, minerals that crystallize at a high temperature tend to be the most susceptible to weathering. Thus, olivine and Ca-plagioclase weather most rapidly, followed by pyroxene and amphibole and then by Na-plagioclase and K-feldspar. Quartz is one of the most highly resistant minerals because it resists alteration by chemical decomposition and alteration by mechanical processes.

In warm, humid climates, the rate of weathering is very rapid. This fact is demonstrated by some of the recent lava flows in Hawaii, which, within a few years, have decomposed sufficiently to support some vegetation. Soils have also formed on the ash deposits resulting from the 1883 eruption of Krakatoa. Measurements made 45 years after the Krakatoa eruption show a soil nearly 35 cm thick, and chemical modifications within the soil are indicated by a loss of silica and an enrichment in aluminum and iron oxides. Another example from the island of St. Vincent in the

West Indies shows that a clay soil 2 m deep has developed in 4,000 years on volcanic material.

It is also clear that the weathering process slows down as the fresh rock is covered by soil and regolith forms upon it. Ultimately a state of equilibrium may be reached. By contrast, in colder climates, the same rock types have remained fresh and essentially unaltered since the Ice Age.

In spite of the complexity of the weathering process, we can get some idea of weathering rates by measuring the amount of decay on rock surfaces of known age. Tombstones, ancient buildings, and monuments, for example, provide rock surfaces that were fresh and unaltered when they were constructed but have since suffered weathering over a known period of time.

Example

How do the pyramids of Egypt provide information on rates of weathering?

The Egyptian pyramids provide an interesting example of rates of weathering. The Great Pyramid of Cheops, near Cairo, was originally covered with polished, well-fitted blocks of limestone. These blocks pro-

tected the rock in the pyramid core from weathering until the outer, polished layers were removed, about 1,000 years ago, to build mosques in Cairo. Since then, without the rock facing, weathering has attacked all four main rock types used in the construction of the pyramid. The most durable (least weathered) rock in the pyramid is a granite, which today remains essentially unweathered. Also resistant is a hard, gray limestone, which still retains marks of the quarry tools. The shaly limestone and fossiliferous limestone used for other blocks, however, have weathered rapidly. Many of these blocks have a zone of decayed minerals as deep as 20 cm. Most of the weathered debris remains as a talus on individual tiers and around the base of the pyramid (Figure 9.12). The volume of weathered debris produced from the pyramid during the last 1,000 years has been calculated to be 50,000 m^3, or an average of 50 m^3 per year. This is a loss of approximately 3 mm per year over the entire surface of the pyramid.

Some of the older stepped pyramids, built 4,500 years ago, show much greater weathering. Many of their large blocks are almost completely decayed or reduced to small, spherical boulders. In addition, high piles of debris have accumulated on each major terrace (Figure 9.13).

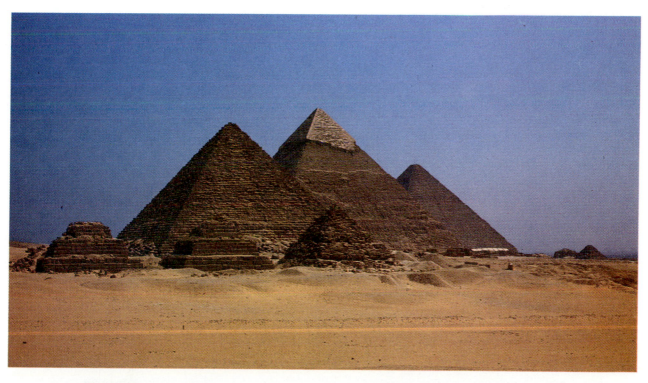

Figure 9.12
The products of weathering can be observed on the Great Pyramids of Egypt. The rectangular building blocks have been modified into ellipsoidal forms by spheroidal weathering, and talus debris has accumulated on each tier.

Figure 9.13
Deep weathering is apparent on the ancient stepped pyramids of Egypt. Each step is completely covered with talus, and many individual blocks are weathered to spherical form.

SUMMARY

1. Weathering is the breakdown and alteration of rocks at Earth's surface through physical and chemical reactions with the atmosphere.
2. Mechanical weathering fragments the rock by various physical stresses. Ice wedging and sheeting are the most important processes.
3. Chemical weathering includes a variety of chemical reactions between elements in the atmosphere and those in the rocks. Hydrolysis, dissolution, and oxidation are the most important reactions.
4. Rock disintegration is greatly influenced by patterns of joints, bedding, and other planes of structural weakness in the parent rock material.
5. The major results of weathering are (a) formation of a discontinuous blanket of desegregated and decomposed rock debris (regolith) and (b) rocks of various sizes modified into spherical forms.
6. Climate is the single most important factor in weathering, the two most important variables being precipitation and temperature.
7. Rates of weathering depend mostly on climate and the composition of the rock. Rates of weathering can be measured on volcanic extrusions and man-made monuments, where the age of original surfaces can be dated.

KEY TERMS

A horizon (p. 137)
B horizon (p. 137)
C horizon (p. 137)
chemical
  weathering (p. 132)
differential
  weathering (p. 136)

dissolution (p. 132)
exfoliation (p. 135)
granular
  disintegration (p. 134)
hydrolysis (p. 132)
ice wedging (p. 130)
joint (p. 133)

joint block
  separation (p. 133)
mechanical
  weathering (p. 130)
oxidation (p. 132)
regolith (p. 136)
shattering (p. 135)

sheeting (p. 130)
soil (p. 136)
soil profile (p. 137)
spheroidal
  weathering (p. 137)
talus (p. 131)

REVIEW QUESTIONS

1. List five ways in which the effects of weathering are expressed in natural outcrops of bedrock.
2. Discuss the processes involved in the most important types of mechanical weathering.
3. Which rock types are most susceptible to chemical weathering?
4. Discuss the chemical reactions involved in hydrolysis.
5. Explain why joints are important in weathering processes.
6. How does soil originate?
7. Explain how climate, rock types, topography, and time influence the types of soil produced by weathering.
8. What is spheroidal weathering?
9. Draw a schematic diagram showing variations in weathering from the arctic to the tropics.
10. How do the Great Pyramids of Egypt provide information on rates of weathering?

Multiple-Choice Questions

1. The most common type of mechanical weathering is
   a. sheeting.
   b. ice wedging.
   c. thermal expansion and contraction.
   d. action of burrowing organisms.
   e. oxidation.
2. Ice wedging would be most effective in
   a. the permafrost areas of the north.
   b. areas where freezing and thawing occur many times a year.
   c. areas where it is cold enough to produce long periods below zero degrees centigrade.
   d. beneath the glaciers in Antarctica and Greenland.
   e. in latitudes where heavy frost occurs.
3. Ice wedging on the jointed bedrock of steep cliffs results in the formation of
   a. an exfoliation dome.
   b. a talus slope.
   c. a thick layer of saprolite.
   d. mudflows.
   e. minor tributaries.
4. Chemical weathering would progress most rapidly in
   a. the Rocky Mountains of Colorado.
   b. Greenland.
   c. the Great Lakes area.
   d. northern South America.
   e. New Mexico.
5. When granite weathers, its quartz grains
   a. dissolve completely.
   b. remain essentially unaltered.
   c. alter into clays.
   d. are replaced with silica.
   e. react with water to form soil.
6. Joints are important in weathering because they
   a. break down the rock into fragments.
   b. increase the surface area available for chemical reactions.
   c. act as channel ways for elements of the atmosphere to penetrate the rock.
   d. act as channels for subsurface water.
   e. all of the above.

7. The general term for the blanket of loose rock debris that covers large areas of Earth's surface is
   a. bedrock.
   b. regolith.
   c. outcrop.
   d. lithic layer.
   e. laterite.
8. Which of the following changes take place during the chemical weathering of granite?
   a. Feldspar increases and clay decreases.
   b. Clay increases and feldspar decreases.
   c. Clay decreases and iron oxide increases.
   d. Quartz and plagioclase decrease and K-feldspar increases.
   e. Clay increases and feldspar increases.
9. Which of the following common rock-forming minerals weathers to form most of the clay found in soils and regolith?
   a. calcite
   b. feldspar
   c. quartz
   d. halite
   e. mica
10. The thickest soils are produced in the
    a. arctic.
    b. low-latitude deserts.
    c. temperate zones.
    d. rain forests.
    e. savannas.

True/False Questions

1. *T F* Frost wedging is most effective in areas where freezing and thawing occur many times a year.
2. *T F* The rate and degree of chemical weathering are influenced greatly by the amount of precipitation.
3. *T F* Chemical weathering is the breakdown of rock into smaller fragments by various physical stresses.
4. *T F* In humid areas limestone commonly weathers into cliffs, but in arid regions limestone formations form valleys.
5. *T F* In deserts, arctic regions, and high mountainous regions, chemical weathering dominates as the means of soil production.

Fill in the Missing Terms

1. _____ weathering is the breakdown of rock into smaller fragments by various physical stresses.
2. The products of ice wedging commonly accumulate at the bases of cliffs in piles of angular rock fragments called _____.
3. In the weathering process, there is an almost universal tendency for exposed rock to assume a _____ shape.
4. The most important type of mechanical weathering is _____ _____.
5. In _____ weathering the internal structures of the original minerals are commonly destroyed, and new minerals are formed with new crystal structures that are stable under conditions at Earth's surface.

10

RIVER SYSTEMS

River systems are not simply channels through which water flows to the sea; they are the major agents by which the surface of Earth is sculptured into an infinite variety of fascinating erosional and depositional landforms. Stream erosion is a fundamental part of the hydrologic system and will continue as long as there is land exposed above the sea.

Humans have always lived by rivers. Most cities, townsites, transportation routes, and farmlands are located near rivers and the valley plains they have formed. Rivers will always be our principal source of water. From the time of the earliest dwellers on the Nile to the present, rivers have been diverted and manipulated for irrigation, power, and transportation. They are also our major garbage disposal system. It is therefore imperative that we understand how rivers operate as a dynamic system.

MAJOR CONCEPTS

1. Running water is part of Earth's hydrologic system and is the most important agent of erosion. Stream valleys are the most abundant and widespread landforms on the continents.

2. A river system consists of a main channel and all of the tributaries that flow into it. It can be divided into three subsystems: (a) a collecting system, (b) a transporting system, and (c) a dispersing system.

3. River systems have a universal tendency to establish equilibrium among the various factors influencing stream flow (velocity, water volume, stream gradient, and volume of sediment).

4. The downcutting of stream channels and headward erosion are the major processes of stream erosion.

5. Slope retreat causes valleys to grow wider and is associated with downcutting and headward erosion. It results from various types of mass movement.

6. As a river develops a low gradient, it deposits part of its load on point bars, on natural levees, and across the surface of its flood plain.

7. Most of a river's sediment is deposited where the river empties into a lake or the ocean. This deposition commonly builds a delta at the river's mouth. In arid regions many streams deposit their load as alluvial fans at the base of steep slopes.

THE MAJOR CHARACTERISTIC OF A RIVER SYSTEM

Statement

A river system, also referred to as a drainage basin, consists of a main channel and all of the tributaries that flow into it. It is bounded by a divide (ridge) beyond which water is drained by another system. Within a river system, the surface of the ground slopes toward the network of tributaries, so that the drainage basin acts as a funneling mechanism for removing precipitation and weathered rock debris. A typical river can be divided into three segments, (1) a collecting system, (2) a transporting system, and (3) a dispersing system. These are shown on an idealized map of a typical river in Figure 10.1

Discussion

Geologic Importance of Running Water

An attempt to appreciate the significance of streams and stream valleys in the regional landscape of Earth presents a problem of perspective, much like trying to appreciate the abundance of craters on the Moon from viewpoints on the lunar surface. To an astronaut on the Moon, the surface appears to be an irregular, broken landscape cluttered with rock debris. The crater systems and terrain patterns, so striking when viewed from space, are not at all apparent from vantage points on the Moon's surface. Indeed, crater rims may appear only as rounded hills, and without the aid of maps or space photographs, some of the larger craters may not even be recognizable as circular land forms.

Viewed from the ground, Earth's stream valleys may appear to be relatively insignificant, irregular depressions between rolling hills, mountain peaks, and broad plains. Viewed from space, however, stream valleys are seen to dominate most continental landscapes of Earth, much as craters dominate the landscape of the Moon.

The ubiquitous stream valleys on Earth's surface and the importance of running water as the major agent of erosion can best be appreciated by considering a broad, regional view of the continents and their major river systems, as seen through high-altitude photography. As the photograph in Figure 3.2 shows, throughout broad regions of the continents, the surface is little more than a complex of valleys created by stream erosion. Even in the desert, where it sometimes does not rain for decades, the network of stream valleys commonly is the dominant landform. No other landform on the continents is as abundant and significant.

The Collecting System

The collecting system of a river, consisting of a network of tributaries in the headwater region, collects and funnels water and sediment to the main stream. It commonly has a dendritic "treelike" drainage pattern, with numerous branches that extend upslope toward the divide. Indeed, one of the most remarkable characteristics of the collecting system is the intricate network of tributaries. The detailed network of tributaries shown in the enlargement in Figure 10.1 was made by plotting all visible streams shown on an aerial photograph. That is not, however, the entire system. Each of the smallest tributaries shown in Figure 10.1 has its own system of smaller and smaller tributaries, so that the total number becomes astronomical. From the details in this figure, it is apparent that most of the land's surface is part of some drainage basin.

The Transporting System

The transporting system is the main trunk stream, which functions as a channelway through which water and sediment move from the collecting area toward the ocean. Although the major process is transportation, this subsystem also collects additional water and sediment; deposition occurs where the channel meanders back and forth and at times when the river overflows its banks during a flood stage. Erosion, deposition, and transportation thus occur in the transporting subsystem of a river.

The Dispersing System

The dispersing system consists of a network of *distributaries* at the mouth of a river, where sediment and water are dispersed into an ocean, a lake, or a dry basin. The major processes are the deposition of the coarse sediment load and the dispersal of fine-grained material and river water into the basin.

It is apparent from Figures 10.1 and 10.2 that a stream does not occur as a separate, independent entity. Every stream, every river, and every gully and ravine are part of a drainage system, with each tributary intimately related to the stream into which it flows and to the streams that flow into it. Every stream has tributaries, and every tributary has smaller tributaries, extending down to the smallest gully.

THE DYNAMICS OF STREAM FLOW

Statement

The flow of water in a river system is complex and is influenced by a number of variables; the most important of them are (1) discharge, (2) gradient, or slope,

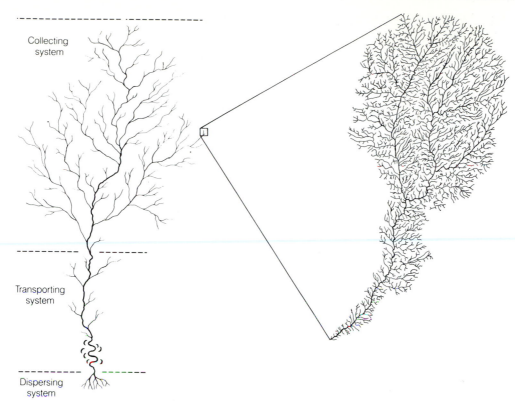

Collecting system

Transporting system

Dispersing system

Figure 10.1
The major parts of a river system are characterized by different geologic processes. The tributaries in the headwaters constitute a subsystem that collects water and sediment and funnels them into a main trunk stream. Erosion is dominant in this headwater area. The main trunk stream is a transporting subsystem. Both erosion and deposition can occur in this area. The lower end of the river is a dispersing subsystem, where most sediment is deposited in a delta or an alluvial fan, and water is dispersed into the ocean. Deposition is the dominant process in this part of the river.

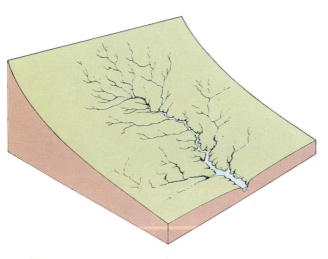

Figure 10.2
The characteristics of a river change systematically downstream. The gradient decreases downstream, and the channel becomes larger. Other downstream changes include an increase in the volume of water and an increase in the size of the valley through which the stream flows.

(3) velocity, (4) sediment load, (5) channel slope, and (6) base level. These factors are interrelated in a systematic way so that a change in one leads to a regular and predictable change in the others.

Discussion

Discharge is the amount of water passing a given point during a specific interval of time. The water in a river system comes from both surface runoff and seepage of groundwater into the stream channels. It varies from season to season and with fluctuations of longer climatic cycles.

The gradient, or slope, of the streambed is steepest in the headwaters and decreases downslope (Figure 10.2). The longitudinal profile (a cross section from the headwaters to the sea) is a smooth, concave upward curve that becomes very flat at the lower end of the stream.

Sediment load is the material transported by the river. Fine particles (silt and mud) are moved downstream in suspension (Figure 10.3) at the velocity of the flowing water. Particles too large to remain in suspension (sand and gravel) collect on the floor of the stream channel and form the bed load, or traction load. These particles move by sliding, rolling, and saltation (short leaps) (Figure 10.3). The bed load can constitute 50 percent of the total load in some rivers, but it usually ranges from 7 to 10 percent. The movement of the bed load is one of the major tools of stream erosion because as the particles of sand and gravel move, they abrade (wear away) the bottom and side of the stream channel. In addition to the suspended and bed load, all streams carry some material in solution as a dissolved load.

The base level of a stream is the lowest level to which the stream can erode. The ultimate base level, for all practical purposes, is sea level, but local base levels occur. For example, a tributary cannot erode lower than the level of the stream into which it flows. Similarly, a lake is a temporary base level for all the streams that flow into it.

Studies of extensive records of stream flow from thousands of rivers throughout the world show that the interaction of these variables is not random but varies in a systematic way: that is, the factors are so interrelated and balanced that a change in one leads to surprisingly regular and predictable changes in the others. For example, if the discharge of a stream doubles because of heavy rainfall, velocity will increase by a factor of 1.3, channel slope will change, and the amount of sediment carried by the stream will increase by a factor of 8. What is astonishing is not that changes in the variables occur, but that the quantity of such changes is regular and predictable. As we will see in subsequent sections, the effect of modifying river systems by dams, levees, locks, and canals can bring on many critical consequences.

EQUILIBRIUM IN RIVER SYSTEMS

What major variables of a river system continually change in order to establish equilibrium?
What happens when a river's equilibrium is disrupted?

Statement

A river system functions as a unified whole, and any change in one part of the system affects other parts. The major variables of stream flow (discharge, gradient, base level sediment load, volume, and channel shape) constantly change toward a balance, or equilibrium, so that neither erosion nor deposition of the sediment load occurs. A change in any of these factors causes compensating adjustments in the other variables in a manner that will tend to restore equilibrium. An understanding of the concept of equilibrium in river systems has practical consequences because as we continually modify rivers to suit our needs, we should know how the river system will respond to artificial modifications.

Discussion

The concept of equilibrium in a river system can best be appreciated by considering a hypothetical stream in which equilibrium has been established. In Figure 10.4A, the variables in the stream system (discharge, velocity, gradient, base level, sediment load, volume, and channel shape) are in balance, so that neither erosion nor sedimentation occurs along the stream's profile. There is just enough water to transport the available sediment down the existing slope. Such a stream is in equilibrium and is known as a graded stream. In Figure 10.4B, the stream's profile is dis-

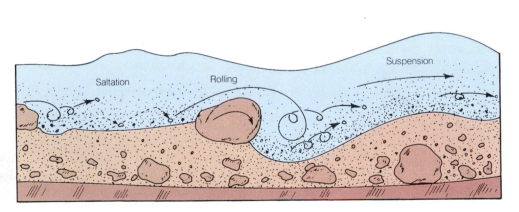

Figure 10.3
Movement of the bedload in a stream is accomplished in a variety of ways. Particles that are too large to remain in suspension are moved by sliding, rolling, and saltation. Increases in discharge, because of heavy rainfall or spring snowmelt, can flush out all of the loose sand and gravel, so that the bedrock is eroded by abrasion.

placed by a fault that creates a waterfall or rapids. The increased gradient across the falls greatly increases the stream's velocity at that point, so that rapid erosion occurs and the waterfall (or the rapids)

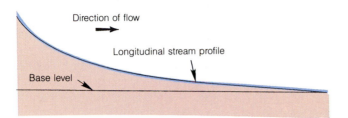

(A) Initially, when the stream profile is at equilibrium, the velocity, load, gradient, and volume of water are in balance. Neither erosion nor deposition occurs.

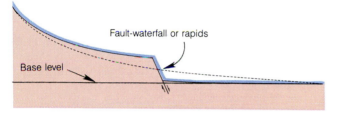

(B) Faulting disrupts equilibrium by decreasing the gradient downstream and increasing the gradient at the fault line.

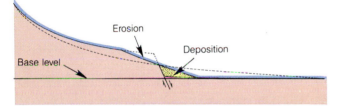

(C) Erosion proceeds upstream from the fault, and deposition occurs downstream.

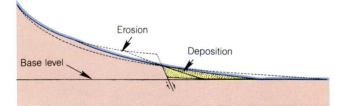

(D) Erosion and deposition continue to develop a new stream profile at which the velocity, load, gradient, and volume of water will be in balance, so that neither erosion nor deposition occurs.

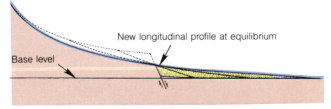

(E) A new profile of equilibrium, in which neither erosion nor deposition occurs, is eventually reestablished.

Figure 10.4
Adjustments of a stream to reestablish equilibrium are illustrated by profile changes after disruption by faulting.

migrates upstream. The eroded sediment added to the stream segment on the dropped fault block is more than the stream can transport because the system was already in equilibrium with the sediment load before faulting occurred. The river therefore deposits part of its load at that point, thus building up the channel gradient (the shaded areas in Figure 10.4C–E) until a new profile of equilibrium is established.

Example

As the population has grown, the artificial modification of the natural water systems of the continents has greatly increased, largely through the construction of dams and canals and through urbanization. Dams are built to store water for industrial and irrigational use, to control floods, and to produce electrical power. When a dam is built, however, the equilibrium of the river system, established over thousands or millions of years, is instantly upset. Many unforeseen, long-term effects occur both upstream and downstream from the construction.

The Aswan Dam

The Aswan High Dam of the Nile provides a good example of the many consequences of modifying a river system. For centuries, the Nile River has been the sole source of life in Egypt. The principal headwaters of the Nile are located in the high plateaus of Ethiopia. Once a year, for approximately a month, the Nile used to rise to flood stage and cover much of the farmland in the Nile Delta area, making it fertile. The Aswan High Dam, completed in the summer of 1970, was to provide Egypt with water to irrigate one million acres of arid land and generate 10 billion kilowatts of power. This, in turn, was to double the national income and permit industrialization. The dam, however, destroyed the equilibrium of the Nile, and many unforeseen adjustments in the river resulted. This is what happened.

The Nile is not only the source of water for the delta, it is also the source of sediment. When the dam was finished and began to trap sediment in a reservoir (Lake Nasser), the physical and biological balance in the delta area was destroyed. Without the annual "gift of the Nile," the delta coastline is now exposed to the full force of marine currents, and wave erosion is eating away at the delta front. Some parts of the delta are receding several meters per year.

The sediment previously carried by the Nile was an important link in the aquatic food chain, nourishing marine life in front of the delta. The recent lack of Nile sediment has reduced plankton and organic carbon to a third of the former levels. This change either killed off or drove away sardines, mackerel, clams, and crustaceans. The annual harvest of 16,000 metric tons of sardines and a fifth of the fish catch have been lost. The sediment of the Nile also naturally fertilized the flood plain. Without this annual addition of soil nutrients, Egypt's one million cultivated acres need artificial fertilizer.

The water discharged from the reservoir is clear, free of most of its sediment load. Without its load, the discharged water flows downstream much faster and is vigorously eroding the channel bank. This scouring process has already destroyed three old barrier dams and more than five hundred bridges built since 1953. Ten new barrier dams between Aswan and the ocean must be built at a cost equal to one-fourth the cost of the Aswan Dam itself.

The annual flood of the Nile was also important to the ecology of the area because it washed away salts that formed in the arid soil. Soil salinity has already increased, not only in the delta, but throughout the middle and upper Nile area. Unless corrective measures are taken (at a cost of over $1 billion), millions of acres will revert to desert within a decade.

The change in the river system has permitted double cropping, but there are no periods of dryness. The dry seasons previously helped limit the population of bilharzia, a blood parasite carried by snails, which infects the intestinal and urinary tracts of humans. One out of every two Egyptians now has the infection, and it causes a tenth of the deaths in the country.

Problems have also occurred in the lake behind the dam. The lake was to have reached a maximum level in 1970, but it might actually take two hundred years to fill. More than 15,000,000 m³ of water annually escape underground into the porous Nubian Sandstone, which lines 480 km of the lake's western bank. The sandstone is capable of absorbing an almost unlimited quantity of water. Moreover, the lake is located in one of the hottest and driest places on Earth, and the rate of evaporation is staggering. A high rate was expected, but additional losses from transpiration by plants growing along the lakeshore and increased evaporation from high winds have brought the total loss of water from the lake to nearly double the expected rate. This loss equals half the total amount of water that once was "wasted," flowing unused to the ocean.

Urbanization

Another way that human activity has modified river systems is through urbanization. The construction of cities may at first seem unrelated to the modification of river systems, but a city significantly changes the surface runoff in a large segment of a drainage basin. The resulting changes in river dynamics are becoming serious and costly. Water that falls to Earth as precipitation usually follows several paths in the hydrologic system. Generally, from 54 to 97 percent of precipitation returns to the air directly by evaporation and transpiration; from 2 to 27 percent collects in stream systems as surface runoff, and from 1 to 20 percent infiltrates the ground and moves slowly through the subsurface toward the ocean. Urbanization disrupts the flow of water in each of these paths. One of the major effects is the striking change in the rates and percentages of runoff and infiltration. Roads, sidewalks, and roofs of buildings render a large percentage of the surface impervious to infiltration. This causes the volume of surface runoff to increase substantially. In addition, surface runoff is much faster because water is channeled through gutters, storm drains, and sewers. As a result, this rapid runoff tends to produce frequent and severe floods.

PROCESSES OF STREAM EROSION

Statement

The basic processes of stream erosion are

1. removal of the regolith.
2. downcutting of the stream channel by abrasion.
3. headward erosion.

Discussion

Removal of the Regolith

One of the most important processes of erosion is the removal and transport of regolith, the loose rock debris produced by weathering. The process is simple but important. The mineral grains produced by weathering (gravel, sand, clay, etc.) are washed downslope into the drainage system and transported as sediment load in streams and rivers. In addition, soluble material is carried in solution. The net result is that the blanket of regolith over Earth's surface is continually being removed and transported to the sea by stream action. As the regolith is removed, fresh rock is exposed and weathered, so the supply of regolith is always being replenished. Thus, a universal process in the erosion of the surface of Earth is the breakdown and decay of solid rock by weathering and removal of this loose material by the drainage system.

Downcutting of Stream Channels

Downcutting is one of the basic processes of erosion in all stream channels, whether small, hillside gullies or great canyons of major rivers. The process is abrasion by sand, cobbles, and boulders moving along the channel floor. In a very real sense, the sediment load acts like sandpaper and is capable of wearing down the channel floor at an astonishing rate. Some dramatic examples of the power of streams to cut downward are the steep, nearly vertical gorges in many

Figure 10.5
The tools of erosion are sand and gravel. Transported by a river, they act as powerful abrasives, cutting through the bedrock as they are moved by flowing water. The abrasive action of sand and gravel cuts this vertical gorge through resistant limestone in the Grand Canyon, Arizona.

Figure 10.6
Potholes are eroded in a streambed by sand, pebbles, and cobbles whirled around by eddies.

canyons in the southwestern United States (Figure 10.5).

An effective and interesting type of stream abrasion is the drilling action of pebbles and cobbles trapped in a depression and swirled around by currents. The rotational movement of the sand, gravel, and boulders acts like a drill and cuts deep, vertical holes called potholes. As the pebbles and cobbles are worn away, new ones take their place and continue to drill into the stream channel. Holes eventually develop that can measure several meters in diameter and more than 5 m in depth (Figure 10.6).

Another important factor in the downcutting of a stream channel is the upstream migration of waterfalls and rapids. Here again, the process is simple but important. It can be appreciated by considering the erosion of Niagara Falls (Figure 10.7). The increased

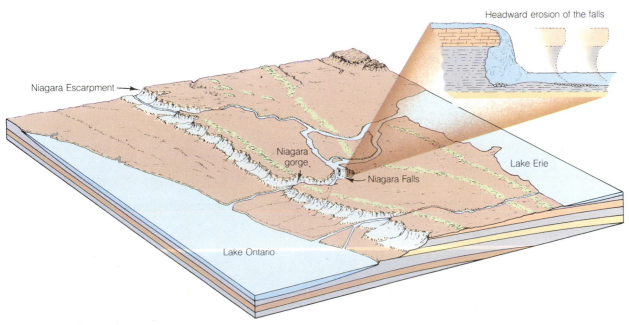

Headward erosion of the falls

Niagara Escarpment

Niagara gorge

Niagara Falls

Lake Erie

Lake Ontario

Figure 10.7
Retreat of Niagara Falls upstream occurs as hydraulic action undercuts the weak shale below the limestone. The Niagara River originated as the last glacier receded from the area and water flowed from Lake Erie to Lake Ontario over the Niagara cliffs. Erosion causes the waterfalls to migrate upstream at an average rate of 1.3 m per year.

velocity of the falling water sets up strong turbulence at the base of the falls, which rapidly erodes the weak, nonresistant rock layers. The cliff is gradually undermined, and the falls retreat upstream.

Headward Erosion

In the process of stream erosion and valley evolution, streams have a universal tendency to erode headward, or upslope, and to increase the length of the valley until they reach the divide. Headward erosion can be analyzed by reference to Figure 10.8. The reason erosion is more vigorous at the head of a valley than on its sides is apparent from the relationship between the valley and the regional slope. Above the head of a valley, water flows down the regional slope as a sheet (sheet flow), but the water starts to converge to a point where a definite stream channel begins. As the water is concentrated into a channel, its velocity and erosive power increase far beyond that of the slower-moving sheet of water on the surrounding, ungullied surface. The additional volume and velocity of the channel water erode the head of the valley much faster than sheet flow erodes ungullied slopes or the valley walls. In addition, groundwater moves toward the valley, so that the head of the valley is a favorable location for the development of springs and seeps. These, in turn, help to undercut overlying resistant rock and cause headward erosion to occur much faster than the retreat of the valley walls. The head of the valley is thus extended upslope.

With the universal tendency for headward erosion, the tributaries of one stream can extend upslope and intersect the middle course of another stream, thus diverting the headwater of one stream to the other. This process, known as stream piracy, or stream capture, is illustrated in Figure 10.9. Stream piracy is most likely to occur if headward erosion of one stream

is favored by a steeper gradient or by a course in more easily eroded rocks. Some of the most spectacular examples occur in the folded Appalachian Mountains, where nonresistant shale and limestone are interbedded with resistant quartzite formations. The process of stream capture and the evolution of the region's drainage system are shown in the series of diagrams in Figure 10.10. The original streams flowed in a dendritic pattern (a branching, treelike pattern) on horizontal sediments that once covered the folds. As uplift occurred, erosion removed the horizontal sediments, and the dendritic drainage pattern became superposed, or placed, upon the folded rock beneath. The stream thus cuts across weak and resistant rocks alike. As the major stream cuts a valley across the folded rocks, new tributaries rapidly extend headward along the nonresistant formations. By headward erosion, these new streams progressively capture the superposed tributaries and change the dendritic drainage pattern to a trellis drainage pattern (a pattern in which the tributaries join the main stream at right angles).

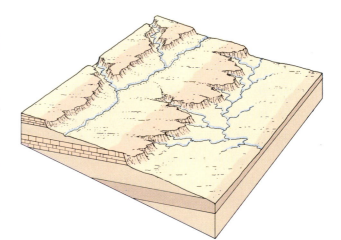

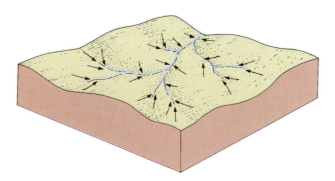

Figure 10.8
Headward erosion is the basic mechanism by which a drainage system is extended upslope. Water flows as a sheet down the regional slope (away from the foreground of the figure). As it converges toward the head of a tributary valley, its velocity is greatly increased, and so its ability to erode also increases. The tributary valley is thus eroded headward, up the regional slope.

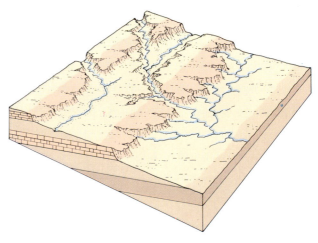

Figure 10.9
Stream piracy occurs where a tributary with a high gradient rapidly erodes headward and captures the tributary of another stream.

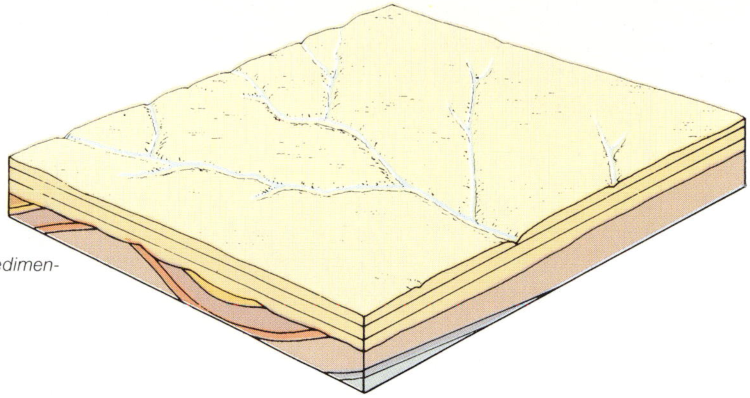

(A) Initially a dendritic pattern formed on horizontal sedimentary rocks, which cover the older, eroded folds.

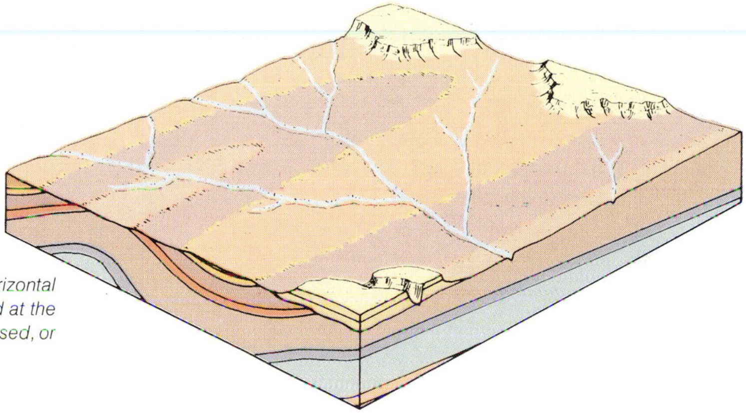

(B) Regional uplift causes erosion to remove the horizontal sediments, so that the older, folded rocks are exposed at the surface. The dendritic drainage pattern is then superposed, or placed upon, the folded rocks.

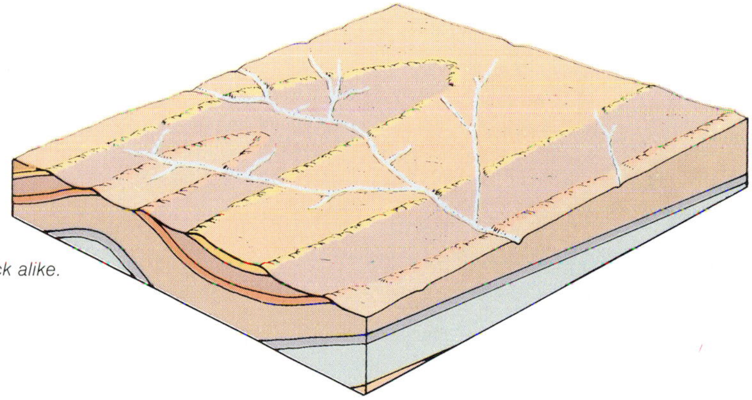

(C) Streams cut across resistant and nonresistant rock alike.

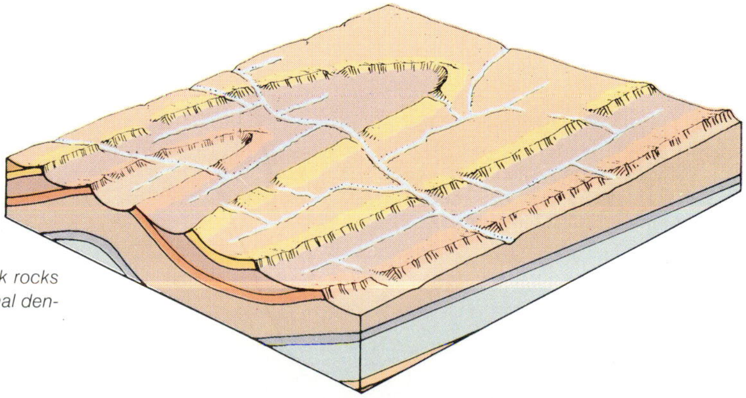

(D) Rapid headward erosion along exposures of weak rocks results in stream capture and modification of the original dendritic pattern to a trellis pattern.

Figure 10.10
A dendritic drainage pattern superposed on a series of folded rocks evolves into a trellis pattern as headward erosion proceeds along nonresistant rock formations.

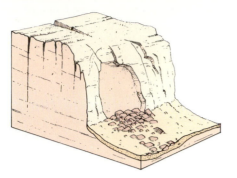

(A) A rockfall is the free fall of rock from steep cliffs.

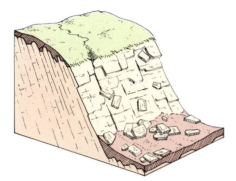

(B) A rockslide is the rapid downslope movement of rock material along a bedding plane, joint, or other plane of structural weakness.

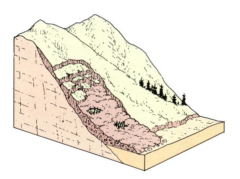

(C) A debris slide is the rapid movement of soil and loose rock fragments. The mass can be dry or moderately wet.

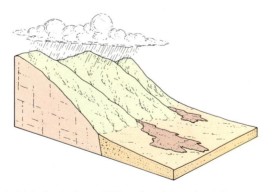

(D) A debris flow is the rapid flow of a mixture of rock fragments, soil, mud, and water. The mixture generally contains a large proportion of mud and water. Mudflows are common.

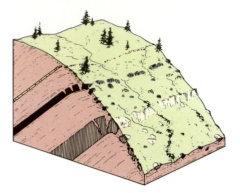

(E) Creep is the slow downslope migration of soil and loose rock fragments that results from a variety of processes, including frost heaving.

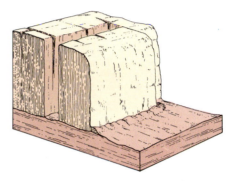

(F) A blockslide is the slow movement of large blocks of material over a layer of weak, plastic material (such as clay or shale).

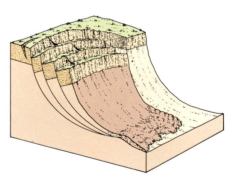

(G) Slump is the slow or moderately rapid movement of a coherent body of rock along a curved rupture surface. Debris flows commonly occur at the end of a slump block.

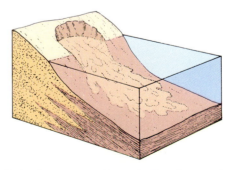

(H) A subaqueous sand flow is the flow of saturated sand or silt beneath the surface of a lake or an ocean.

Figure 10.11

Mass movement takes various forms, all of which produce slope retreat and enlarge valleys. Examples of various types of mass movement are illustrated in the diagrams.

Mass Movement

Erosion of the land surface takes place as loose rock debris produced by weathering is moved downward, by gravity, toward the sea. As we have seen, the movement usually involves flowing water—river systems—but in many areas movement may begin under the influence of gravity alone, as soil slumps down a steep hillside, or where a large block of rock, loosened by weathering, falls from the face of a cliff. These direct movements of material by the force of gravity are called mass movements, or mass wasting.

The mechanics of mass movement can be very complicated. The processes can be rapid and devastating, such as catastrophic landslides, or imperceptibility slow, such as creep down the gentle slope of a grass-covered field. In most types of mass movement, some triggering mechanism is important. The force of gravity is continuous, of course, but it can move material only if it can overcome inertia. Any factor that tends to reduce resistance to motion aids mass movement. Such factors include the saturation of the material by water, oversteepening of slopes through undercutting by streams or waves, alternating freezing and thawing, and vibrations from earthquakes. Water is an important factor in mass movement because it lubricates and adds weight to the mass of material. The various types of mass movement generally fall into two major groups: (1) predominantly slow movement and (2) predominantly rapid movement. Some examples of each are shown in Figure 10.11.

Creep is the most universal type of mass movement and is active on all slopes that are covered with soil. The motion is so slow that it generally is difficult to observe directly, but it is expressed in a variety of ways. On weakly consolidated, grass-covered slopes, evidence of creep can be seen as bulges or low, wavelike swells in the soil. In road cuts and stream banks,

creep can be expressed by the bending of steeply dipping strata in a downslope direction or by the movement of blocks of a distinctive rock type downslope from their outcrop. Additional signs of creep include tilted trees and posts, displaced monuments, deformed roads and fence lines, and tilted retaining walls (Figure 10.12). The slow movement of large blocks (blockslides) can be considered a type of creep.

Many factors combine to cause creep. Undoubtedly, soil moisture is important because it weakens the soil's resistance to movement. In colder regions creep is produced by frost heaving. Water percolates into the pore spaces of the rocks, and then it freezes and expands, lifting the ground surface at a right angle to the slope. When the ground thaws, each particle drops vertically, coming to rest slightly downslope from its original position (Figure 10.13). Repeated freezing and thawing cause the particles to move downslope in a zigzag path. Wetting and drying result in a similar motion of loose particles because moisture causes the expansion of clay minerals.

Many other factors contribute to creep. Growing plants exert a wedgelike pressure between rock particles in the soil, causing them to be displaced downslope. Burrowing organisms also displace particles, and the force of gravity moves them downslope.

Slopes may therefore be considered dynamic systems in which the effects of weathering, mass movement, and headward erosion of small gully tributaries combine to transport material down to a main stream. The net result is slope retreat. The diagram in Figure 10.14 illustrates the major components of a slope system and shows how they function.

The resistant rock units break up into blocks and fragments, which accumulate at the base of the cliff. This coarse talus then moves farther downslope, chiefly by rockslides and by rolling. Farther downslope, the debris moves slowly by creep. In the pro-

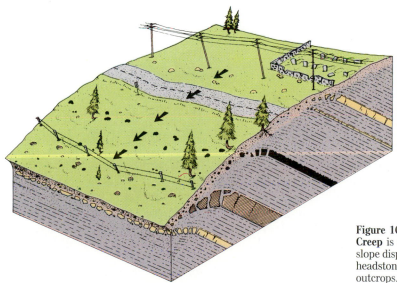

Figure 10.12
Creep is expressed in various ways, such as the downslope displacement of fence lines, roads, telephone poles, headstones in cemeteries, and rock debris from original outcrops.

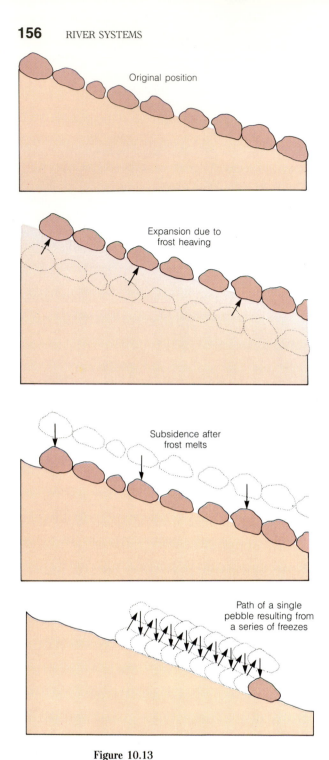

Original position

(A) Water seeps into the pore spaces between fragments of loose rock debris.

Expansion due to frost heaving

(B) As the water freezes and expands, the soil and rock fragments are lifted perpendicular to the ground surface.

Subsidence after frost melts

(C) As the ice melts, gravity pulls the particles down vertically, displacing them slightly downhill.

Path of a single pebble resulting from a series of freezes

(D) Repeated freezing and thawing cause a significant net displacement downslope.

Figure 10.13
Creep can result from repeated frost heaving. With each freeze, there is a net downslope displacement of all loose material.

cess the rock particles are weathered and broken into smaller fragments and continue on a downslope course. The fine-grained rock debris is eventually fed into a stream where it is carried away by running water. As the system operates, the valley slope gradually retreats from the stream channel, although the shape of the profile may remain essentially constant.

Remember that headward erosion is also active. Numerous tributaries and small gullies operate on essentially all slopes, as they collect and funnel debris to the main stream. These small tributaries work headward, undercutting the resistant cliff, and thus making it more susceptible to rockfalls and slumping.

To appreciate the importance of mass movement

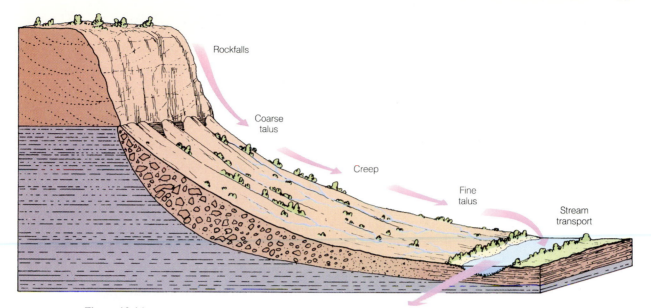

Figure 10.14
Slopes are dynamic systems in which material is continually moving downslope and into a drainage channel. Talus produced by frost action and rockfalls accumulates at the base of the cliff. It is then transformed into smaller and smaller particles by mechanical and chemical weathering and moves downslope by creep and other types of mass movement. Some of the debris may be collected by minor tributaries and moved to the main stream by running water. The remaining fine-grained debris continues to move downslope by gravity. It is eventually fed into a stream, which carries it out through the drainage system.

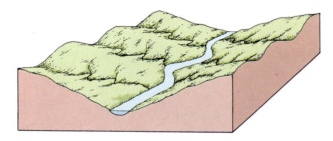

(A) Where rocks are very resistant, mass movement cannot keep pace with downcutting. As a result, a vertical-walled canyon develops.

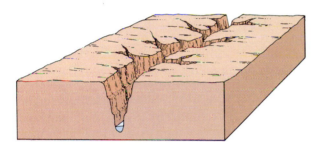

(B) If slope processes can keep pace with downcutting, smooth rolling hills and valleys develop.

Figure 10.15
Slope processes affect the development of landscapes in important ways.

and headward erosion of minor tributaries in the process of valley development, try to visualize what a valley would look like without these processes. If downcutting of the stream channel were the only process in operation, a deep, vertical-walled canyon would result (Figure 10.15A). This does occur where rocks are strong enough to resist the force of gravity, such as in the deep canyons cut in resistant sandstones and limestones in the Colorado Plateau (Figure 10.5). More commonly, slope retreat occurs contemporaneously with downcutting, to produce sloping valleys walls (Figure 10.15B).

One of the most important features of slopes is that, although most of them may appear to be stable and static, they actually are dynamic systems in which surface material is constantly moving downslope in a variety of ways. Large, catastrophic landslides and debris flows are the most hazardous, but imperceptible creep and mass movements of moderate magnitude probably move the most material. Downslope movements are natural events that occur everywhere. The artificial modification of slopes by the construction of dams, roads, and buildings can also cause an increase in mass movement or, where stabilizing structures are built, a temporary reduction in movement. Whether catastrophic or slow moving, the mass movement of slope material may represent a major environmental hazard and a continual challenge for effective land-use planning.

Examples

Many catastrophic landslides are the natural result of high, steep slopes with weak surface material, and they occur regardless of human activity. They are commonly triggered by exceptionally heavy snowfalls or rainfalls, or by earthquakes. For example, during the earthquake in Peru on May 31, 1970, 4,000 m³ of debris roared down the slope of Mount Huascara at a speed of more than 300 km per hour. The debris killed approximately 40,000 people as it covered the town of Yungay with as much as 14 m of mud and rock. Similar types of mass movement can be expected to occur in high mountainous areas in the future.

Many landslides, however, are actually caused by artificial manipulation of the landscape, though the event may be initiated by excess moisture. An example is the landslide that occurred at Vaiont Dam in northern Italy. This, the worst dam disaster in history, resulted from a huge landslide into the Vaiont Reservoir on October 9, 1963. The landslide began as a relatively slow creep over a three-year period. The rate of creep ranged to as much as 7 cm per week, until a month before the catastrophe, when it increased to 25 cm per day. On October 1, animals grazing on the slopes sensed the danger and moved away. Finally, on the day before the slide, the rate of creep was about 40 cm per day. Engineers expected a small landslide and did not realize, until the day before the disaster, that a large area of the mountain slope was moving en masse at a uniform rate.

When the slide broke loose, more than 240,000,000 m³ of rock rushed down the hillslope and splashed into the reservoir. It produced a wave of water over 100 m high, which swept over the dam and rushed down the valley, completely destroying everything in its path for many kilometers downstream. The entire catastrophic event, including the slide and flood, lasted only seven minutes, but it took approximately 2,600 lives and caused untold property damage.

DIFFERENTIAL EROSION

Statement

Different rock types erode at different rates, a phenomenon referred to as differential erosion. Some rocks, such as shale, are soft and weak and typically erode into gentle slopes and valleys. Others, such as sandstone and quartzite, are firm and hard and form prominent ledges and cliffs. Differential erosion occurs on all scales, from ridges and valleys in a mountain belt to delicate paper-thin layers within a single bed, and gives rise to a variety of fascinating landforms.

Discussion

Numerous examples of differential erosion can be seen in every part of the world. In areas of horizontal rock, alternating layers of sandstone, shale, and limestone (a common sequence of rock produced on the stable platforms of all continents) typically erode into alternating cliffs and slopes. This is probably the most widespread example of differential erosion and is superbly expressed in arid regions where plateaus, mesas, and buttes are formed on resistant sandstone or limestone and the underlying shale forms gentle slopes (Figure 10.16). In some areas differential erosion may produce a fascinating variety of landforms, such as the pinnacles of Bryce Canyon (Figure 10.17) and the natural arches of the Colorado Plateau (Figure 10.18).

Where a sequence of alternating resistant and nonresistant strata is tilted or folded, the nonresistant layers are eroded into long lowlands or valleys parallel to the trend of the rock. Ridges form on the resistant strata (Figure 10.19). The Valley and Ridge Province of the eastern United States is a classic example (Figure 10.20).

Figure 10.16
Differential erosion produces plateaus, mesas, buttes, and pinnacles in Monument Valley, Utah, from resistant sandstone overlying nonresistant shale.

Figure 10.17
The columns in Bryce Canyon National Park are produced by differential erosion along receeding cliffs where zones of weakness occur along fractures and bedding planes.

Figure 10.18
Natural arches develop in massive sandstone layers in arid regions where groundwater out near the base of a cliff and dissolves the cement that holds the sand grains together.

Figure 10.19
Differential erosion of a sequence of tilted strata produces ridges on resistant formations and long, narrow valleys in nonresistant rocks.

Figure 10.20
Erosion of folded strata in the Appalachian Mountains, in the eastern United States, produces a series of zigzag ridges and valleys.

PROCESSES OF STREAM DEPOSITION

Statement

In the lower part of a river system, the surface of the land slopes gently toward the sea and the stream gradient is very low. As a result the river is unable to transport all of its load, and a significant amount of sediment may be deposited across the floodplain. Most of the sediment load, however, is carried to the sea, where it is deposited as a delta. In arid regions where rivers enter dry basins, the sediment load is deposited as an alluvial fan.

Discussion

Floodplain Deposits

Erosion is the dominant process in the high headwaters of a drainage system, but on the more gentle slopes of the stable platform or shield, most stream valleys are floored by large quantities of sediment that make up a flat surface over which the stream flows. This surface is called the floodplain, and during high floods it may be completely covered with water. Sediment is deposited across the floodplain by a variety of processes, foremost among which form point bars, natural levees, and backswamps. A schematic diagram showing features commonly developed on a river floodplain is shown in Figure 10.21. It serves as a simple graphic model of floodplain sedimentation.

Meanders and Point Bars

All rivers naturally tend to flow in a sinuous pattern, even if the slope is relatively steep. This is because water flow is turbulent, and any bend or irregularity in the channel deflects the flow of water to the opposite bank. The force of the water striking the stream bank causes erosion and undercutting, which initiate a small bend in the river channel. In time, as the current continues to impinge on the outside of the channel, the bend grows larger and is accentuated; a small bend ultimately grows into a large meander bend (Figure 10.22). On the side of the meander, velocity is at a minimum, so that some of the sediment load is deposited. These deposits occur on the point of the meander bend and are called a point bar.

As a meander bend becomes accentuated, it develops an almost complete circle. Eventually the river channel cuts across the meander loop and follows a more direct course downslope. The meander cutoff forms a short, but sharp increase in stream gradient, causing the river to abandon completely the old meander loop, which remains as a crescent-shaped lake called an oxbow lake. The point bar sediment inside the meander thus remains river deposits.

Natural Levees

Another key process operating on a floodplain is the development of natural levees. If a river overflows its banks during the flood stage, the water is no longer confined to a channel but flows over the land surface

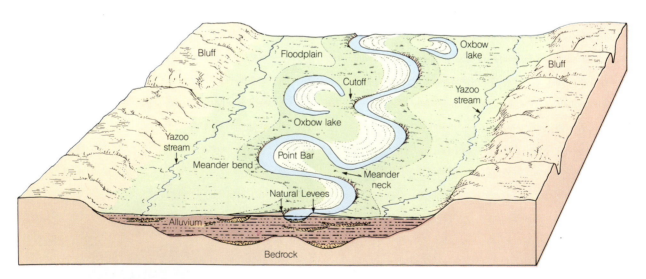

Figure 10.21
The major features of a floodplain include meanders, point bars, oxbow lakes, natural levees, backswamps, and streams. A stream flowing around a meander bend erodes the outside curve and deposits sediment on the inside curve to form a point bar. The meander bend migrates laterally and is ultimately cut off, to form an oxbow lake. Natural levees build up the banks of the stream, and backswamps develop on the lower surfaces of the floodplain. Yazoo streams have difficulty entering the main stream because of the high natural levees and thus flow parallel to it for considerable distances before becoming tributaries. Slope retreat continues to widen the low valley, which is partly filled with river sediment.

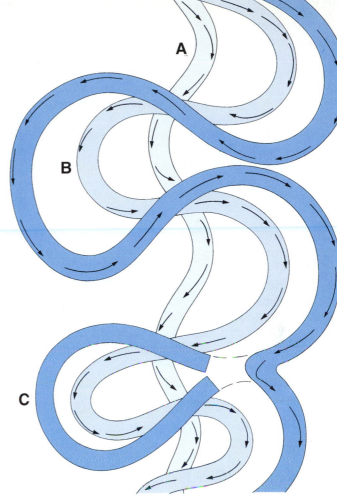

Figure 10.22

The evolution of stream meanders results because erosion occurs on the outside of a curve in the stream channel, where velocity is greatest, and deposition occurs on the inside of the curve, where velocity is least.

(A) (light blue) Stream flow is deflected by an irregularity and moves to the opposite bank, where erosion begins.

(B) (darker blue) Once the bend begins to form, the flow of water continues to impinge on its outside curve, so that a meander loop develops. At the same time, deposition occurs on the inside of the bend as a result of the lower stream velocities in that area.

(C) (darkest blue) The meander is enlarged and migrates laterally, with the contemporaneous growth of a point bar. There is a general downslope migration of meanders as they grow larger and ultimately cut themselves off to form oxbow lakes.

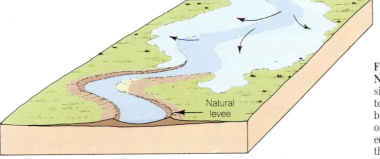

Figure 10.23

Natural levees are wedge-shaped deposits of fine sand, silt, and mud that taper away from the stream banks toward the backswamp. They form during flood stages because, as the stream overflows its banks, the velocity of the water is reduced, and silt is deposited. As the levees grow higher, the stream channel also rises, and thus the river can be higher than the surrounding floodplain.

in a broad sheet. This flow pattern significantly reduces the water's velocity, and some of the suspended sediment settles out. The coarsest material is deposited close to the channel, where it builds up an embankment known as a natural levee. Natural levees grow with each flood. They soon form high embankments, and thus a river can actually build its channel higher than the surrounding area (Figure 10.23).

Backswamps

As a result of the growth and development of natural levees, much of the floodplain may be below river level. This area, known as the backswamp, is poorly drained and is commonly the site of marshes and swamps. Fine-grained mud carried by the floodwater ultimately settles out, adding to the accumulation of floodplain sediment.

Braided Streams

If streams are supplied with more coarse-grained sediment than they can carry, they deposit the excess material on the channel floor as sand and gravel bars. This deposition forces a stream to split into two or more channels, so the stream pattern forms an interlacing network of braided channels and islands (Figure 10.24). The braided pattern is best developed in rivers that are heavily laden with coarse sand and gravel and fluctuate greatly in the volume of water that they discharge.

Stream Terraces

Many streams may fill part of their valleys with sediment during one period of their history and then cut through the sediment fill during a subsequent period. The factors that may cause a stream to change from deposition to erosion include (1) changes in the volume of discharge as a result of climatic fluctuations, (2) changes in gradient caused by regional uplift, and (3) changes in the amount of sediment load. These fluctuations commonly produce stream terraces.

The basic steps in the evolution of stream terraces are shown in the diagrams in Figure 10.25. In block A, a stream cuts a valley by downcutting and slope retreat. In block B, changes such as regional tilting of the land or a rise of base level cause the stream to deposit part of its sediment load and build up a floodplain, which forms a broad, flat valley floor. In block C, subsequent changes (such as uplift or increased runoff) cause renewed downcutting into the easily eroded floodplain deposits, so that a single set of terraces develops on both sides of the river. Further erosion can produce additional terraces (block D) by the lateral shifting of the meandering stream.

During the last Ice Age, the hydrology of most rivers changed significantly and produced numerous stream terraces. Stream runoff was increased greatly by the melting ice, and large quantities of sediment deposited by the glaciers were reworked by the streams, causing many to become overloaded. In addition, the climatic changes accompanying the Ice Age caused a general worldwide increase in precipitation. As a result, many streams that filled part of their valleys with sediment during the Ice Age are now cutting through that sediment fill to form stream terraces.

Deltas

As a river enters a lake or the ocean, its velocity suddenly diminishes, and most of its sediment load is deposited to form a delta. The growth of a delta can be complex, especially for large rivers depositing huge volumes of sediment. Two major processes, however,

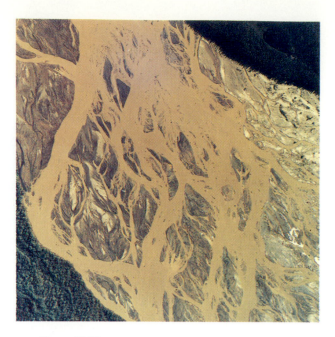

Figure 10.24
A braided stream pattern commonly results if a river is supplied with more sediment than it can carry. Deposition occurs, causing the river to develop new channels.

are fundamental to the formation and growth of a delta:

1. The splitting of a stream into a distributary channel system, which extends into the open water in a branching pattern
2. The development of local breaks, called crevasses, in natural levees, through which sediment is diverted and deposited as splays in the area between the distributaries

The diagrams in Figure 10.26 illustrate the development of distributaries. As a river enters the ocean (or a lake) and the flowing water is no longer confined to a channel, the currents flare out, rapidly losing velocity and flow energy. The coarse material carried by the stream is deposited in two specific areas: (1) along the margin of quiet water on either side of the main channel (these deposits build up subaqueous natural levees) and (2) in the channel at the river mouth, where there is a sudden loss of velocity (these deposits build a bar at the mouth of the channel). These two deposits effectively create two smaller channels (distributary channels), which extend seaward for some distance. The process is then repeated, and each new distributary is divided into two smaller distributaries. In this manner a system of branching distributaries builds seaward in a fan-shaped pattern.

Figure 10.27 shows how the area between distributaries is filled with sediment. A local break in the levee, called a crevasse, forms during periods of high runoff and diverts a significant volume of water and sediment from the main stream. The escaping water

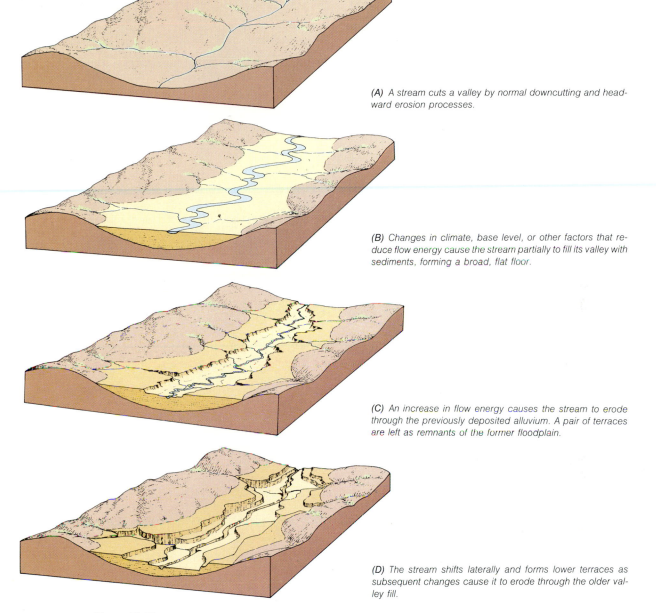

(A) A stream cuts a valley by normal downcutting and headward erosion processes.

(B) Changes in climate, base level, or other factors that reduce flow energy cause the stream partially to fill its valley with sediments, forming a broad, flat floor.

(C) An increase in flow energy causes the stream to erode through the previously deposited alluvium. A pair of terraces are left as remnants of the former floodplain.

(D) The stream shifts laterally and forms lower terraces as subsequent changes cause it to erode through the older valley fill.

Figure 10.25
The evolution of stream terraces involves the deposition of sediment in a stream valley, subsequent change in the stream's gradient, and renewed downcutting. These changes can be initiated by various factors that affect a stream's capacity to transport sediment, such as changes in climate, changes in base level, or regional uplift.

spreads out and deposits its sediment to form a splay, which is essentially a small delta with small distributaries and systems of subsplays.

A major phenomenon in the construction of deltas is the shifting of the entire course of a river. Distributaries cannot extend indefinitely into the ocean because the gradient and capacity of the river to flow gradually decrease. The river, therefore, is eventually diverted to a new course, which has a higher gradient. This diversion generally happens during a flood. The river breaks through its natural levee, far inland from

the active distributaries of the delta, and develops a new course to the ocean. The new course shifts the site of sedimentation to a different area, and the abandoned segment of the delta is attacked by wave and current action. The new, active delta builds seaward, developing distributaries and splays, until it also is eventually abandoned, and another site of active sedimentation is formed. The shifting back and forth of the main river channel is thus a major way in which sediment is dispersed and a delta grows (Figure 10.28).

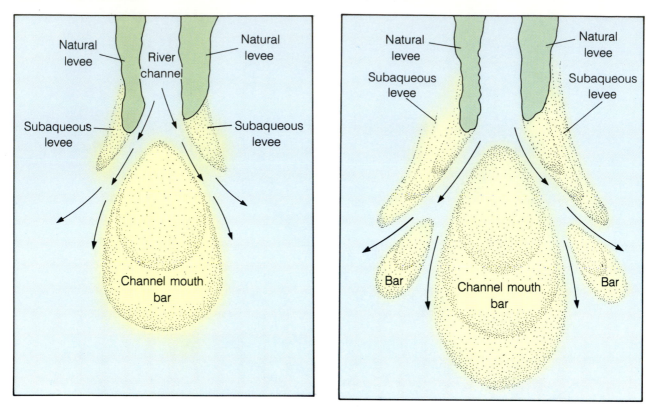

Figure 10.26
Distributaries develop where a stream enters a lake or the sea. A bar forms at the mouth of a river channel where water that was previously confined to the channel loses velocity, resulting in deposition as the stream enters the ocean. Subaqueous natural levees also form below water level. The bar then diverts the water coming from the main stream into two distributary channels, which grow seaward. This process is repeated, forming branching distributaries.

Figure 10.27
The formation of splays. Splays form where a break in a natural levee permits part of the stream to be diverted to the backswamp. This diversion reduces the velocity and causes deposition of sediment in a fan-shaped splay. Like the main river, a splay has distributaries and a series of smaller subsplays.

Alluvial Fans

Alluvial fans are stream deposits that accumulate in dry basins at the bases of mountain fronts. Deposition results from the sudden decrease in velocity as a stream emerges from the steep slopes of the upland and flows across the adjacent basin with its gentle gradient. Alluvial fans form mostly in arid regions, where streams flow intermittently. Such areas usually have a large quantity of loose, weathered rock debris on the surface; thus, when rain falls, the streams transport huge volumes of sediment. Where a river emerges from a mountain front into the adjacent

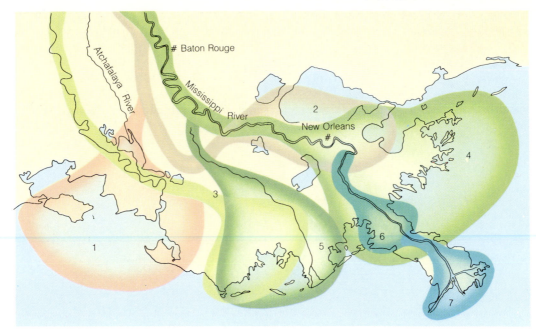

Figure 10.28
The history of the Mississippi Delta involves repeated shifting of the main channel, which has formed seven subdeltas. The flow of the Mississippi River is relatively constant throughout most of the year, and the water is confined to the main distributary. Most of the sediment is deposited in a small sector of the delta front. The gradient of the river where it enters the sea is very low, so the extension of the delta seaward cannot continue indefinitely. A major break in the natural levee upstream eventually diverts the entire flow to some other sector, and the process is repeated. Wave action then erodes the inactive bird-foot deltas. Previous subdeltas are indicated by numbers (1 through 6) according to age. Number 7 is the present subdelta. The active distributary system (6 and 7) has built a major bird-foot delta during the last five hundred years.

Figure 10.29
Alluvial fans form in arid regions, where a stream enters a dry basin and deposits its load of sediment.

basin, its gradient is sharply reduced and deposition occurs. The channel soon becomes clogged with sediment and the stream is forced to seek a new course. In this manner the stream shifts from side to side and builds up an arcuate, fan-shaped deposit (Figure 10.29). As several fans build basinward at the mouths of adjacent canyons, they eventually merge to form broad slopes of alluvium at the base of the mountain range.

Although alluvial fans and deltas are somewhat similar, they differ in mode of origin and internal structure. In deltas stream flow is checked by standing water, and sediment is deposited in a body of water. The level of the ocean or lake effectively forms the upper limit to which the delta can be built. In contrast a fan is deposited in a dry basin, and its upper surface is not limited by water level. The coarse, unweathered, poorly sorted sands and gravels of an alluvial fan also contrast with the fine sand, silt, and mud that predominate in a delta.

SUMMARY

1. A river system is a network of connecting channels that collect surface water and funnel it back to the ocean. It can be divided into three subsystems: (a) tributaries, which collect and transport water and sediment into the main stream, (b) a main trunk stream, which is largely a transporting system, and (c) a dispersing system at the river's mouth.

2. The major variables in a river system are (a) discharge, (b) gradient, (c) base level, (d) sediment load, (e) volume, (f) channel shape, and (g) base level.

3. A river system functions as a unified whole, adjusting its profile to establish equilibrium among the factors that influence flow. If one factor changes, the river system adjusts to reestablish equilibrium.

4. Streams erode the landscape by (a) removal of the regolith, (b) downcutting stream channels, (c) headward erosion, and (d) mass movement.

5. The most important types of mass movement are creep, debris flows, and slumping. Mass movement is especially important in erosion of valley slopes.

6. In the transporting segment of a river system, sediment is deposited on the floodplain to form (a) point bars, (b) natural levees, and (c) backswamps.

7. Stream terraces form by deposition of sediment in a stream valley, followed by rejuvenation of erosion. The renewed erosion may result from uplift, climatic changes, or changes in base level.

8. Most of the sediment load carried by a river is deposited as a delta where the river enters a lake or an ocean. The major delta-forming processes are (a) development of distributaries and (b) development of splays.

9. Where a stream enters a dry basin and is unable to transport, deposition occurs to form an alluvial fan.

KEY TERMS

abrasion (p. 151)
alluvial fan (p. 164)
backswamp (p. 161)
base level (p. 148)
bed load (p. 148)
braided stream (p. 162)
collecting system (p. 146)
creep (p. 155)
delta (p. 162)
dendritic
  drainage (p. 146)

discharge (p. 147)
dispersing
  system (p. 146)
distributary (p. 146)
divide (p. 146)
drainage basin (p. 146)
fan (p. 164)
floodplain (p. 160)
gradient (p. 147)
headward
  erosion (p. 152)

longitudinal
  profile (p. 149)
meander (p. 160)
natural levee (p. 160)
mass movement (p. 155)
oxbow lake (p. 160)
point bar (p. 160)
pothole (p. 151)
saltation (p. 148)
splay (p. 163)

stream load (p. 147)
stream piracy (p. 152)
stream terrace (p. 162)
superposed stream (p. 152)
suspended load (p. 148)
transporting
  system (p. 146)
trellis drainage
  pattern (p. 153)
tributary (p. 146)

REVIEW QUESTIONS

1. Explain how stream action is able to cut a valley through solid bedrock.
2. What is headward erosion? Why does it occur?
3. Explain the process of stream piracy and cite examples of how it modifies a drainage system.
4. Name and describe the important landforms associated with floodplain deposits.
5. Describe the steps involved in the growth of a stream meander and the formation of an oxbow lake.
6. Explain the origin of natural levees.
7. Describe and illustrate the steps in the development of stream terraces.
8. Explain how a delta is built where a stream enters a lake or the sea.
9. Explain how an alluvial fan is built.
10. Explain the reasons for concluding that stream action (running water) is the most important process of erosion on Earth.
11. Describe and illustrate the three major subsystems of a river.
12. Explain the concept of equilibrium in river systems and cite several examples of how streams adjust to attain equilibrium.
13. How does urbanization affect surface runoff?

Multiple-Choice Questions

1. Most stream tributaries occur in
   a. a collecting system.
   b. a transporting system.
   c. a dispersal system.
   d. a differentiation system.
   e. a merging system.
2. The lowest level to which a stream can erode its channel is called the stream
   a. gradient.
   b. base level.
   c. capacity.
   d. competence.
   e. degradation.
3. A stream transports sediment load
   a. in suspension.
   b. in solution.
   c. by traction.
   d. All of the above.
   e. Only (a) and (b) are correct.
4. Urbanization causes problems in stream flow because
   a. it causes more rapid runoff.
   b. it prevents infiltration into the groundwater system.
   c. it changes the nature of the terrain.
   d. it upsets stream equilibrium.
   e. All of the above.
5. The main tool used by a stream to erode its channel is
   a. bank deposit.
   b. sediment load.
   c. rapids.
   d. base level.
   e. material in solution.
6. The development of a drainage system may involve all of the following but
   a. superposition.
   b. stream piracy.
   c. antiposition.
   d. downcutting.
   e. headward erosion.

7. A dendritic drainage pattern, superposed upon a series of folded rocks, will commonly
   a. retain its basic pattern, but downcut its channel.
   b. evolve into a larger dendritic system.
   c. evolve into a trellis pattern.
   d. evolve into a longitudinal pattern by stream piracy.
   e. downcut its channel and enlarge the dendritic pattern.
8. Which of the following types of mass movement is most universal, regardless of terrain, rock type, and climate?
   a. rock falls
   b. debris flows
   c. solifluction
   d. creep
   e. slump
9. As a river overflows its banks during floods, it drops much of its suspended load right next to the channel to form
   a. natural levees.
   b. alluvial fans.
   c. point bars.
   d. flood plains.
   e. backswamps.
10. Point bars along a meandering stream are deposited
    a. on the inside of the meanders.
    b. on the outside of the meanders.
    c. along straight segments between meanders.
    d. in the central part of the channel.
    e. as a natural levee.

True/False Questions

1. *T F* The Mississippi River and its tributaries, the Missouri and Ohio rivers, function as independent systems.
2. *T F* By far the greatest amount of material carried by a stream is derived from erosion of the stream channel.
3. *T F* A river is in equilibrium if its channel form, gradient, volume, and velocity are balanced, so that neither erosion nor deposition occurs.
4. *T F* Extensive stream piracy results in the development of a dendritic drainage pattern.
5. *T F* The development of distributaries in a delta is enhanced if waves or tides are strong.

Fill in the Missing Terms

1. _____ is the major process occurring on the outside of a meander.
2. A _____ stream pattern is best developed in rivers that are heavily laden with erosional debris and that fluctuate in water discharge.
3. The two main processes in the growth of a delta are the development of _____ and _____.
4. A dendritic drainage pattern superposed on a series of folded rocks typically evolves into a _____ pattern.
5. The _____ _____ of a stream is the lowest level to which the stream can erode its channel.

11

GROUNDWATER SYSTEMS

The movement of water in the pore spaces of rocks and soil beneath Earth's surface is a geologic process not easily observed and therefore not readily appreciated; yet, groundwater is an integral part of the hydrologic system and a vital natural resource. Groundwater is not rare or unusual. It is distributed everywhere beneath the surface. It occurs not only in humid areas, but also in desert regions, under the frozen Arctic, and in high mountain ranges. In many areas the amount of water seeping into the ground equals or exceeds the surface runoff, but in arid regions it may be less than 5 percent.

In limestone formations groundwater commonly dissolves the rock to form a network of caves and subterranean passageways. The caves may enlarge and their roofs commonly collapse, so that circular depressions called sinkholes are formed. In areas where sinkholes dominate, a unique landscape called karst evolves. There is no integrated surface drainage in the area, and most of the water falling on the surface as rain becomes part of the groundwater system.

In this chapter we will study the groundwater system, how water moves through various types of pore spaces in the rock, and how groundwater forms karst topography. We will then consider how this most precious natural resource is used, and how we attempt to cope with the environmental problems that result when this part of the hydrologic system is manipulated and modified.

MAJOR CONCEPTS

1. The movement of groundwater is controlled largely by the porosity and permeability of the rocks through which it flows.
2. The water table is the upper surface of the zone of saturation.
3. Groundwater moves slowly (percolates) through the pore spaces in the rocks by the pull of gravity. In artesian systems, it is moved by hydrostatic pressure.
4. The natural discharge of groundwater is generally into streams, marshes, and lakes.
5. Artesian water is water that is confined under pressure, like water in a pipe. It occurs in permeable beds bounded by impermeable beds.
6. Erosion by groundwater produces karst topography, which is characterized by sinkholes, solution valleys, and disappearing streams.

CHARACTERISTICS OF THE GROUNDWATER SYSTEM

Statement

Water can infiltrate the subsurface because solid bedrock, as well as loose soil, sand, and gravel, contains pore spaces. As water seeps into the ground, gravity pulls it downward, and below a certain level all of the openings in the rocks are saturated. The upper surface of this saturated zone is called the water table. Groundwater moves from areas of high pressure to areas of low pressure (toward streams, lakes, and swamps), and under various geologic conditions, it will discharge as springs. The pores, or open spaces, within a rock can be spaces between mineral grains, cracks, solution cavities, or vesicles. Two physical properties of a rock largely control the amount and movement of groundwater. One is porosity, the percentage of the total volume of the rock consisting of voids. Porosity determines how much water a rock body can hold. The second property is permeability, the capacity of a rock to transmit fluids. Permeability depends on such factors as the size of the voids and the degree to which they are interconnected. Porosity and permeability are not identical. Some very porous rocks (such as shale) have such small pore spaces that it is difficult for water to move through them. Even though they have high porosity, such rocks are impermeable.

Contrary to popular belief, underground rivers are uncommon and occur only in cavernous limestone and in some lava tunnels in volcanic terrain. Most groundwater occurs in the pore space between grains, in fractures, and in other small voids.

Discussion

Porosity

There are four main types of pore spaces, or voids, in rocks (Figure 11.1): (1) spaces between mineral grains, (2) fractures, (3) solution cavities, and (4) vesicles. In sand and gravel deposits, pore space can constitute from 12 to 45 percent of the total volume. If several grain sizes are abundant and the smaller grains fill in the space between larger grains, or if a significant amount of cementing material fills in the space between grains, the porosity is greatly reduced. In addition, all rocks are cut by fractures, and in some dense rocks (such as granite), fractures constitute the only significant pore spaces. Solution activity, especially in limestone, commonly removes soluble material, forming pits and holes. Some limestones thus have high porosity. As water moves along joints and bedding planes in limestone, solution enlarges fractures in the rock and develops passageways, which

may grow to become caves. In basalts and other volcanic rocks, vesicles formed by trapped gas bubbles significantly increased porosity. Vesicles commonly are concentrated near the top of a lava flow and form zones of very high porosity, which can be interconnected by columnar joints or by cinders and rubble at the top and base of the flow.

Permeability

Permeability, the capacity of a rock to transmit a fluid, varies with the fluid's viscosity, the size of openings, and particularly, the degree to which the openings are interconnected. Rocks that commonly have high permeability are conglomerates, sandstones, basalt, and certain limestones. Permeability in sandstones and conglomerates is high because of the relatively large, interconnected pore spaces between the grains. Basalt is permeable because it often is extensively fractured with columnar jointing and because the tops of most flows are highly vesicular. Fractured limestones are also permeable, as are limestones in which solution activity has created many small solution cavities. Rocks that have low permeability are shale, unfractured granite, quartzite, and other dense, crystalline metamorphic rocks.

Water moves through the available pore spaces, twisting and turning through the tiny voids. Regardless of the degree of permeability, groundwater flows extremely slowly in comparison to the turbulent flow of rivers. Whereas the velocity of water in rivers is measured in kilometers per hour, the velocity of groundwater commonly ranges from 1 m per day to 1 m per year. The highest rate of percolation measured in the United States, in exceptionally permeable material, is only 250 m per day. Only in special cases, such as the flow of water in caves, does the movement of groundwater even approach the velocity of slow-moving surface streams.

The Water Table

As water seeps into the ground, gravity pulls it downward through two zones of soil and rock. In the upper zone, the pore spaces in the rocks are only partly saturated, and the water forms a thin film, clinging to grains by surface tension. This zone, in which pore space is filled partly with air and partly with water, is called the zone of aeration. Below a certain level, all of the openings in the rock are completely filled with water (Figure 11.2) (the zone of saturation). The water table, which is the upper surface of the zone of saturation, is an important element in the groundwater system. It may be only a meter or less deep in humid regions, but hundreds or even thousands of meters below the surface in deserts. In swamps and lakes, the water table is essentially at the land surface.

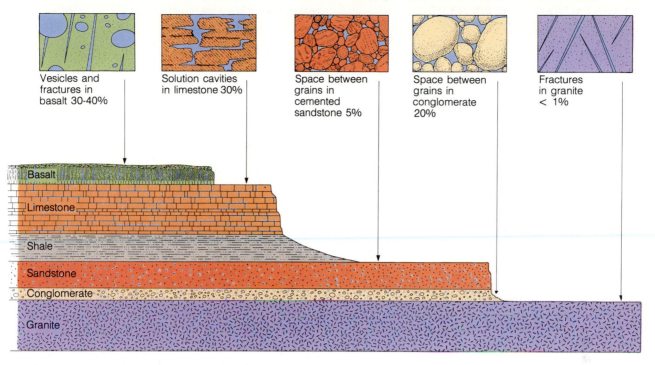

Figure 11.1
Various types of pore spaces in rocks permit the flow of groundwater. Porosity resulting from vesicles is common in basalt. Porosity resulting from solution activity is common in limestones. Porosity resulting from spaces between grains is exemplified in sandstone and conglomerate. Porosity resulting from fractures occurs in almost all rocks.

Although the water table cannot be observed directly, it has been studied and mapped from data collected from wells, springs, and surface drainage, and the movement of groundwater has been studied by means of radioactive isotopes, dyes, and other tracers, so that extensive knowledge of this largely invisible body of water has been acquired.

Several important generalizations can be made about the water table and its relation to surface topography and surface drainage. These are diagrammed in Figure 11.3. In general the water table tends to mimic the surface topography. In flat country the water table is flat. In areas of rolling hills, it rises and falls with the surface of the land. The reason for this is that groundwater moves very slowly, so that the water table rises in the areas beneath the hills during periods of greater precipitation, and during droughts it tends to flatten out. Where impermeable layers (such as shale) occur within the zone of aeration, the groundwater is trapped above the regional water table, forming a local perched water table. If a perched water table extends to the side of a valley,

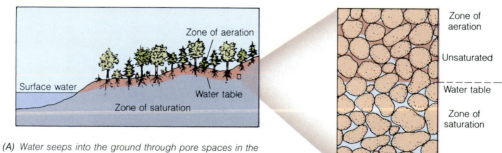

(A) Water seeps into the ground through pore spaces in the rock and soil. It passes first through the zone of aeration, in which the pore spaces are occupied by both air and water, and then into the zone of saturation, in which all of the pore spaces are filled with water. The depth of the water table varies with climate and amount of precipitation.

(B) Zones of aeration and saturation and the water table are shown in microscopic view.

Figure 11.2
The water table is the upper surface of the zone of saturation.

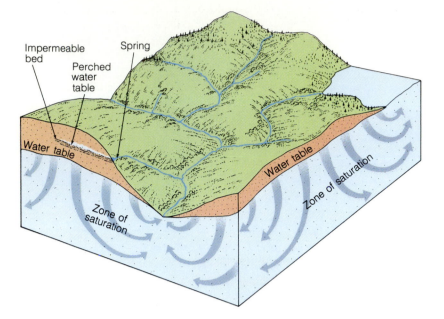

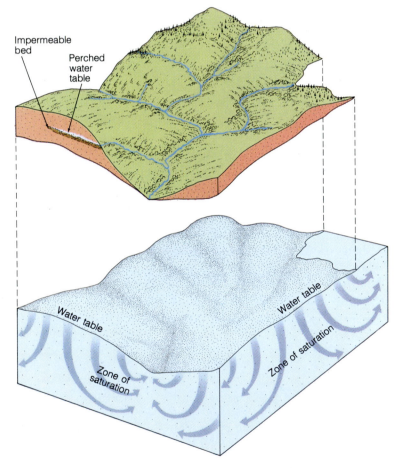

Figure 11.3
The movement of groundwater is directed toward areas of least pressure. In the idealized system depicted here, groundwater moves downward to the water table (by the pull of gravity) and then moves toward areas of least pressure. The configuration of the water table has a strong influence on the direction of movement. In most areas the water table is a subdued replica of the topography: the exploded view (B) shows high and low areas in the water table, much like the hills and valleys on the surface. Differences in the height of the water table cause differences in the pressure on water in the saturation zone at a given point. Water thus moves downward beneath the high areas of the water table (because of the higher pressure) and upward beneath the low areas. It commonly seeps into streams, lakes, and swamps, where the table is near the surface. In areas where the water table is low, water from streams and lakes moves downward toward the zone of saturation. A line of springs and seeps commonly occurs where an impermeable rock layer that has formed a perched water table is exposed at the surface.

springs and seeps occur. The water table is at the surface in lakes, swamps, and most streams, and water moves through the subsurface material toward these areas following the general paths shown in Figure 11.3. In arid regions, however, most streams lie above the water table, so that they lose much of their water through seepage into the subsurface.

The Movement of Groundwater

The difference in elevation between parts of the water table is called the hydraulic head. It causes the water to follow the paths illustrated in Figure 11.3. If we could trace the path of a particle of water, we would find that gravity slowly pulls it through the zone of aeration to the water table. When a particle encounters the water table, it continues to move downward by the pull of gravity along curved paths from areas where the water table is high toward areas where it is low (lakes, streams, and swamps). The path of groundwater movement is not down the slope of the water table, as one might first suspect. The explanation for this seemingly indirect flow is that the water table is not a solid surface like the ground surface. It is a surface of liquid, which in some ways resembles the surface of a wave. Water at any given point below the water table beneath a hill is under greater pressure than water at the same elevation below the lower water table in a valley. Groundwater therefore moves downward and toward points of lesser pressure.

Although these paths of groundwater movement may seem indirect, they conform to the laws of fluid physics and have been mapped in many areas by tracing the movement of dye injected into the system. The movement of the dye reveals a continual slow circulation of groundwater from infiltration at the surface to seepage into steams, rivers, and lakes. Groundwater is not stagnant and motionless. Rather, it is a dynamic part of the hydrologic system, in constant motion, and is intimately related to surface drainage. Like other parts of the hydrologic system (rivers and glaciers), the groundwater system is an open system into which water enters via surface water infiltrating into the ground. Water then moves through the system by percolating through the pore spaces of the rock and ultimately leaves the system by seeping into streams or lakes. At considerable depths all pore spaces in the rocks are closed by high pressure and there is no free water. This is the lower limit, or base, of the groundwater system.

Natural and Artificial Discharge

Groundwater reservoirs discharge naturally wherever the water table intersects the surface of the ground. Generally, such places are inconspicuous, typically occurring in the channels of streams and on the floors and banks of marshes and lakes. Under various special geologic conditions, the water table intersects the ground surface and discharges as springs and seeps (Figure 11.4).

The natural discharge of groundwater reservoirs into the drainage system introduces a significant volume of water into streams. This discharge is the major link between groundwater reservoirs and other parts of the hydrologic system. If it were not for groundwater discharge, many permanent streams would be dry during parts of the year.

Artificial discharge results from the extraction of water from wells, which are made by simply digging or drilling holes into the zone of saturation. Many millions of wells have been drilled, so that in some areas artificial discharge has modified the groundwater system. Indeed, in some areas more water is removed from the groundwater system by artificial discharge than by natural recharge and the level of the water table drops.

Ordinary wells are made simply by digging or drilling holes through the zone of aeration into the zone of saturation, as shown in Figure 11.5. Water then flows out of the pores into the well, filling it to the level of the water table. When a well is pumped, the water table is drawn down around the well in the shape of a cone, called the cone of depression. If water is withdrawn faster than it can be replenished, the cone of depression continues to grow, and the well ultimately goes dry.

Example

Although groundwater is largely invisible and only isolated glimpses of its presence can be seen in one place, a dramatic example of its movement can be seen in the Snake River Plain in Idaho. This region is a vast lava plain extending across the entire southern part of the state. It was built up by innumerable floods of basaltic lava with some interbedded coarse sand and gravel deposited in streams and lakes, which occupied the region during the intervals between volcanic eruptions. The porosity and permeability of the basaltic rocks are remarkably high. Porosity is produced by columnar joints and vesicular texture in the basalt and by spore space in layers of ash and rubble between the basalt flows. In addition, porosity is naturally high in lava tubes and in layers of unconsolidated coarse sand and gravel between some of the flows. In terms of permeability, the rock sequence is almost like a sieve.

The Snake River is located near the southern margin of the plain, so that tributary streams coming from the mountains to the north are forced to flow across the plain before they can join the Snake River (Figure 11.6). Only one river actually completes the short journey! The rest terminate after flowing a short distance across the plain; they lose their entire vol-

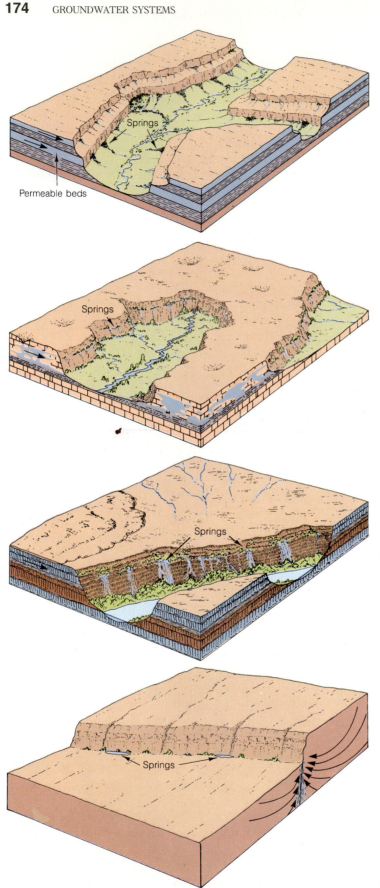

(A) A spring line develops on valley walls where impermeable beds cause groundwater in permeable layers to migrate laterally and eventually seep out at the surface.

Springs

Permeable beds

(B) Springs form along valley slopes where cavernous limestone permits the free flow of groundwater to the surface.

Springs

(C) Surface water readily seeps into vesicular and jointed basalt flows. It then migrates laterally and forms springs where basalt units are exposed in canyon walls.

Springs

(D) Many faults displace rocks so that impermeable beds are placed next to permeable beds. A spring line commonly results as groundwater migrates upward along a fault line.

Springs

Figure 11.4
Springs can be produced under a variety of geologic conditions, some of which are illustrated in the block diagrams here. They are natural discharges of the groundwater reservoir and introduce a significant volume of water to surface runoff.

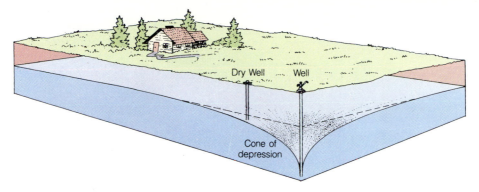

Figure 11.5
A cone of depression in a water table results if water is withdrawn from a well faster than it can be replenished. The cone can extend out for hundreds of meters around large deep wells and effectively lower the water table over a large area. Shallow wells nearby then run dry because they lie above the lowered water table.

ume of water by seeping into the subsurface. Two of the major, would-be tributaries are known, respectively, as Big Lost and Little Lost rivers.

However, the groundwater returns to the surface in a series of spectacular springs, approximately 400 km downstream. The largest and best known are the Thousand Springs just west of Twin Falls, Idaho. These springs clearly show the tremendous movement of groundwater as they discharge about 5,000 cubic feet per second (nearly 37,000 gallons). The visible springs issue from a layer of vesicular basalt 50 m above the river. However, the volume of water that seeps into the Snake River in a less spectacular fashion below the banks is no doubt many times as great. So much water comes from the Thousand Springs area that an electric power plant has been built on the site to utilize the energy (Figure 11.7).

Figure 11.6
Drainage in the Snake River Plain, Idaho, is influenced to a considerable extent by the high porosity and permeability of the basaltic bedrock. The major tributaries coming from the north lose their entire volume of water by seepage into the subsurface, and the rivers simply terminate. Much of the groundwater reappears to form the spectacular Thousand Springs, approximately 400 km downstream.

Figure 11.7
The Thousand Springs, Snake River Canyon, Idaho, issue from the north wall of the canyon and are fed by water from the mountains, 400 km to the northeast.

ARTESIAN WATER

What produces flowing wells?

Statement

Artesian water is confined groundwater that builds up an abnormally high hydrostatic pressure. The necessary geologic conditions for artesian water, include the following:

1. The rock sequence must contain interbedded permeable and impermeable strata. This sequence occurs commonly in nature as interbedded sandstone and shale. Permeable beds usually are called aquifers.
2. The rocks must be tilted and exposed in an elevated area where water can infiltrate into the aquifer.
3. Sufficient precipitation and surface drainage must occur in the outcrop area to keep the aquifer filled.

Discussion

Figure 11.8 illustrates the general geologic conditions necessary for artesian water. The height to which artesian water rises above the aquifer is shown by the dashed colored line. The surface defined by this line is called the artesian-pressure surface. You might expect it to be a horizontal surface, but actually an artesian-pressure surface slopes away from the recharge area (the area where water is absorbed into the ground and into the aquifer, usually where a stream or river crosses an exposed aquifer). Pores in the aquifer provide resistance to flow, and pressure is lost through fractures (leaks) in the underground plumbing system. If a well were drilled at location A or C in

Figure 11.8, water would rise in the well, but it would not flow to the surface because the artesian-pressure surface is below the ground surface. Water in the well at location B or D, where the artesian-pressure surface is above the ground surface, would flow to the surface. Nonetheless, all of these wells are artesian wells—that is, the water is under artesian pressure and rises above the top of the aquifer.

THERMAL SPRINGS AND GEYSERS

Why do geysers erupt in cycles?

Statement

In areas of recent igneous activity, rocks associated with old magma chambers can remain hot for hundreds of thousands of years. Groundwater migrating through these areas of hot rocks becomes heated and, when discharged to the surface, produces thermal springs and geysers.

Discussion

The three most famous regions of hot springs and geysers (a hot spring that intermittently erupts jets of hot water and steam) are Yellowstone National Park, Iceland, and New Zealand. All are regions of recent volcanic activity, so the rock temperatures just below the surface are quite high. Although no two geysers are alike, all require certain conditions for their development:

1. A body of hot rocks must lie relatively close to the surface.

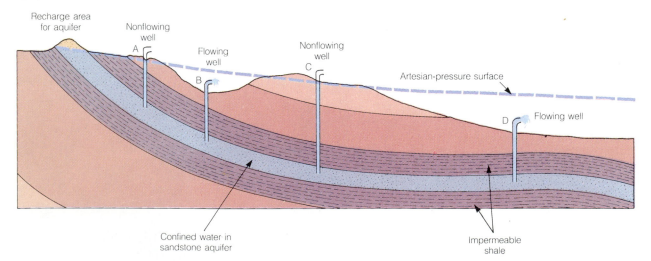

Figure 11.8
Necessary geologic conditions for an artesian system include (1) a permeable bed (aquifer) confined between impermeable layers, (2) rocks tilted so the aquifer can receive infiltration from surface waters, and (3) adequate infiltration to fill the aquifer and create hydrostatic pressure. The diagram shows an idealized artesian system. All wells are artesian wells (that is, water rises in them under pressure). Flowing wells occur only when the top of the well is below the artesian-pressure surface.

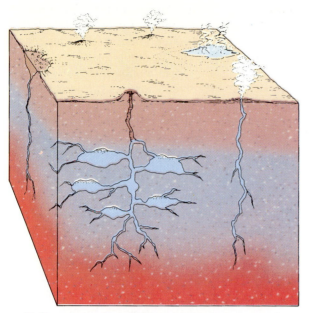

(A) Groundwater circulating through hot rocks in an area of recent volcanic activity collects in caverns and fractures. As steam bubbles rise, they grow in size and number and tend to accumulate in restricted parts of the geyser tube.

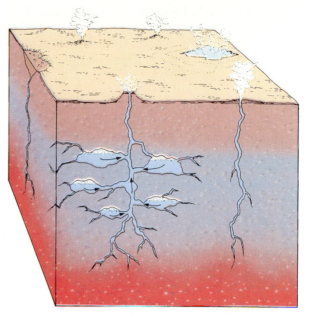

(B) The expanding steam forces water upward until it is discharged at the surface vent. The deeper part of the geyser system then becomes ready for the major eruption.

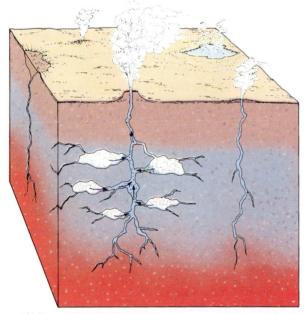

(C) The preliminary discharge of water reduces the pressure on the water lower down. Water from the side chambers and pore spaces begins to flash into steam, forcing the water in the geyser system to erupt.

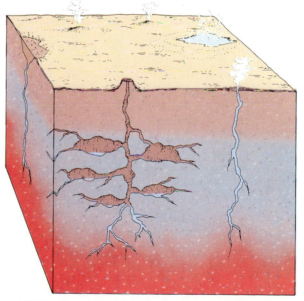

(D) Eruption ceases when the pressure from the steam is spent and the geyser tubes are empty. The system then begins to fill with water again, and the eruption cycle starts anew.

Figure 11.9
The origin of geysers is depicted in this series of diagrams. A geyser can develop only if (1) a body of hot rock lies relatively close to the surface, (2) a system of irregular fractures extends down from the surface, and (3) there is a constant supply of groundwater.

2. A system of irregular fractures must extend downward from the surface.
3. A relatively large and constant supply of groundwater must be present.

Geyser eruptions occur when groundwater pressure in fractures, caverns, or layers of porous rock builds to a critical point where the temperature-pressure balance is such that any increase in temperature will cause the water to change instantly into steam (Figure 11.9). Because the water at the base of the fracture is under greater pressure than the water above, it must be heated to a higher temperature before it boils. Eventually, a slight increase in temperature or a decrease in pressure (resulting from the liberation of dissolved gases) causes the deeper water to boil. The expanding steam throws water from the underground chambers high into the air. After the pressure is released, the caverns refill with water and the process is repeated.

This process accounts for the periodic eruption of many geysers. The interval between eruptions is the amount of time required for water to percolate into the fracture and be heated to the critical temperature. Geysers like Old Faithful in Yellowstone National Park erupt at definite intervals because the rocks are permeable and the "plumbing system" refills rapidly. Other geysers, which require more time for water to percolate into the chambers, erupt at irregular intervals because the water supply over a longer period of time can fluctuate.

Geothermal Energy

The thermal energy of groundwater, called geothermal energy, offers an attractive source of energy for human uses. Presently it is used in various ways in local areas of the United States, Mexico, Italy, Japan, and Iceland. Recent estimates show that from 1 to 2 percent of our current energy needs could be met by geothermal sources.

In Iceland geothermal energy has been used successfully since 1928. The plan is simple. Wells are drilled in geothermal areas, and the steam and hot water are piped to storage tanks and then pumped to homes and municipal buildings for heating and hot water. The cost of this direct heating is only about 60 percent of that of fuel-oil heating and about 75 percent of the cost of the cheapest method of electrical heating. Steam from geothermal energy is also used to run electric generators, producing an easily transported form of energy. Corrosion is a problem, however, because most geothermal waters are acidic and contain undesirable dissolved salts. As a result, the expense of periodically replacing the plumbing system may make geothermal energy uneconomical in some areas.

EROSION BY GROUNDWATER

What unique landforms are produced by groundwater activity?

Statement

Slow-moving groundwater cannot erode rock by abrasion, as a surface stream does, but it can dissolve great quantities of soluble rock such as limestone and carry it off in solution. In some limestone terrains, groundwater is the dominant agent of erosion and is responsible for the development of the major features of the landscape. Groundwater erosion starts with water percolating through joints, faults, and bedding planes and dissolving the soluble rock. In time the fractures enlarge to form a subterranean network of caves, which can extend for many kilometers. The caves grow larger until ultimately the roof collapses, and a craterlike depression called a sinkhole forms. Solution activity then enlarges the sinkhole to form a solution valley, which continues to grow until the soluble rock is removed completely.

Discussion

Karst topography is a distinctive type of terrain resulting largely from erosion by groundwater. In contrast to a landscape formed by surface streams, which is characterized by an intricate network of stream valleys, karst topography lacks a well-integrated drainage system. Sinkholes are generally numerous and, in many karst regions, they dominate the landscape. Where sinkholes grow and enlarge, they merge and form elongate or irregular closed depressions (Figure 11.10) called solution valleys. Small streams commonly flow on the surface for only a short distance and then disappear down a sinkhole. There the water moves slowly through a system of caverns and caves, sometimes as sluggish underground streams. Springs, which are common in karst areas, return water to the surface drainage.

In tropical areas where dissolution is at a maximum because of the abundance of water from heavy rainfall, a particular type of karst topography, called tower karst, develops. Tower karst is characterized by steep, cone-shaped hills rather than sinkholes and solution valleys (Figure 11.11). The towers are largely residual landforms left after most of the rock has been removed by collapse of caverns and enlargement of solution valleys. They are the remnants of a once continuous layer of rock that covered the area.

Thus, in detail karst topography is highly diverse, ranging from fantastic landscapes, such as that shown in Figure 11.11, to the low relief of a plane pitted with small depressions, as shown in Figure 11.10. What is common to all karst terrains is that their landforms are caused by the unusually great solubility of certain rock types. Humid climate is a very important factor in developing karst topography. The more water that is moving through the system, the more solution activity will occur. Karst topography is, therefore, largely restricted to humid and temperate climatic zones. In desert regions, where little rain falls, karst topography will not develop.

A model of how karst topography evolves is shown in the block diagrams in Figure 11.12. Initially, water follows surface drainage until a large river cuts a deep valley below the limestone layers. Groundwater then moves through the joints and bedding surfaces in the

Figure 11.10
Karst topography in the limestone region of Kentucky is dominated by sinkholes and solution valleys.
Note the absence of an apparent drainage system in this region.

(A) Tower karst near Guilin, China.

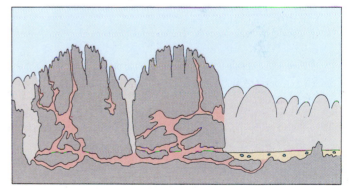

(B) Diagram showing the general nature of the structure of the towers.

Figure 11.11
The tower karst topography of central China forms some of the most spectacular landforms in the world.
The towers are largely residual landforms. Many are laced with caves and caverns.

limestone and emerges at the river banks. As time goes on, the passageways become larger, and caverns develop. Surface waters disappear into solution depressions. The roofs of caves collapse, so that numerous sinkholes are produced (Figure 11.12A). Springs commonly occur along the margins of major stream valleys. Sinkholes proliferate and grow as the limestone formation is dissolved away. The cavernous terrain of central Kentucky, for example, is marked by over 60,000 holes. As solution processes continue, sinkholes increase in number and size. Some merge to form larger depressions with irregular outlines. This process ultimately develops solution valleys (Figure 11.12B). Most of the original surface is finally dissolved, with only scattered hills remaining (Figure 11.12C). When the soluble bedrock has been removed by groundwater solution, normal surface drainage patterns reappear.

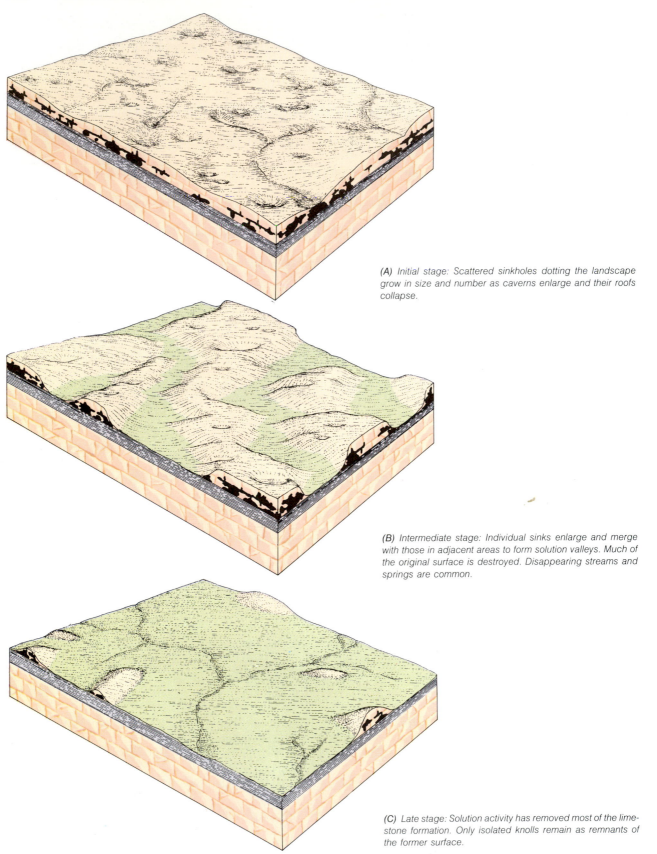

(A) Initial stage: Scattered sinkholes dotting the landscape grow in size and number as caverns enlarge and their roofs collapse.

(B) Intermediate stage: Individual sinks enlarge and merge with those in adjacent areas to form solution valleys. Much of the original surface is destroyed. Disappearing streams and springs are common.

(C) Late stage: Solution activity has removed most of the limestone formation. Only isolated knolls remain as remnants of the former surface.

Figure 11.12
The evolution of karst topography involves these major processes: (1) the enlargement of caves and the development of sinkholes, (2) the enlargement of sinkholes and the development of solution valleys, and (3) the enlargement of solution valleys until the original limestone terrain is completely destroyed.

Examples

The best-known karst topography in the United States is the sinkhole country of Florida, Kentucky, and southern Indiana (Figure 11.10). Surface streams of any significant length are extremely rare. In Kentucky the Green River is the only major stream to cross the limestone terrain. In southern Indiana the White River is the sole river to cross the karst area. Statistical studies of "sinkhole plains" indicate that the size and frequency of the sinks are closely related to the characteristics of specific types of limestone.

In striking contrast to the sinkhole plains of Indiana and Kentucky, the tower karst topography of southern China presents some of the most spectacular limestone scenery on Earth (Figure 11.11). Here an area of thousands of square kilometers, once covered by thick limestone formations, is in an advanced stage of dissection by groundwater. Only remnants between sinkholes and solution basins remain, standing like clusters of towers above the surrounding terrain. These strange mountains, shaped like upended loaves of French bread, form an intricate system of precipitous slopes and overhanging cliffs, with caves, arches, and strange landforms made by solution activity.

The small-scale features in this area are almost as impressive as the larger features. Evidence of solution activity is everywhere. Most outcrops look like Swiss cheese because of the maze of interconnected cavities dissolved and enlarged by groundwater.

Classical Chinese art is noted for portraying these bizarre and exotic landforms, which appear unreal to the foreign eye. Western artists believed that the Chinese masters who painted these landforms were impressionists, but anyone fortunate enough to visit the region realizes that the artists were not visionaries; the shapes they painted were nature's own.

DEPOSITION BY GROUNDWATER

What are the major features formed by the mineral matter deposited by groundwater?

Statement

The mineral matter dissolved by groundwater can be deposited in a variety of ways. The most spectacular deposits are stalactites and stalagmites, frequently found in caves as dripstone. Less obvious are the deposits in permeable rocks, such as sandstone and conglomerates, where groundwater commonly deposits mineral matter as a cement between grains. The precipitation of minerals by groundwater is also responsible for the formation of certain mineral deposits, such as the uranium deposits found in the Colorado Plateau.

Discussion

Cave deposits are familiar to almost everyone, and a variety of forms have been named. Most originate in a similar way, however, and they are referred to collectively as dripstone. The process of dripstone formation is shown in Figure 11.13. As water enters the cave (usually from a fracture in the ceiling), part of it evaporates, and a small amount of calcium carbonate is left behind. Each succeeding drop adds more calcium carbonate, so that eventually a cylindrical or cone-shaped projection is built downward from the ceiling. Many beautiful and strange forms result, some of which are shown in Figure 11.14. Icicle-shaped forms growing down from the ceiling are called stalactites. These commonly are matched by deposits growing up from the floor, called stalagmites, because the water dripping from a stalactite precipitates additional calcium carbonate on the floor directly below. Many stalactites and stalagmites eventually unite to form columns. Water percolating from a fracture in the roof may form a thin, vertical sheet of rock called a drip curtain. Pools of water on the cave floor flow from one place to another, and as they evaporate, calcium carbonate is deposited on the floor, forming travertine terraces.

Example

Although cave deposits are a spectacular expression of deposition by groundwater, they are trivial compared to the amount of material deposited in the pore spaces of rock. In sandstones and conglomerates, precipitation of silica and calcium carbonate cement the loose grains into a hard, strong rock body. In some formations the cementing minerals deposited by groundwater may exceed 20 percent of the volume of the original rock.

Mineral precipitation by groundwater action is a slow process, in some cases involving the slow removal—one at a time—of atoms, or molecules of organic matter, and their simultaneous replacement by other mineral ions carried by the groundwater. One example of this process is petrified wood. Perhaps the best-known deposit of petrified wood is the Petrified Forest National Park in eastern Arizona. There, great accumulations of petrified logs buried in ancient river sediments are now being uncovered by weathering and erosion (Figure 11.15).

Another expression of deposition by groundwater is the mineral deposits formed around springs. Mammoth Hot Springs in Yellowstone National Park is one of the most spectacular examples (Figure 11.16).

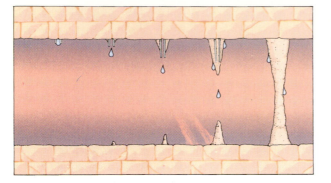

(A) Diagram showing the evolution of stalactites → stalagmites → columns.

Figure 11.13
Dripstone originates on the ceilings of caves. Water seeps through a crack and partially evaporates. This causes a small ring of calcite to be deposited around the crack. The ring grows into a tube, which commonly acquires a tapering shape, as water seeps from adjacent areas and flows down its outer surface.

(B) Photograph of long, slender stalactites (soda straws), which grow as the drop of water suspended at the end evaporates.

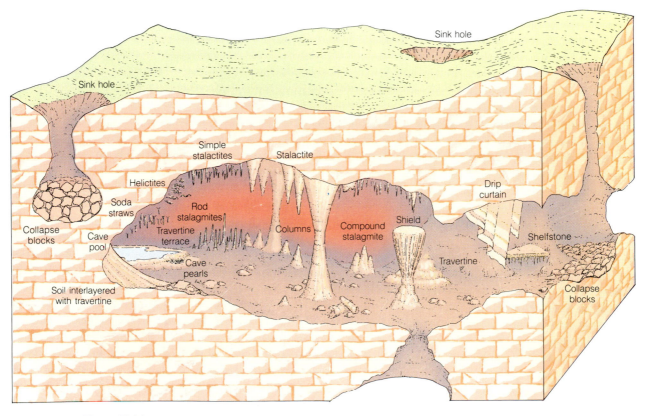

Figure 11.14
Many varieties of cave deposits have been recognized. All of them are composed of calcite deposited by water that seeps into a cave and then evaporates.

Figure 11.15
Petrified trees litter the area, piled like giant jackstraws about a rolling landscape on the Petrified Forest Member of the Chinle Formation, Arizona.

Figure 11.16
Mammoth Hot Springs, Yellowstone National Park, was formed by the deposition of travertine ($CaCO_3$) as the warm spring water cooled and evaporated.

ALTERATION OF GROUNDWATER SYSTEMS

How does human activity alter the groundwater systems?

Statement

Groundwater constitutes a valuable resource that is being exploited at an ever-increasing rate. A variety of problems results from human activities that alter the groundwater system. Some of the more important problems are:

1. Changes in the chemical composition of groundwater
2. Saltwater encroachment
3. Changes in the position of the water table
4. Subsidence

Discussion

Groundwater is an integral part of the hydrologic system and is intimately related to other parts of the system. As we have seen in this chapter, its sources are precipitation and infiltration from surface runoff. Its natural discharge is into streams and lakes and ultimately the sea. With time a balance, or equilibrium, among precipitation, surface runoff, infiltration, and discharge is established. These, in turn, approach an equilibrium with surface conditions such as slope angles, soil cover, and vegetation. When any of these interrelated factors is changed or modified, the others respond to reestablish equilibrium.

Examples

Changes in Composition. The composition of groundwater can be changed by increases in the concentration of dissolved solids in surface water. The soil is like a filtration system through which groundwater moves. Obviously, any concentration of chemicals or waste creates local pockets that potentially can contaminate the groundwater reservoir. Material that is leached (dissolved by percolating groundwater) from waste-disposal sites, for example, includes both chemical and biological contaminants. Upon entering the groundwater flow system, the contaminants move according to the hydraulics of that system. The character and strength of the leachates (pollutants) depend on the length of time the infiltrated water is in contact with the waste deposit, the volume of infiltrated water, and the kinds of waste involved.

Figure 11.17 illustrates four geologic environments in which waste disposal affects the groundwater system. In the environment shown in Figure 11.17A,

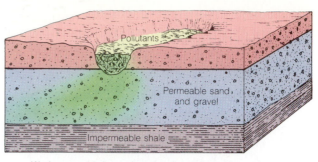

(A) A permeable layer of sand and gravel overlying an impermeable shale creates a potential pollution problem because contaminants are free to move with groundwater.

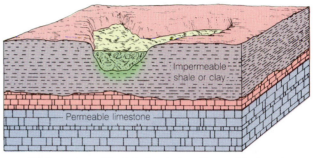

(B) An impermeable shale (or clay) confines pollutants and prevents significant infiltration into the groundwater system in the limestone below.

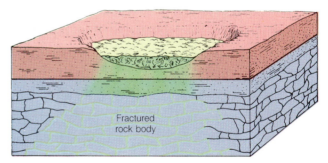

(C) A fractured rock body provides a zone where pollutants can move readily in the general direction of groundwater flow.

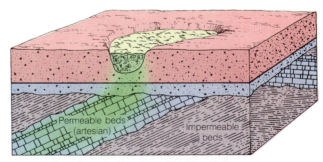

(D) An inclined, permeable aquifer below a disposal site permits pollutants to enter an artesian system and move down the dip of the beds, so that they contaminate the system.

Figure 11.17
The effects of solid-waste disposal on a groundwater system depend on the geologic setting. In many cases, water seeping through the disposal site enters and pollutes the groundwater system.

the near-surface material is permeable and essentially homogeneous. Leachates percolate downward through the zone of aeration and, upon reaching the water table, enter the groundwater flow system. The flowing leachates ultimately become part of the surface drainage system. As is shown in Figure 11.17B, an impermeable layer of shale confines pollutants and prevents their free movement in the groundwater system. As a result, the pollutants are restricted and inhibited from moving freely through the groundwater system. Figure 11.17C illustrates a disposal site above a fractured rock body. Upon reaching the fractured rock, the contaminants can move more readily in the general direction of groundwater flow. Dispersion of the contaminants is limited, however, because of the restriction of flow to the fractures. Figure 11.17D illustrates a critical condition in which a waste-disposal site is located in highly permeable sand and gravel above an inclined aquifer. There, leachates move down past the water table and enter the aquifer as recharge. If the waste-disposal site were located directly above the aquifer, as is shown in the diagram, most of the leachates would enter the aquifer and contaminate the groundwater system.

Saltwater Encroachment. On an island or a peninsula, where permeable rocks are in contact with the ocean, a lens-shaped body of fresh groundwater is buoyed up by the denser saltwater below, as is illustrated in Figure 11.18A. The fresh water literally floats on the salt water and is in a state of balance with it. If excessive pumping develops a large cone of depression in the water table, the pressure of the fresh water on the salt water directly below the well is decreased, and a large cone of saltwater encroachment develops below the well, as is shown in Figure 11.18B. Continued excessive pumping causes the cone of salt water to extend up the well and contaminate the fresh water. It is then necessary to stop pumping for a long time, to allow the water table to rise to its former position and depress the cone of salt water. Restoration of the balance between the freshwater lens and the underlying salt water can be hastened if fresh water is pumped down an adjacent well (Figure 11.18C).

Changes in the Position of the Water Table. The water table is intimately related to surface runoff, the configuration of the landscape, and the ecological conditions at the surface. The balance between the water table and surface conditions, established over thousands or millions of years, can be completely upset by changes in the position of the water table. Two examples illustrate some of the many potential ecological problems.

In southern Florida fresh water from Lake Okeechobee has flowed for the past 5,000 years as an al-

most imperceptible "river," only a few centimeters deep and 64 km wide. This sheet of shallow water created the swampy Everglades. The movement of the water was not confined to channels. It flowed as a

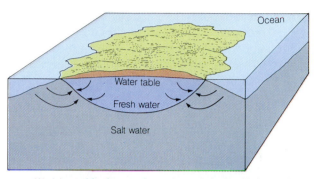

(A) A lens of fresh groundwater beneath the land is buoyed up by denser salt water below.

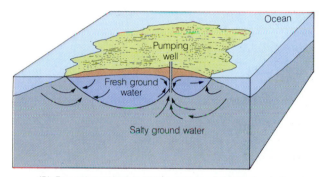

(B) Excessive pumping causes a cone of depression in the water table on top of the fresh water lens, and a cone of salt water encroachment at the base of the freshwater lens.

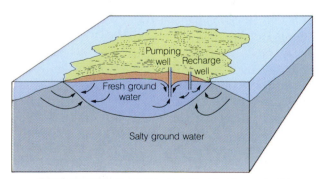

(C) Fresh water pumped down an adjacent well can raise the water table around the well and lower the interface between the fresh water and the salt water.

Figure 11.18
The relationship between fresh water and salt water on an island or a peninsula is affected by the withdrawal of water from wells. Excessive pumping causes a cone of salt water encroachment, which limits the usefulness of the well.

sheet in a great curving swath for more than 160 km (Figure 11.19). The surface of the Everglades slopes southward only 2 cm per km, but this gradient was enough to keep the water moving slowly to the coast and to prevent salt water from invading the Everglades and the subsurface aquifers along the coast. In effect the water table in the swamp was at the surface, and the ecology of the Everglades was in balance with the water table.

Today many canals have been constructed to drain swamp areas for farmland, to help control flooding, and to supply fresh water to the coastal megalopolis (Figure 11.20). The canals diverted the natural flow of water across the swamp, in effect lowering the water table, in some places as much as 0.5 m below sea level. This change in position of the water table produced many unforeseen and often unfortunate results. As the water table was lowered, saltwater encroachment occurred in wells all along the coast. Some cities had to move their wells far inland to obtain fresh water.

The most visible effects, however, involve the ecology of the swamp. In the past the high water table could maintain a marsh during periods of natural drought. Now the surface is dry during droughts. Forest fires ignite the dry organic muck, which burns like peat and smolders long after the surface fires die out. This effectively destroys the ecology of the swamp. The lowering of the water table also caused the muck to compact, so that it subsided in places as much as 2 m. In addition, muck exposed to the air oxidizes and disappears at a rate of about 2.5 cm per year. Once the muck is gone from the swamp, it can be replaced only by nature.

Raising the water table can also modify many surface processes. An example is found in the environmental changes caused by irrigation in the Pasco basin of Washington. This area, which lies in the rain shadow of the Cascade Mountains, receives only 15 to 25 cm of precipitation a year. The basin's surface conditions (the soil cover, slope of land, drainage network, and so on) developed in response to an overall increase in aridity over the last several million years. The surface material, slope, and vegetation developed a balance with an arid climate and a low water table.

In recent years extensive irrigation has caused the water table to rise, introducing many changes in the surface conditions. Today, from 100 to 150 cm of water are applied each year to the ground by irrigation, which simulates the effect of a climatic change of considerable magnitude. The higher water table has rapidly developed large springs along the sides of river valleys. The springs are now permanent, reflecting saturation of much of the ground. Erosion is accelerated, and many farms and roads have been damaged severely. Landslides present the most serious problems. Slopes that were stable under arid conditions are now unstable, because they are partly satu-

rated from the high water table and from the formation of perched water bodies.

In many areas it is imperative that people modify the environment by reclaiming land or by irrigation, but unless we are careful, the detrimental effects of our modifications may outweigh the advantages. Before we seriously modify an environment, we must attempt to understand the many consequences of altering the natural systems.

Subsidence. Surface subsidence related to groundwater can result from natural Earth processes, such as the development of sinkholes in a karst area, or from the artificial withdrawal of fluids.

An ever-present hazard in limestone terrains is the collapse of subterranean caverns and sinkhole formations. Buildings and roads have frequently been damaged by sudden collapses into previously undiscovered caverns below. In the United States important karst regions occur in central Tennessee, Kentucky, southern Indiana, Alabama, Florida, and Texas. The problem of potential collapse is difficult to solve. Important construction in karst regions should be preceded by test borings to determine whether subterranean cavernous zones are present. Wet concrete can be pumped down into caves and solution cavities, but such remedies can be very expensive and can alter subsurface flow patterns.

Compaction and subsidence also present serious problems in areas of recently deposited sediments. In New Orleans, for example, large areas of the city are now 4 m below sea level, a drop due largely to the pumping of groundwater. As a result, the Mississippi River flows some 5 m above parts of the city, and rainwater must be pumped out of the city at considerable cost. Also, as the Earth subsides, water lines and sewers are damaged.

Where groundwater, oil, or gas is withdrawn from the subsurface, significant subsidence can also occur, damaging construction, water supply lines, sewers, and roads. Long Beach, California, has subsided 9 m as the result of forty years of oil production from the Wilmington oil field. This subsidence resulted in almost $100 million worth of damage to wells, pipelines, transportation facilities, and harbor installations. Parts of Houston, Texas, have subsided as much as 1.5 m, as a result of the withdrawal of groundwater.

Probably the most spectacular example of subsidence is Mexico City, which is built on a former lake bed. The subsurface formations are water-saturated clay, sand, and volcanic ash. The sediment compacts as groundwater is pumped for domestic and industrial use, and slow subsidence is widespread. The opera house (weighing 54,000 metric tons) has settled more than 3 m, and half of the first floor is now below ground level. Other large structures are noticeably tilted.

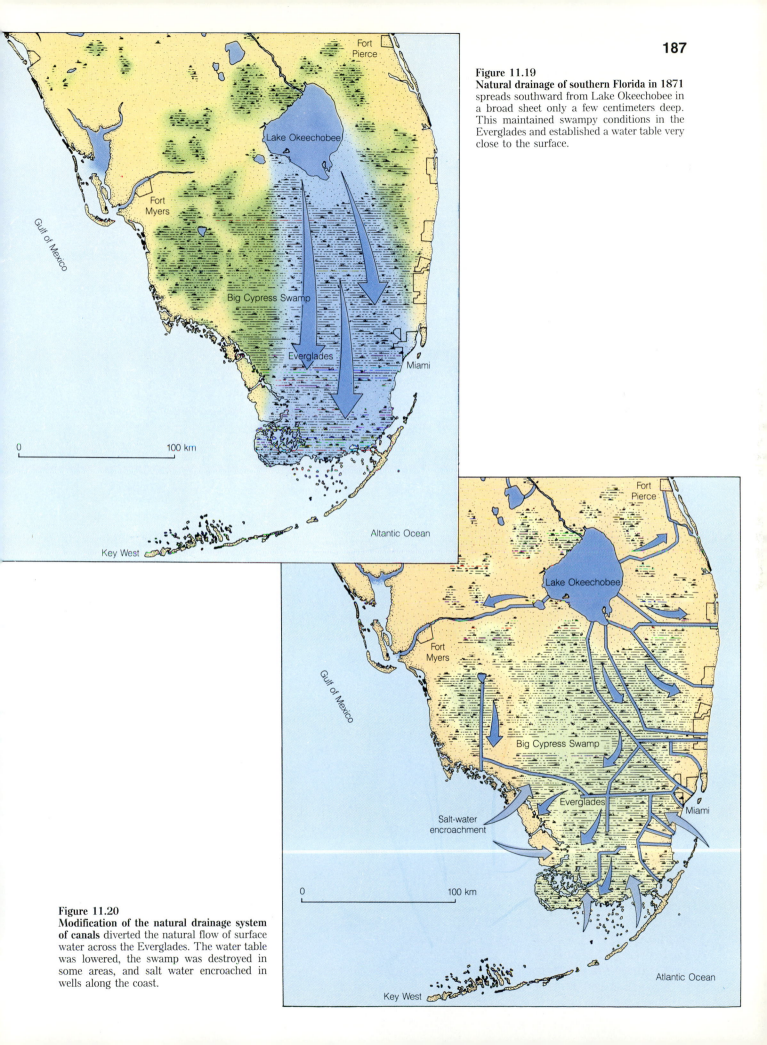

Figure 11.19
Natural drainage of southern Florida in 1871
spreads southward from Lake Okeechobee in
a broad sheet only a few centimeters deep.
This maintained swampy conditions in the
Everglades and established a water table very
close to the surface.

Figure 11.20
**Modification of the natural drainage system
of canals** diverted the natural flow of surface
water across the Everglades. The water table
was lowered, the swamp was destroyed in
some areas, and salt water encroached in
wells along the coast.

SUMMARY

1. Groundwater is an integral part of the hydrologic system, and it is intimately related to surface water drainage.
2. The movement of groundwater is very slow, controlled largely by the porosity and permeability of the rock. At some depth below the surface, all pore spaces are filled with water.
3. The upper surface of this saturated zone is called the water table.
4. Movement of groundwater below the water table is closely associated with surface runoff, and groundwater commonly discharges into streams, lakes, and swamps.
5. Artesian water is water confined between impermeable beds and is under hydrostatic pressure.
6. Groundwater erodes by dissolving soluble rocks, such as limestone, rock salt, and gypsum. Such erosion produces karst topography, characterized by sinkholes, solution valleys, and disappearing streams.
7. Under normal conditions, groundwater is essentially in balance, or in equilibrium with surface runoff, surface topography, and salt water in the ocean.
8. Alteration of the groundwater system can produce many unforeseen problems, as does alteration of any segment of the hydrologic system.

KEY TERMS

Artesian water (p. 176)
cave (p. 178)
cone of depression (p. 175)
disappearing
  stream (p. 178)
dripstone (p. 181)
geothermal energy (p. 178)

geyser (p. 176)
hydraulic head (p. 176)
hydrostatic
  pressure (p. 176)
karst topography (p. 178)
perched water
  table (p. 172)

permeability (p. 170)
pore spaces (p. 170)
porosity (p. 170)
saltwater
  encroachment (p. 185)
sinkhole (p. 178)
solution valley (p. 178)

stalactite (p. 181)
stalagmite (p. 181)
tower karst (p. 179)
travertine terrace (p. 181)
water table (p. 170)
zone of aeration (p. 170)
zone of saturation (p. 170)

REVIEW QUESTIONS

1. Define porosity and permeability.
2. Describe and illustrate the major types of pores, or voids, in rocks.
3. What rock types are generally impermeable or nearly impermeable?
4. Describe the major zones of subsurface water and explain how water moves through each zone.
5. Explain some of the ways in which springs originate.
6. Explain how artesian flow occurs.
7. Explain the origin of geysers.
8. Describe the evolution of a landscape in which groundwater is the dominant agent of erosion.
9. Explain how stalagmites and stalactites originate.
10. Describe the forms and processes of groundwater pollution.
11. Explain how the alteration of the natural drainage system in southern Florida has affected the Everglades.

Multiple-Choice Questions

1. The ability of a porous solid to transmit a fluid is called
   a. permeability.
   b. transmissibility.
   c. porosity.
   d. viscosity.
   e. vesicularity.
2. In most places the water table has a form similar to that of
   a. sine waves.
   b. meanders.
   c. dunes.
   d. sinkholes.
   e. the land surface.
3. The water table is
   a. the level of water in an artesian well.
   b. the level of downward percolating water.
   c. the upper limit in which pore space in the rock is completely saturated by water.
   d. the lower limit in which pore space in the rock is completely saturated by water.
   e. within the part of the regolith or bedrock that is saturated with water.
4. If a 25-foot well is drilled at a place where the water table is 20 feet below the surface, how many feet of water will be in the well?
   a. 5 feet
   b. 0 feet
   c. 45 feet
   d. 20 feet
   e. 25 feet
5. Artesian wells flow because the water in them is
   a. hard and resists gravity pull.
   b. soft and fluid.
   c. under free gravity flow.
   d. confined and under pressure.
   e. under capillary pull.
6. Conditions necessary for artesian water include
   a. impermeable cap rock.
   b. inclined aquifer.
   c. recharge area.
   d. permeable aquifer.
   e. all of the above.
7. Thermal springs and geysers commonly form in areas of
   a. normal faulting.
   b. recent igneous activity.
   c. normal geothermal gradient.
   d. interior of small plates.
   e. deep geothermal energy in the atmosphere.
8. Good examples of karst topography can be found in
   a. Kentucky.
   b. Florida.
   c. southern Illinois.
   d. the area of the Texas-New Mexico state line.
   e. all of the above.
9. Which of the following rocks is known to have low permeability?
   a. sandstone
   b. shale
   c. vesicular basalt
   d. conglomerates
   e. fractured limestone
10. Caves are most commonly formed in
    a. limestone.
    b. sandstone.
    c. shale.
    d. rock salt.
    e. gypsum.

True/False Questions

1. *T F* In swamps and lakes the water table is essentially at the surface.
2. *T F* The path of groundwater movement is down the slope of the water table.
3. *T F* Groundwater moves downward and toward points of lesser pressure.
4. *T F* On entering the groundwater system, pollutants move independently of the hydraulics of that system.
5. *T F* On an island or on a peninsula, where permeable rock is in contact with the ocean, the entire groundwater system is salty.

Fill in the Missing Terms

1. The zone of _____ is characterized by pore spaces in the rock that are only partly saturated, so the water forms a thin film, clinging to the grains by surface tension.
2. The _____ _____ is the upper surface of the zone of saturation.
3. When a well is pumped, the water table is drawn down around the well in the shape of a cone, called the _____ _____ _____.
4. _____ topography is characterized by a surface pitted with sinkholes, solution valleys, and disappearing streams.
5. Icicle-shaped forms growing down from the ceiling of a cave are called _____.

12

GLACIAL SYSTEMS

No event in recent geologic history has had such a profound effect upon Earth as the last, great Ice Age. Its impact extended far beyond the margins of the ice itself and influenced almost every aspect of the physical and biological world. For example, the present sites of many northern cities, such as Chicago, Detroit, Montreal, and Toronto, were buried beneath thousands of feet of glacial ice as recently as 15,000 to 20,000 years ago.

When glaciation occurs, many geologic processes are interrupted or modified significantly. Much precipitation becomes trapped in glaciers, instead of flowing immediately back to the ocean. Consequently, sea level drops and the hydrology of streams is greatly altered. As the great ice sheets advance over the continents, they obliterate preexisting drainage networks. The moving ice scours and erodes the landscape and deposits the debris near its margins, covering the preexisting topography. The crust of the Earth is pushed down by the weight of the ice, and meltwater commonly collects and forms lakes along the ice margins. As the glaciers melt, new drainage systems are established to accommodate the large volume of meltwater. Far beyond the margins of the glaciers, stream systems are modified by changing climatic patterns. Even in arid regions, the imprint of climatic changes associated with glaciation is seen in the development of large lakes in closed basins.

The subject of glaciation is of serious concern to every person. We are just emerging from the Ice Age and may be living in an interglacial period. Major climatic changes lie ahead. One probable change might be a return to another major expansion of glaciers. Another would be a continued warming trend, which could melt the ice on Greenland and Antarctica and flood most coastal areas of the world. Earth has passed through similar glacial and interglacial periods many times in the past. Human welfare depends on how we adjust.

MAJOR CONCEPTS

1. Glaciers are systems of flowing ice that form where more snow accumulates each year than melts.
2. As ice flows, it erodes the surface by abrasion and plucking. Sediment is transported by the glacier and deposited where the ice melts.
3. The two major types of glaciers—continental and valley glaciers—produce distinctive erosional and depositional landforms.
4. The major effects of an ice age include the rise and fall of sea level, isostatic adjustments of land, modification of drainage systems, creation of numerous lakes, and the migration and selective extinction of plant and animal species.
5. The cause of glaciation is not completely understood, but it may be related to several simultaneously occurring factors, such as astronomical cycles, plate tectonics, and ocean currents.

GLACIAL SYSTEM

How does solid ice flow?

Statement

A glacier is a system of flowing ice that originates on land through the accumulation and recrystallization of snow. It is an open system and has much in common with other gravity flow systems, such as rivers and groundwater. Water enters the system primarily in the upper parts of the glacier, where snow accumulates and is transformed into ice. The ice then flows out of the zone of accumulation, generally moving a few centimeters per day. At the lower end (or terminus) of the glacier, ice leaves the system by melting and evaporating. As glacial ice flows over the land, it erodes, transports, and deposits vast amounts of rock material and greatly modifies the preexisting landscape. A generalized block diagram showing how glaciers erode, transport, and deposit material is show in Figure 12.1.

Discussion

Erosion

Glaciers erode bedrock in two ways: (1) by glacial plucking (a process by which large blocks of rock are frozen to the bottom surface of a glacier and are torn out of the bedrock and transported with the moving ice), and (2) by abrasion (grinding). The eroded material is then carried in suspension in the ice and deposited near the margins of the glacier, where melting dominates. Glacial plucking is the lifting out and removal of fragments of bedrock by a glacier. It is one of the most effective ways in which a glacier erodes the land. The process involves ice wedging. Beneath the glacier, meltwater seeps into joints or fractures, where it freezes and expands, wedging blocks of rock loose. The loosened blocks freeze to the bottom of the glacier and are plucked, or quarried, from the bedrock, becoming incorporated in the moving ice. The process is especially effective where the bedrock is cut by numerous joints and where the surface is unsupported on the downstream side.

Abrasion is essentially a filing process. The angular blocks plucked and quarried by the moving ice freeze firmly into the glacier, where they act as tools that grind and scrape the bedrock. Aided by the pressure of the overlying ice, the angular blocks are very effective agents of erosion, capable of wearing away large quantities of bedrock. The rock fragments incorporated in the glacial ice are themselves abraded and worn down as they grind against the bedrock surface. As a result, they usually develop flat surfaces that are deeply scratched.

Evidence of the distinctive abrasive and quarrying action of glaciers can be seen on most bedrock surfaces over which glacial ice has moved. Small hills of bedrock commonly are streamlined by glacial abrasion. Their upstream side typically is rounded off, while the downstream side is made steep and rugged by glacial plucking (Figure 12.2). Glacial striations, such as those illustrated in Figure 12.3, range from hairline scratches to large, deep furrows more than a kilometer long.

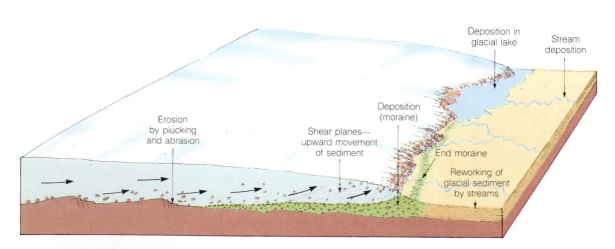

Figure 12.1
Glaciers erode, transport, and deposit rock material in the manner illustrated in this diagram. Erosion by glacial plucking occurs as blocks of rock, separated by joints, freeze to the base or sides of the glacier and are lifted from the outcrop by the moving ice. These fragments, frozen in the ice, then act as abrasives and wear down the surface by grinding. Near the end of the glacier, the ice is commonly stagnant and no longer moves. Much of the sediment load is thus forced upward along shear planes and is concentrated at the surface of the glacier. As the ice melts, this sediment accumulates in an end moraine.

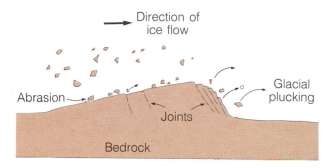

Figure 12.2
Roche moutonnée is an erosional feature that forms as ice moves over bedrock, eroding it into a streamlined shape. Glacial plucking commonly produces a ragged edge on the downcurrent side.

Figure 12.3
Glacial striations result from the abrasive action of a glacier and clearly show the direction in which the ice moved.

Transportation

The rock fragments transported by a glacier are collectively referred to as its load. The manner in which a glacier transports its load, however, differs from stream transport in a very important way. The load in a glacier is carried in suspension, and large blocks are transported side by side with small grains, without the sorting or separation of the material according to size that occurs in streams. As a result, the deposits of a glacier are unsorted and unstratified, and thus they differ markedly from stream deposits. The load of a glacier is concentrated near the contact between the ice and the bedrock, from which it was derived.

Most of the particles carried by a glacier are fresh and unweathered and have angular, jagged surfaces. The grinding action of the moving ice further crushes the grains to produce an abundance of fine particles known as rock flour.

Deposition

Most of the particles transported by a glacier are deposited near the terminus, where melting dominates. There, the ice may become stagnant. This commonly forces the active upstream ice to move up and over the "dead ice" at the terminus, so that much of the load is carried upward along shear planes and is concentrated on the surface of the glacier (Figure 12.1). When melting is completed, this material accumulates as a ridge, marking the former margins of the glacier.

Some of the sediment deposited by a glacier is picked up and reworked by meltwater and redeposited beyond the margins of ice by streams. The sediment, deposited directly by the glacier or indirectly in glacial lakes and streams, is referred to as glacial drift (such deposits were once believed to have "drifted" to their present resting place during the flood of Noah). Both stratified and unstratified glacial drift occurs. The term till is used to designate unsorted, unstratified drift, deposited directly by the glacier.

Rates of Ice Flow

Ice flow in a glacier may seem extremely slow in comparison to the flow of water in rivers, but the movement is continuous, and over the years, vast quantities of ice can move through a glacier. Measurements show that some of Switzerland's large valley glaciers move as much as 180 m per year. Smaller glaciers move from 90 to 150 m per year. Some of the most rapid rates have been measured on Greenland, where ice is funneled through mountain passes at a speed of 8 km per year. From these and other measurements, flow rates of a few centimeters per day appear common, and velocities of 3 m per day are exceptional. The most rapid movement occurs in glaciers, where pressure built up behind a mountain range helps to force the ice through a mountain pass.

Recent close observations of the movements of valley glaciers show that occasionally a glacier can surge forward more than several hundred meters per day. Now that movements can be monitored by satellite photography, these surges are known to be fairly common. (In the past only a few surges had been observed because the flow is so short-lived.) A glacier in the Himalayas, for instance, advanced 11 km in three months. In 1966 the Steele Glacier, in the Yukon Territory, advanced approximately 8 km within a matter of weeks. Glacial surges apparently result from sudden slippages along the base of glaciers, caused by the buildup of extreme stress upstream. Stagnant or slow-moving ice near the terminus can act as a dam for the faster-moving ice upstream. If this happens, stress builds up behind the slow-moving ice, and a surge occurs when a critical point is reached. Surges can also be caused by a sudden addition of mass to the glacier, such as a large avalanche or landslide on its surface. Lubrication of the base of the glacier, by meltwater, may also be an important factor.

In considering the flow of ice in a glacier and the erosion it can cause, remember that a glacier is an open system. Material enters the system in the zone of accumulation, flows through, and then leaves the system at the distal margins by melting or evaporation. Ice within a glacier flows continuously through the system, regardless of whether the terminal margin is advancing, retreating, or stationary. The length of a valley glacier is therefore no indication of the amount of erosion it can accomplish or has accomplished. Erosion by a glacier is a function of the duration of the process, the thickness of the ice, and the velocity at which it flows.

VALLEY GLACIERS

Statement

Valley glaciers are long, narrow rivers of ice that originate in the snowfields of high mountain ranges and flow down preexisting stream valleys. They range from a few hundred meters to more than a hundred kilometers in length. In many ways they resemble river systems. They receive an input of water (in the form of snow) in the higher reaches of the mountains. They have a system of tributaries leading to a main trunk system. The flow direction is controlled by the valley the glacier occupies, and as the ice moves, it erodes and modifies the landscape over which it flows.

Discussion

The System

The essential parts of a valley glacial system are shown in Figures 12.4 and 12.5: the zone of accumulation, where there is a net gain of ice, and the zone of ablation, where ice leaves the system by melting and evaporating. In the zone of accumulation, snow is transformed into glacial ice through a process identical to rock metamorphism. Freshly fallen snow consists of delicate hexagonal ice crystals or needles, with as much as 90 percent of the total volume as empty space. As snow accumulates, the ice at the points of the snowflake melts from the pressure of snow buildup and migrates toward the center of the flake, eventually forming an elliptical granule of recrystallized ice approximately 1 mm thick. The accumulation of these particles packed together is called firn, or neve. With repeated annual deposits, the loosely packed neve granules are compressed by the weight of the overlying snow. Meltwater, which results from daily temperature fluctuations and the pressure of the overlying snow, seeps through the pore spaces between the grains; when it freezes, it adds to the recrystallization process. Air in the pore spaces is driven out. When the ice reaches a thickness of approximately 30 or 40 m, it can no longer support its own weight and yields to plastic flow. The upper part of a glacier is thus rigid and tends to fracture, but the ice beneath moves by plastic flow.

As is shown in Figures 12.4 and 12.5, the boundary between the zone of accumulation and the zone of ablation is approximated by the snowline. Above the snowline, the surface of the glacier is smooth and white—because more snow accumulates than is lost by melting—and any irregularities are soon covered and filled with freshly fallen snow. Below the snowline, melting and evaporation exceed snowfall. There, the surface of the ice is rough and pitted and commonly is broken by open crevasses.

At the terminal margin of a glacier, the loss of ice by melting and evaporation exceeds the rate of accumulation. This margin is the major exit boundary of the system.

It is important for you to understand that the margins of a glacier constitute the boundaries of a system of flowing ice, much as the banks and mouth of a river constitute the boundaries of a river system. If more snow is added in the zone of accumulation than is lost by melting or evaporation at the end of the glacier, the ice mass increases, and the glacial system expands. If the accumulation of ice is less than ablation, there is a net loss of mass and the size of the glacial system is reduced. If accumulation and ablation are in balance, the mass of ice remains constant, the size of the system remains constant, and the terminus of the ice remains stationary. Nevertheless, ice within the

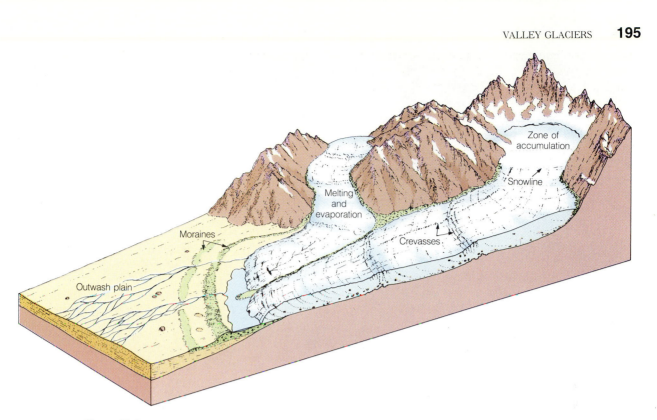

Figure 12.4
A valley glacial system is an open system of ice that flows under the pull of gravity. Snow enters the system by precipitation and is transformed into ice. The ice flows outward from the zone of accumulation under the pressure of its own weight. The ice leaves the system by evaporating and melting in the zone of ablation. The balance between the rate of accumulation and the rate of melting determines the size of the glacial system. The major parts of a valley glacial system are the zone of accumulation, where snow enters the system, and the zone of ablation, where ice leaves the system by melting and evaporation. The boundary between these zones is approximated by the snow line. As the ice moves through the system it erodes the preexisting stream valley and deposits sediment at the end of the glacier. Meltwater can rework much of the glacial sediment and redeposit it downstream.

glacier continually flows toward the terminus, or terminal margins, regardless of whether the terminal margins are advancing, retreating, or stationary.

The behavior of a glacial system is determined by the balance between the rate of input and the rate of output of ice. The two major variables in this balance are temperature and precipitation. A glacier can grow or shrink with an unchanging rate of precipitation if the temperature varies enough to increase or decrease the rate of melting (rate of output). The length of a glacier in no way represents the amount of ice that has moved through the system, just as the length of a river does not represent the volume of water that has flowed through it. Length simply shows the amount of ice currently in the system.

An example from the last Ice Age illustrates this point. A glacial valley 20 km long in the Rocky Mountains was eroded 600 m deeper than the original stream valley. This large amount of erosion was not accomplished by 20 km of ice moving down the valley. It was the result of many thousands of kilometers of ice flowing through the valley. If the ice occupied the valley during each glacial epoch and moved 0.3 m per day, a total of approximately 72,000 km of

ice would have moved down the valley. The enormous abrasion caused by such a long stream of ice would be able to wear down the valley 600 m.

Landforms Developed by Valley Glaciers

The idealized diagram in Figure 12.6 illustrates the major erosional landforms resulting from valley glaciation. This figure permits a comparison and contrast of landscapes formed only by running water with those that have been modified by valley glaciers. Figure 12.6A shows the typical topography of a mountain region being eroded by streams. A relatively thick mantle of soil and weathered rock debris covers the slopes. The valleys are V-shaped in cross section and have many bends at tributary junctions, so that ridges and divides between tributaries appear to overlap if you look up the valley. In Figure 12.6B the area is shown occupied by glaciers. The growing glaciers expand down the tributary valleys and merge to form a major glacier. During glaciation several thousand kilometers of ice might flow down the valley. The enormous quantity of moving ice can erode the valley as much as 600 m below its original level.

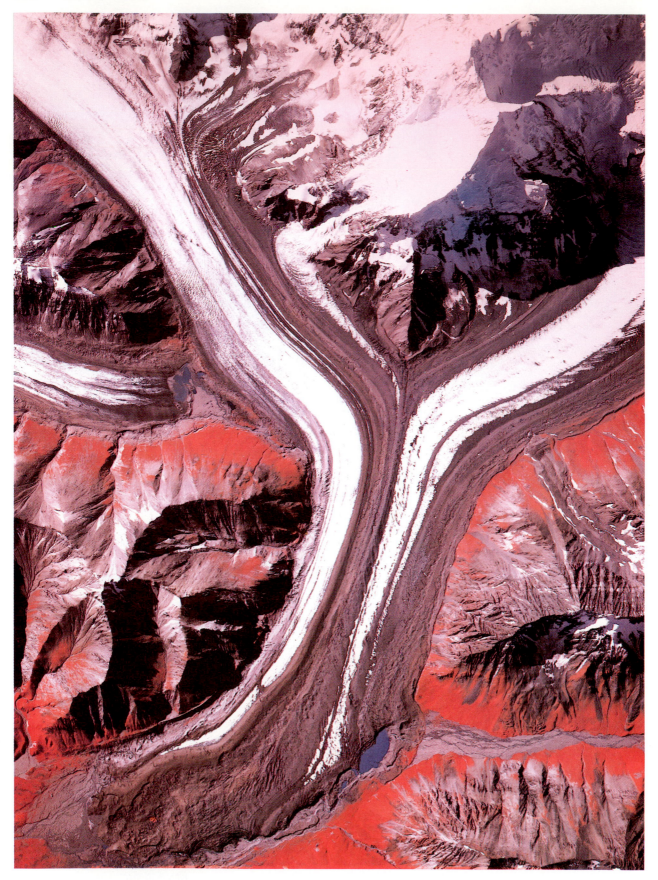

Figure 12.5
Valley glaciers in Alaska show many of the features of a glacial system illustrated in the diagram of Figure 12.4. Note the snow-covered zone of accumulation, the rough ice in the zone of ablation, the flow structure in the glacier, and the sediment transported and deposited at the end of the glacier.

(A) The topography before glaciation is shaped by running water. Valleys typically are V-shaped and have many curves and irregularities in map views. Hills are rounded.

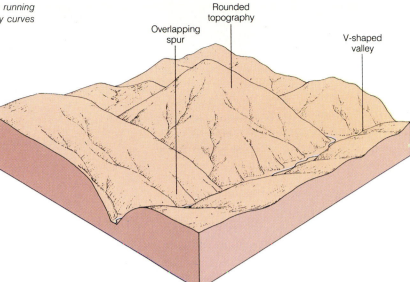

(B) Valley glaciers form in high areas and move down major stream valleys. A network of tributaries delivers ice to the main glacier. Frost action on the valley walls produced abundant rock fragments, which accumulate as lateral moraines on the flanks of the glacier.

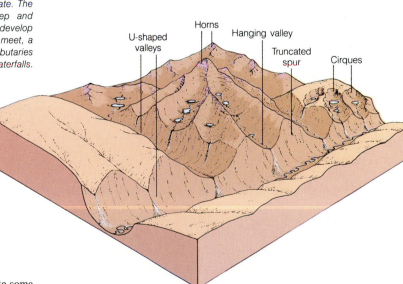

(C) When the glacier recedes, the topography has been significantly modified. Sharp, angular landforms dominate. The valleys previously occupied by glaciers are deep and U-shaped. Bowl-shaped depressions called cirques develop at the heads of the valleys. Where several cirques meet, a sharp, pyramidlike peak called a horn is formed. Tributaries form hanging valleys, which can have spectacular waterfalls.

Figure 12.6
Landforms produced by valley glaciers constitute some of the most spectacular scenery in the world. In these diagrams an idealized landscape formed by stream erosion is shown as it might appear before, during, and after glaciation.

Ice wedging is a major process in the cold regions of glaciers. It effectively sharpens the mountain summits and causes glacial plucking. The heads of the glaciers are enlarged by plucking and grow headward toward the mountain crest to form a cirque. Where two or more cirques approach the summit crest, they sculpt the mountain peak into a horn.

Note that where tributaries enter the main glacier, the upper surfaces of the glaciers are at the same level. The main glacier, however, is much thicker, and it therefore erodes its valley to a greater depth than that of the tributary valleys. When the glaciers recede from the area, the floors of the tributary valley will be higher than the floor of the main valley; for this reason the tributary valleys are called hanging valleys.

Figure 12.6C shows the region after the glaciers have disappeared. The most conspicuous and magnificent landforms developed by valley glaciers are the long, straight, U-shaped valleys, or troughs. Many are several hundred meters deep and tens of kilometers long. The heads of glacial valleys terminate in large amphitheater-shaped or bowl-like cirques, which commonly contain small lakes. The divides between glacial valleys are sharp ridges called aretes, which have a rugged and sharp topography because of the severe effects of frost action.

The diagrams in Figure 12.6 do not show the landforms that develop at the terminus of a valley glacier. These are illustrated in the photograph in Figure 12.5. Three types of moraine are immediately obvious: (1) lateral moraines, along the margins of the glacier; (2) medial moraines, formed where two lateral moraines join; and (3) a terminal moraine, or end moraine. The terminal moraine characteristically extends in a broad arc, conforming to the shape of the terminus of the ice. It commonly traps meltwater and forms a temporary lake. If periods of stabilization occur during the recession of ice, recessional moraines may form behind the terminal moraine.

Much of the sediment forming the moraine is derived from abrasion and plucking as the glacial ice moves through the valley; however, a large volume of the glacier's load is derived from rock fragments that avalanche down the steep valley walls and accumulate along the glacier margins. Ice wedging is especially efficient in the cold climate of valley glaciers and produces large quantities of angular rock fragments. This material is transported along the surface of the ice near the glaciers' margins, forming lateral moraine (Figure 12.7). Where tributary glaciers enter the main valley, the lateral moraine merges to form a medial moraine in the central part of the main glacier.

The great volume of meltwater released at the terminus of a glacier reworks much of the previously deposited moraine and redeposits the material in an outwash plain beyond the glacier. Outwash sediment has all of the characteristics of stream deposits, and the sediment is typically rounded, sorted, and stratified.

Figure 12.7
Lateral moraines are accumulations of debris derived from erosion of the valley walls. Much of the debris is produced by ice wedging on the high valley slopes. This material moves downslope by mass movement and accumulates along the margins of the glacier. It is then transported downstream by the moving ice. Where tributary glaciers merge with the main stream of the glacier, the lateral moraines also merge to form a medial moraine. Terminal moraines form at the end of a glacier, where sediment transported by the glacier is deposited as the ice melts.

CONTINENTAL GLACIERS

Statement

In terms of their effect on the landscape and the Earth's hydrologic system, continental glaciers are by far the most important type of glacial system because of their immense size and huge volume. The basic elements of a continental glacier, however, are much the same as those of a valley glacier. Both systems have a zone of accumulation, where there is a net gain of ice from snowfall. The ice flows out from the zone of accumulation to the zone of ablation, where it leaves the system through melting evaporation.

The weight of such a huge ice mass causes the Earth's crust to subside, so that the surface of the land commonly slopes toward the glacier. Preexisting drainage systems are modified or completely obliterated. Thus, when the ice melts, no established, integrated drainage system exists, so numerous lakes form in the natural depressions. At the margins of continental glaciers the sediment is deposited as terminal moraine.

Discussion

The System

A continental glacier is a thick ice sheet that completely covers a large part of a continent. Such a large mass of ice (commonly 2 m thick) causes a number of significant changes in the regional physical setting. Some of these are shown in Figure 12.8. The weight of the ice depresses the ground surface, so the land commonly slopes toward the glacier. Consequently, glacier lakes form in the depression along ice margins, or an arm of the ocean may invade the depression. The original drainage system is greatly modified, as the streams that flow toward the ice margins are impounded to form lakes. Because the glacier advances more rapidly into lowlands, the margins are not smooth but are typically irregular or lobate. As the system expands and contracts, ridges of sediment are deposited along the margins, and a variety of ero-

sional and depositional landforms develop beneath the ice. The balance between the rate of accumulation and the rate of melting determines the size of the glacial system. The Barnes Ice Cap in Canada is a good example (Figure 12.9).

The glacier that covers nearly 80 percent of Greenland is another excellent example of a continental glacier. In cross section the glacier is shaped like a drop of water on a table (Figure 12.10). Its upper surface is a broad, almost flat-topped arch and is typically smooth and featureless. The base of the glacier is relatively flat. The Greenland glacier is more than 3,000 m thick in its central part, but it thins toward the margins. The zone of accumulation is in the central part of the island, where the ice sheet is nourished by snowstorms moving from west to east. The snow line lies from 50 to 250 km inland; thus, the area of ablation constitutes only a narrow belt along the glacial margins.

In rugged terrain, especially in areas close to the margins, the direction of ice movement is greatly influenced by mountain ranges, and the ice moves through mountain passes in large streams called outlet glaciers. These resemble valley glaciers in that they are confined by the topography. Pressure builds up in the ice behind a mountain range and forces outlet glaciers through mountain passes at relatively high speeds. Measurements in Greenland show that the main ice mass advances from approximately 10 to

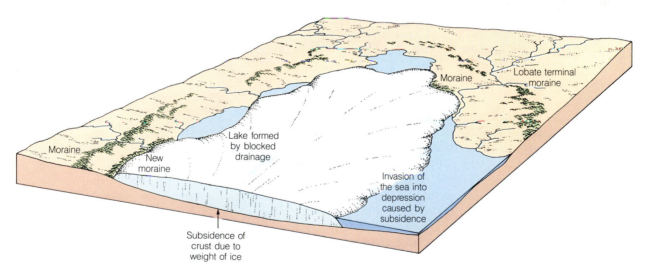

Figure 12.8
A continental glacier system covers a large part of a continent and causes a number of significant changes in the regional physical setting. The weight of the ice depresses the ground surface, so the land commonly slopes toward the glacier. Consequently, glacial lakes form in the depression along ice margins, or an arm of the ocean may invade the depression. The original drainage system is greatly modified, as the streams that flow toward the ice margins are impounded to form lakes. The glacier advances more rapidly into lowlands, so the margins are not smooth but are typically irregular or lobate. As the system expands and contracts, ridges of sediment are deposited along the margins, and a variety of erosional and depositional landforms develop beneath the ice. The balance between the rate of accumulation and the rate of melting determines the size of the glacial system.

Figure 12.9
The margins of the Barnes Ice Cap are marked by ridges of sediment deposited as the ice melts. The glacier's upper surface is gently arched, and meltwater has formed small, meandering streams. The landscape of the Great Lakes region must have appeared something like this 20,000 years ago.

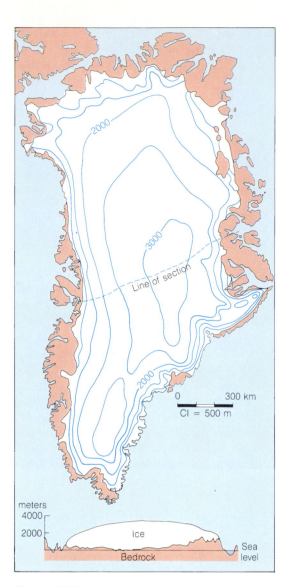

Figure 12.10
The Greenland Glacier covers nearly 80 percent of the island. In this diagram, the thickness of the glacier is shown by contour lines. Note from the cross section that the central part of Greenland has been depressed below sea level by the weight of the ice.

30 cm per day. Outlet glaciers, however, can move as fast as 1 m per hour. In some places you can actually see the ice move.

The glacier of Antarctica is similar to that of Greenland in that it covers essentially the entire landmass. Antarctica, however, is much larger than Greenland, and its glacier contains more than 90 percent of the Earth's ice. Much of the ice is more than 3,000 m thick; its weight has depressed large parts of the continent's surface below sea level. Parts of Antarctica (mostly near the continental margins) are mountainous, with the high peaks and ranges protruding above the ice. In the mountains outlet glaciers funnel ice from the interior to the coast. Two large ice shelves and numerous small ones occur along the coast. These break up in a process called calving, to form large icebergs in the South Atlantic Ocean.

Both Greenland and Antarctica are surrounded by water, so there is an ample supply of moisture to feed their glaciers. In contrast, although Siberia is cold enough for glaciers to exist, there is insufficient precipitation for ice to accumulate.

Landforms Developed by Continental Glaciers

From viewpoints on the ground, the landforms developed by continental glaciers are not nearly as spectacular as those produced by valley glaciers. On a re-

gional basis, however, continental glaciation modifies the entire landscape, producing many important and distinctive surface features. During the last Ice Age, the great, thick continental ice sheet moved over the flat lowlands of the Canadian Shield and the stable platform, removing existing soil and eroding up to several meters into the bedrock. As a result, many thousands of square kilometers of North America and northern Europe have little or no soil cover, and the effects of glaciation are seen everywhere in the polished and grooved fresh bedrock.

Landforms produced by continental glaciers can best be understood by studying the block diagrams in Figure 12.11 and the photographs in Figure 12.12. Figure 12.11A shows the margins of an ice sheet. Debris (till) transported by the glacier accumulates at

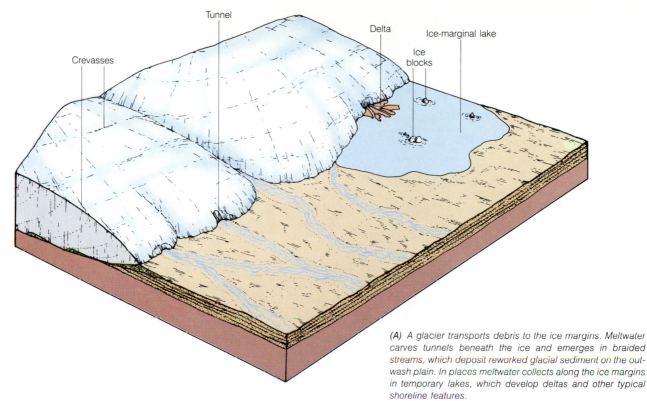

Crevasses

Tunnel

Delta

Ice-marginal lake

Ice blocks

(A) *A glacier transports debris to the ice margins. Meltwater carves tunnels beneath the ice and emerges in braided streams, which deposit reworked glacial sediment on the outwash plain. In places meltwater collects along the ice margins in temporary lakes, which develop deltas and other typical shoreline features.*

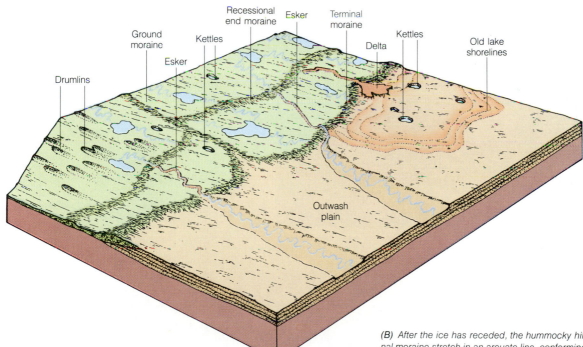

Drumlins

Ground moraine

Esker

Kettles

Recessional end moraine

Esker

Terminal moraine

Delta

Kettles

Old lake shorelines

Outwash plain

(B) *After the ice has receded, the hummocky hills of a terminal moraine stretch in an arcuate line, conforming to the original shape of the ice margins at the farthest advance of the glacier. The retreating glacier leaves behind unsorted debris in ground moraines, and recessional moraines mark the position of the ice margin where the glacier paused during its retreat. Hills of ground moraine can be reshaped by a subsequent advance of ice, forming drumlins. Sinuous eskers remain where sediment was deposited by subglacial streams, and sediment reworked by meltwater forms outwash-plain and lake deposits. Where ice blocks were stranded by the receding glacier and partly buried under debris, the melting of the ice produces kettles.*

Figure 12.11
Landforms developed by continental glaciers commonly are related to the position of the ice margins or the direction of the flow.

(A) Eskers *on the Canadian Shield form long, sinuous ridges composed of sand and gravel deposited by streams that flowed beneath the glacier.*

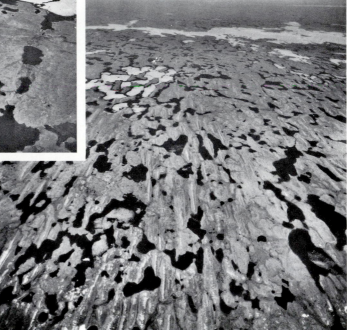

(B) Drumlin *fields in the Canadian Shield are stream-lined hills shaped by the movement of the glacier; they show the direction in which the ice flowed.*

(C) Moraines *in Alberta, Canada, form a distinctive topography of rolling hills and numerous closed depressions.*

(D) Glacial till *resting on horizontal limestone in Iowa is responsible for much of the rich farmland in that area.*

Figure 12.12
Features formed by continental glaciers in various parts of Canada, photographed from an altitude of approximately 9 km, and till deposited on bedrock in Iowa show the profound imprint the ice made on the landscape during the Pleistocene Ice Age.

the ice margin as a terminal moraine. Beneath the ice is a variable thickness of till, which is dragged forward by the glacier and then deposited as a ground moraine. This material, together with outwash-plain sediment, can be reshaped by subsequent advances of ice to produce streamlined hills called drumlins.

Streams of meltwater flow in tunnels within, and beneath, the ice and carry a large bed load, which is deposited to form a long, sinuous ridge called an esker. Debris-laden meltwater forms braided streams that flow from the glacier over the outwash plain, where they deposit much of their load. During the retreat of the glaciers, meltwater forms subglacial channels and tunnels, which open into the outwash plain. Temporary lakes can develop where meltwater is trapped along edges of the glacier, and deltas and other shoreline features form along the lake margins. Deposits on the lake bottom typically are stratified in a series of alternating light and dark layers called varves (Figure 12.13). The coarse, light-colored material accumulates during spring and summer runoff. During the winter, when the lake is frozen over, no new sediment enters the lake, and the fine mud settles out of suspension to form the thin, dark layers.

Ice blocks left behind by the retreating glacier front can be partly or completely buried in the outwash plain or in moraines. Where an isolated block of debris-covered ice melts, a depression called a kettle is formed.

Figure 12.11B shows the area after the glacier has disappeared completely. The end moraine appears as a belt of hummocky hills that mark the former position of the ice. An end moraine can be several kilometers wide, with local relief from 100 to 200 m. From the ground it probably would not be recognized by an untrained observer as anything more than a series of hills. Mapped over a large area, however, it can be seen to have an arcuate pattern, conforming to the lobate margin of the glacier. Many small depressions occur throughout the moraine, some of which may be filled with water, forming small lakes and ponds.

PLEISTOCENE GLACIATION

What evidence suggests multiple advances and retreats of glaciers during the Ice Age?
How did glaciation affect the desert regions of the world?
How did glaciation produce catastrophic flooding in Washington?

Statement

The great Ice Age that existed during the period from two or three million to 10,000 years ago constitutes one of the most significant events in the recent history of Earth. During that time the normal hydrologic system was interrupted completely throughout large areas of the world and was considerably modified in others.

The presence of so much ice on the continents had a profound effect on the Earth's entire hydrologic system. In many areas glacial erosion and deposition replaced running water as the dominant geologic processes. Sea level dropped more than 100 m; river patterns beneath the ice were obliterated; and elsewhere, river discharge and sediment loads were modified. Even areas far removed from the ice were affected by the modification of drainage systems.

Discussion

The evidence of the last Ice Age is overwhelmingly abundant. Over the last century, field observations have provided uncontestable evidence that continental glaciers covered large parts of Europe, North America, and Siberia. These ice sheets disappeared only between 10,000 and 20,000 years ago (Figure 12.14). The general extent of glaciation in the Northern Hemisphere is shown in Figure 12.15.

Four major periods of Pleistocene glaciation in the United States are recorded by broad sheets of till and complex moraines separated by ancient soils and layers of windblown silt. Striations, drumlins, eskers, and other glacier features show that all of Canada, the mountain areas of Alaska, and the eastern and central United States down to the Missouri and Ohio rivers were covered with ice (Figure 12.14). There were three main zones of accumulation. The largest was centered over Hudson Bay. Ice advanced radially from there, northward to the Arctic islands and

Figure 12.13
Varves are thin, alternating layers of light and dark sediment deposited in a glacial lake. A layer of relatively coarse-grained, light-colored sediment accumulates during the spring and summer runoff. During the winter, when the lake is frozen over, fine, dark mud settles to form a dark layer. Each set of light and dark layers therefore represents a year's accumulation.

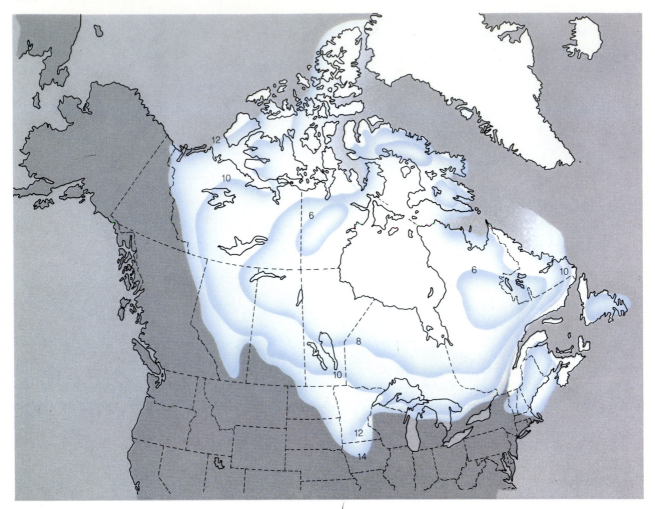

Figure 12.14
Successive position of the ice front during the recession of the last ice sheet have been mapped from data collected by many geologists in Canada and the United States. Contours indicate the position and age of the ice front in thousands of years before the present.

southward into the Great Lakes area. A smaller center was located in the Labrador Peninsula. Ice spread southward from this center into what are now the New England states. In the Canadian Rockies to the west, valley glaciers coalesced into ice caps. These grew into a single ice sheet, which then moved westward to the Pacific shores and eastward down the Rocky Mountain foothills until it merged with the large sheet from Hudson Bay.

Throughout much of central Canada, the glaciers eroded from 15 to 25 m of regolith and solid bedrock. This material was transported to the glacial margins and accumulated as ground moraine, end moraine, and outwash, in a broad belt from Ohio to Montana. In places the glacial debris is more than 300 m thick, but the average thickness is about 15 m. Meltwater carried sediment down the Mississippi River, and much of the fine-grained sediment was transported and redeposited by wind.

Effects of the Ice Age

Isostatic Adjustment. Major isostatic adjustment of the Earth's crust resulted from the weight of the ice, which depressed the continents during Pleistocene glaciation. In Canada a large area around Hudson Bay was depressed below sea level, as was the area around the Baltic Sea in Europe. The land has been rebounding from these depressions ever since the ice melted. The former seafloor around Hudson Bay has risen almost 300 m, and is still rising at a rate of about 2 cm per year. The land must rise an additional 80 m before it regains it preglacial level and reestablishes isostatic balance.

Changes of Sea Level. One of the most important effects of Pleistocene glaciation was the repeated worldwide rise and fall of sea level, a phenomenon that corresponded to the retreat and advance of the

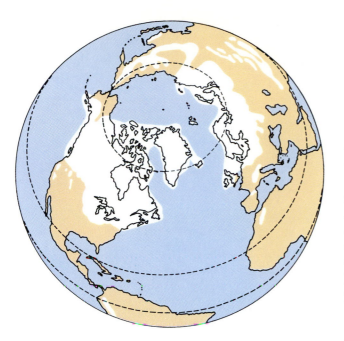

Figure 12.15
Pleistocene glaciers covered large areas in North America, Europe, and Asia, as well as many high mountain regions. Part of Alaska and Siberia were not glaciated because those areas were too dry. They were cold enough, but not enough precipitation fell for glaciers to develop.

glaciers. During a glacial period, water that normally returned to the ocean by runoff became locked on the land as ice, and sea level was lowered. When the glaciers melted, sea level rose again. The amount of change in sea level can be calculated because the area of maximum ice coverage is known in considerable detail, and the thickness of the ice can be estimated from the known volumes of ice in the glaciers of Antarctica and Greenland. The Antarctic ice sheet alone contains enough water to raise sea level throughout the world by about 70 m.

Modification of Drainage Systems. Before glaciation, the landscape of North America was eroded mainly by running water. Well-integrated drainage systems collected runoff and transported it to the ocean. Much of North America was drained by rivers flowing northeastward into Canada because the regional slope throughout the north-central part of the continent was to the northeast. The preglacial drainage patterns are not known in detail. Various features of the present systems, however, together with segments of ancient stream-channels now mostly buried by glacier sediments, suggest a pattern similar to that shown in Figure 12.16A. Before glaciation, the major tributaries of the upper Missouri and Ohio rivers were part of a northeastward-flowing drainage system.

This system also included the major rivers draining the Canadian Rockies, such as the Saskatchewan, Athabasca, Peace, and Liard rivers. It emptied into the Arctic Ocean, probably through Lancaster Sound and Baffin Bay, and an eastern drainage out of the Saint Lawrence River.

As the glaciers spread over the northern part of the continent, they effectively buried the trunk streams of the major drainage systems, damming up the northward-flowing tributaries along the ice front. This damming created a series of lakes along the glacial margins. As the lakes overflowed, the water drained along the ice front and established the present course of the Missouri and Ohio rivers. A similar situation created Lake Athabasca, Great Slave Lake, and Great Bear Lake and their drainage through the Mackenzie River. This established the present drainage patterns over much of North America (Figure 12.16B).

Beyond the margins of the ice, the hydrology of many streams and rivers was profoundly affected by the increased flow from meltwater or by the greater precipitation associated with the glacial epoch. With the appearance of the modern Ohio and Missouri rivers, water that formerly emptied into the Arctic and Atlantic oceans was diverted to the Gulf of Mexico through the Mississippi River.

Other streams became overloaded and partly filled their valleys with sediment. Still others became more effective agents of downcutting as a result of glacial sediment, and they deepened their valleys. Although the history of each river is complex, the general effect of glaciation on rivers was to produce thick alluvial fill in their valleys, which is now being eroded to form stream terraces.

Lakes. Pleistocene glaciation created more lakes than all other geologic processes combined. The reason is obvious if we recall that a continental glacier completely disrupts the preglacial drainage system. The surface over which the glacier moved was scoured and eroded by the ice, which left a myriad of closed, undrained depressions in the bedrock. These depressions filled with water and became lakes (Figure 12.17).

Farther south in the north-central United States, lakes formed in a different manner. There, the surface was covered by glacial deposits of ground moraine and end moraines. These deposits formed closed depressions throughout Michigan, Wisconsin, and Minnesota, which soon filled with water to form tens of thousands of lakes. Many of these lakes still exist. Others have been drained or filled with sediment, leaving a record of their former existence in peat bogs, lake silts, and abandoned shorelines.

During the retreat of both the European and the North American ice sheets, several conditions com-

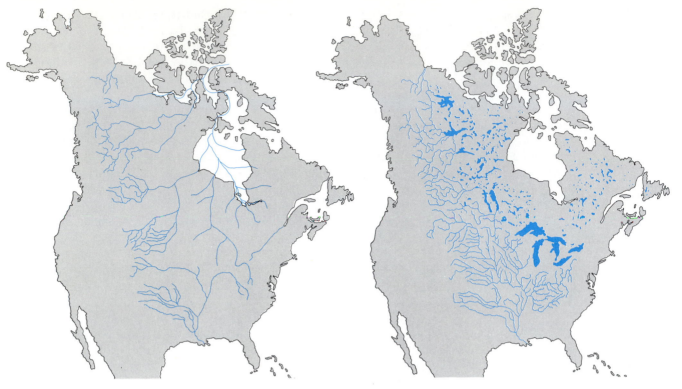

Figure 12.16A
Drainage of central North America before the Ice Age was northeastward, from the northern and central Rocky Mountains, into the St. Lawrence Bay, Hudson Bay, and the Arctic area.

Figure 12.16B
The present drainage patterns show major modifications. Preglacial drainage was impounded against the glacial margins and developed new outlets to the ocean through the Missouri, Ohio, and Mackenzie rivers. The drainage system beneath the ice was obliterated. The present drainage in most of Canada is deranged, consisting of numerous lakes, swamps, and unintegrated meandering streams.

Figure 12.17
Lakes created by continental glaciation in the shield area of North America were photographed from a height of approximately 9 km. More lakes were created by glaciation than by all other geologic processes combined.

bined to create large lakes along the glacial margins. We can envision their formation with the help of the basic model of continental glaciation shown in Figure 12.8. The ice on both continents was about 3,000 m thick near the centers of maximum accumulation, but it tapered toward the glacier margins. Crustal subsidence was greatest beneath the thickest accumulation of ice. In parts of Canada and Scandinavia, the crust was depressed more than 600 m. As the ice melted, rebound of the crust lagged behind, producing a regional slope toward the ice. This slope formed basins that have lasted for thousands of years. These basins became lakes or were invaded by the ocean. The Great Lakes of North America and the Baltic Sea of northern Europe were formed primarily in this way. Although the origin of the Great Lakes is extremely complex, the major elements of their history are known and are illustrated in the four diagrams of Figure 12.18.

To the northwest another group of lakes formed in much the same way, but they have since been reduced to small remnants of their former selves (Figure

12.16B). The largest of these marginal lakes, known as Lake Agassiz, covered the broad, flat region of Manitoba, northwestern Minnesota, and the eastern part of North Dakota (Figure 12.19). It drained into the Mississippi River and then, at lower stages, developed outlets into Lake Superior. Later, when the ice dam retreated, it drained into Hudson Bay. Remnants of this vast lake include Lake Winnipeg, Lake Manitoba, and Lake of the Woods. The sediments deposited on the floor of Lake Agassiz provided much of the rich soil for the wheatlands of North Dakota,

Manitoba, and the Red River Valley of Minnesota. Even now, ancient shorelines of Lake Agassiz remain, marking its former margins.

Northward, along the margin of the Canadian Shield, Lake Athabasca, Great Slave Lake, and Great Bear Lake are remnants of the other great ice-marginal lakes. In northern Europe the recession of the Scandinavian ice sheet caused similar depressions along the ice margins, and the large lakes that were thus produced ultimately connected with the ocean to form the Baltic Sea.

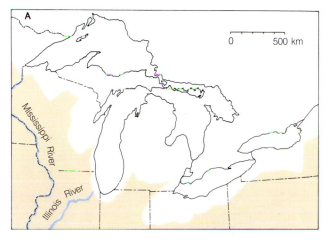

(A) Approximately 16,000 years ago, the ice front extended beyond the present Great Lakes. The ice advanced into lowlands surrounding the Michigan basin, with large lobes extending down from the present sites of lakes Erie and Michigan.

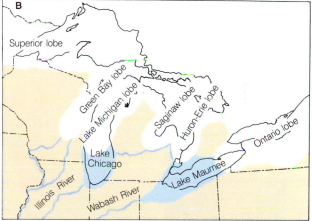

(B) The ancestral Great Lakes appeared about 14,000 years ago, as the ice receded. The northern margins of the lakes were against the retreating ice. Drainage was to the south, to the Mississippi River.

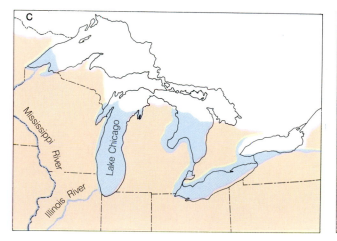

(C) As the ice front continued to retreat, an eastern outlet developed to the Hudson River, but the western lakes still drained into the Mississippi. The lakes began to assume their present outlines about 10,500 years ago.

(D) Niagara Falls originated about 8,000 years ago, when the glacier receded past the Lake Ontario basin, and water from Lake Erie flowed over the Niagara Escarpment into Lake Ontario.

Figure 12.18
The evolution of the Great Lakes can be traced from their origin along the ice margins about 16,000 years ago. The sequence of events and modifications of the landscape are inferred from numerous studies of glacial features in the Great Lakes area.

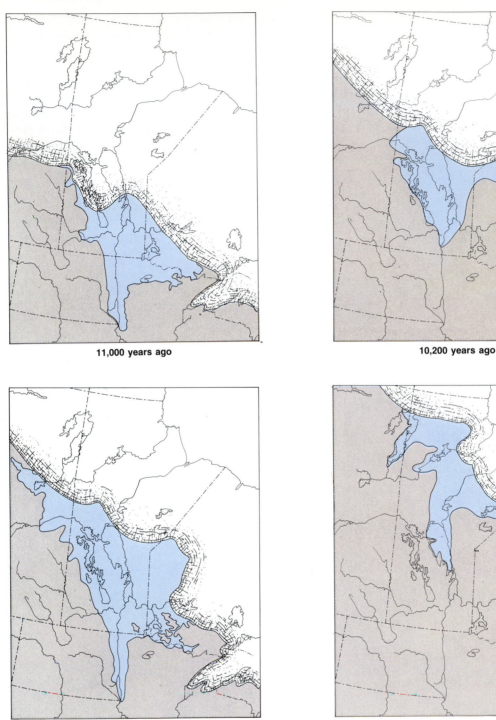

11,000 years ago

10,200 years ago

9,900 years ago

8,300 years ago

Figure 12.19
Lake Agassiz was the largest glacial lake in North America. Its former shorelines are now marked by beach ridges, spits, and bars. The dry lakebed now forms the fertile soil of Manitoba and North Dakota. Remnants of this former glacial lake include Lake Winnipeg, Lake Manitoba, and Lake of the Woods.

Pluvial Lakes. The climatic conditions that caused glaciation had an indirect effect on arid and semiarid regions far removed from the large ice sheets. The increased precipitation that fed the gla-

ciers also increased the runoff of major rivers and intermittent streams, resulting in the growth and development of large pluvial lakes (Latin *pluvia*, "rain") in numerous isolated basins in nonglaciated areas

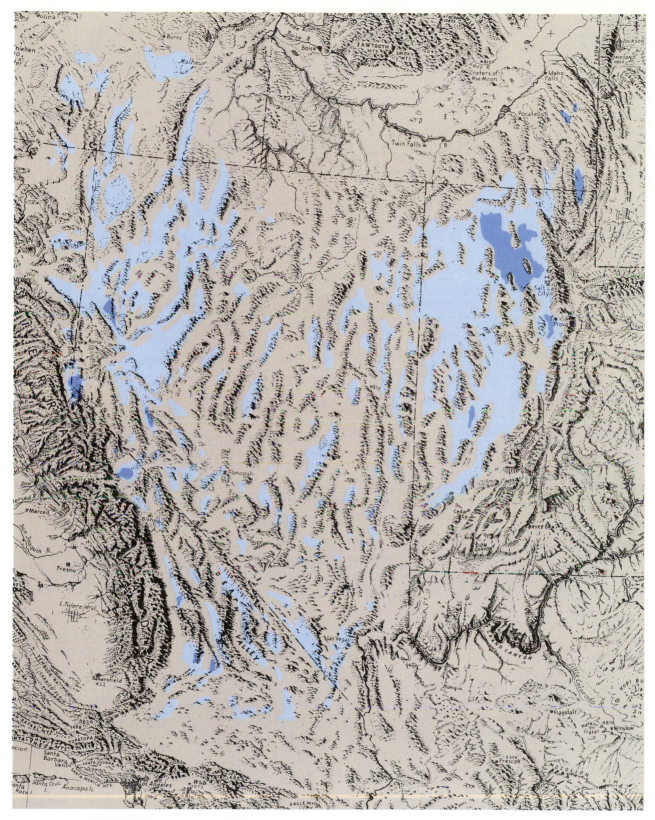

Figure 12.20
Pluvial lakes were formed in the closed basins of the western United States as a result of climatic changes associated with the glacial epoch. Most are now dry lake beds because of the arid climate. Former shorelines of the pluvial lakes are well marked along the basin margins. Lake Bonneville, in western Utah, was the largest. Its present remnants are Great Salt Lake, Utah Lake, and Sevier Lake.

throughout the world. Most pluvial lakes developed in relatively arid regions where, prior to the glacial epoch, there was insufficient rain to establish an integrated, through-flowing drainage system to the sea. Instead, stream runoff in those areas flowed into closed basins and formed playa lakes. With increased rainfall, the playa lakes enlarged and sometimes overflowed. They developed a variety of shoreline features—wave-built terraces, bars, spits, and deltas—now recognized as high-water marks in many desert basins. Pluvial lakes were most extensive during glacial intervals. During interglacial stages, when less precipitation fell, the pluvial lakes shrank to form small salt flats or dry, dusty playas.

The greatest concentration of pluvial lakes in North America was in the northern part of the Basin and Range province of western Utah and Nevada. The fault-block structure there has produced more than 140 closed basins, many of which show evidence of former lakes or former high-water levels of existing lakes. The distribution of the former lakes is shown in Figure 12.20. Lake Bonneville was the largest by far and occupied a number of coalescent intermontane basins. Remnants of this great body of fresh water are Great Salt Lake, Utah Lake, and Sevier Lake. At its maximum extent, Lake Bonneville was approximately the size of Lake Michigan, covering an area of 50,000 km^2, and was 300 m deep.

Biological Effects of the Ice Age. The severe climatic changes during the Ice Age had a drastic impact on most life forms. With each advance of the ice, large areas of the continents (the areas beneath the ice) became totally depopulated, and plants and animals retreating southward in front of the advancing glacier were under tremendous stress. The most severe stresses resulted from drastic climatic changes, reduced living space, and a curtailed food supply. As the glaciers advanced, many species were displaced, along with their environments, across distances of up to 3,200 km. As the ice retreated, some new living space became available in deglaciated areas, but the formerly exposed continental shelves were inundated by the rising sea. During the major glacial advances, when sea level was lower, new routes of migration opened from Asia to North America—because much of Alaska and Siberia were not glaciated (see Figure 12.14)—and from Southeast Asia to the islands of Indonesia. Land plants were forced to migrate with the climatic zones in front of the glaciers. As the glaciers pushed cold-weather belts southward, displaced storm tracks and changes in precipitation affected even the tropics.

Many life forms could not cope with the repeated and overwhelming environmental changes brought about by the cycles of advancing and retreating ice. Numerous species, particularly giant mammals, became extinct. During glaciation, the now-extinct imperial mammoth, 4.2 m high at the shoulders, roamed much of North America. The saber-toothed tiger became extinct about 14,000 years ago. Fossils of the giant beaver, as large as a black bear, and the giant ground sloth, which measured 6 m tall standing on its hind legs, have been found in Pleistocene sediments. In Africa fossil sheep 2 m tall have been found, in addition to pigs as big as the present-day rhinoceros. In Australia giant kangaroos and other marsupials thrived during the Pleistocene.

Effect of Winds. The presence of ice over so much of the continents greatly modified patterns of atmospheric circulation. Winds near the glacial margins were strong and unusually persistent because of the abundance of dense, cold air coming off the glacier fields. These winds picked up and transported large quantities of loose, fine-grained sediment brought down by the glaciers. This dust accumulated as loess (windblown silt), sometimes hundreds of meters thick, forming an irregular blanket over much of the Missouri River valley, central Europe, and northern China.

Sand dunes were much more widespread and active in many areas during the Pleistocene. A good example is the Sand Hills region in western Nebraska, which covers an area of about 60,000 km^2. This region was a large, active dune field during the Pleistocene, but today the dunes are largely stabilized by a cover of grass.

The Oceans. Pleistocene glaciation affected the waters of all of the oceans to some extent. Besides changing the sea level so that shorelines were altered and much of the continental shelves were exposed, the glacial periods cooled the ocean waters. The lower temperatures affected the kind and distribution of marine life and also influenced seawater chemistry. Furthermore, patterns and strengths of oceanic currents were changed. Circulation was significantly restricted by glacially formed features such as the Bering Strait, extensive pack ice, and exposed shelves.

Even the deep-ocean basins did not escape the influence of glaciation. Where glaciers entered the oceans, icebergs broke off and rafted their enclosed load of sediment out into the ocean. As the ice melted, debris—ranging from huge boulders to fine clay—settled on the deep-ocean floor, resulting in an unusual accumulation of coarse glacier boulders in fine oceanic mud. Ice-rafted sediment is most common in the Arctic, the Antarctic, the North Atlantic, and the northeastern Pacific.

CAUSES OF GLACIATION

Is Earth entering a new Ice Age?

Statement

Although the history of Pleistocene glaciation is well established and the many effects of glaciation are clearly recognized, we do not know with complete certainty why climates change and why glaciation takes place. For more than a century, geologists and climatologists have struggled with this problem, but it remains unsolved. An adequate theory of glaciation must account for the following facts:

1. During the last Ice Age, repeated advances of the ice in North America and northern Europe were separated by interglacial periods of warm climate. Glaciation, therefore, is not related to a slow process involving long-term cooling.
2. Glaciation is an unusual event in Earth's history. Widespread glaciation also occurred at the end of the Paleozoic era, 200 to 300 million years ago, and during late Precambrian time, approximately 700 million years ago.
3. Throughout most of Earth's history, the climate has been milder and more uniform than it is now. A period of glaciation would require a lowering of Earth's present average surface temperature by about 5° C and, perhaps, an increase in precipitation.
4. Continental glaciers grow on elevated or polar landmasses that are situated so that storms bring moist, cold air to them. Glaciers can move into lower latitudes, but they originate in highlands or in high latitudes. Greenland and Antarctica provide favorable topographic conditions today, as do the Labrador Peninsula, the northern Rocky Mountains, Scandinavia, and the Andes Mountains.
5. Precipitation is critical to the growth of glaciers. A number of areas are presently cold enough to produce glaciers but do not have sufficient snowfall to develop glacial systems.

It has been known for some time that the Earth's orbit around the sun changes periodically, forcing Earth to be slightly closer to the sun during some epochs than others. M. Milankovitch, a Yugoslavian geophysicist, convincingly calculated that these irregularities in the Earth's orbit could cause climatic cycles. The main period of the cycle is about 100,000 years. In addition, the inclination of the Earth's axis varies periodically between 22° and 24°. The tilt of Earth's axis, of course, causes the seasons: the greater the tilt, the greater the contrasts between summer and winter temperatures. The changes in the angle of inclination occur every 41,000 years. Also, Earth wobbles on its axis like a giant top and completes one wobble every 21,000 years. According to the Milankovitch theory, these astronomical factors cause a periodic cooling of Earth, with the coldest part in the cycle occurring about every 40,000 years.

Milankovitch worked out the ideas of climatic cycles in the 1920s and 1930s, but it was not until the 1970s that a sufficiently long and detailed chronology of the Pleistocene was worked out to test the theory adequately. A correspondence between astronomical cycles and late Cenozoic climatic fluctuations now seems clear. Furthermore, studies of deep-sea cores indicate that the fluctuation of climatic cycles is remarkably close to that predicted by Milankovitch.

A problem with this theory is that the astronomical cycles have been in existence for billions of years. One might expect that glaciation would have been a cyclic event throughout geologic time, instead of a rare occurrence. Other factors must also be involved. Some scholars have proposed that variations in solar energy may possibly be related to sunspots. Others have suggested variations in atmospheric carbon dioxide. Still others argue that volcanic dust injected into the atmosphere would shield Earth from the sun's rays and initiate an ice age.

A more plausible explanation for the erratic occurrence of the conditions necessary for glaciation relates to the position of the continents relative to the poles and to the circulation of the oceans and atmosphere. Here again the theory of plate tectonics may help to explain how Earth's systems operate. Throughout most of geologic time, the polar regions appear to have been broad, open oceans that allowed major ocean currents to move unrestricted. Equatorial waters were spread over the polar regions, warming them with water from more temperate latitudes. This unrestricted circulation produced mild, uniform climates that persisted throughout most of geologic time.

The large North and South American continental plates moved westward from the Eurasian plate throughout Tertiary time. This drift culminated in the development of the Atlantic Ocean, trending north-south, with the North Pole in the small, nearly landlocked basin of the Arctic Ocean. Meanwhile, by late Miocene time, the Antarctic continent had drifted over the South Pole, and glaciation began on that continent. Evidence from deep-sea cores in the southern oceans strongly suggests that glaciation in the Antarctic began long before the Pleistocene and has continued ever since. By the beginning of Pleistocene time, the present location and configuration of the continents and ocean basins had been established, and areas near the polar regions that had adequate precipitation were at the glacial threshold.

It appears that in order for an ice age to occur, there must be

1. special geographic conditions—that is, continents must be in polar regions.
2. adequate precipitation.

3. temperatures low enough so that more snow falls in the winter than melts in the summer.

An important point concerning the question of the origin of the Ice Age is that there are many variables in the interactions of the climate and the hydrologic system. Thus, no single causative agent can be identified. Apparently, an ice age occurs as a result of several simultaneously occurring factors.

SUMMARY

1. A glacier is a system of flowing ice that originates on land through the accumulation and recrystallization of snow.
2. Glaciers erode, transport, and deposit vast amounts of rock material and greatly modify the preexisting landscape.
3. There are two principal types of glaciers: (1) valley glaciers, which originate in snowfields of high mountain ranges and are confined within valley walls, and (2) continental glaciers, which are sheets of ice thousands of meters thick that cover large parts of continents.
4. The major type of landforms resulting from valley glaciation are (a) U-shaped valleys, (b) cirques, (c) horns, (d) hanging valleys, (e) moraines, and (f) outwash plains.
5. The major landforms resulting from continental

glaciation are (a) moraines, (b) drumlins, (c) eskers, (d) outwash plains, (e) kettles, and (f) lake deposits.
6. The Pleistocene Ice Age began two to three million years ago and terminated in most areas about 15,000 years ago. During the Ice Age there were a number of glacial and interglacial epochs.
7. The major effects of the Ice Age were (a) glacial erosion and deposition over large parts of the continents, (b) isostatic adjustment of the crust, (c) changes in sea level, (d) modification of drainage systems, (e) creation of numerous lakes, and (f) stress on most life forms.
8. The exact causes of glaciation are not fully understood. In order for glaciation to occur, there must be adequate precipitation and cool temperatures. Many aspects of glaciation can be explained in terms of plate tectonics, oceanic circulation, and astronomical cycles.

KEY TERMS

abrasion (p. 192)
cirque (p. 197)
continental glacier (p. 198)
crevasse (p. 195)
drumlin (p. 203)
end moraine (p. 198)
esker (p. 201)

glacial plucking (p. 192)
ground moraine (p. 201)
hanging valley (p. 198)
horn (p. 198)
kettle (p. 201)
lateral moraine (p. 198)
loess (p. 210)

medial moraine (p. 198)
moraine (p. 198)
outwash (p. 198)
Pleistocene epoch (p. 203)
pluvial lake (p. 208)
recessional
   moraine (p. 198)

striation (p. 192)
terminal moraine (p. 198)
valley glacier (p. 194)
varve (p. 203)
zone of ablation (p. 194)
zone of
   accumulation (p. 194)

REVIEW QUESTIONS

1. Explain the processes by which glaciers erode the surface over which they flow.
2. Draw a cross section of a typical valley glacier and explain how a valley glacial system operates.
3. Sketch a model of a continental glacial system and explain how it operates.
4. Name and describe the landforms produced by valley glaciers.
5. Briefly describe the major effects, both direct and indirect, of Pleistocene glaciation.
6. List several hypotheses to explain the cause of continental glaciation.
7. Why did sea level change during each period of advance and retreat of the ice?
8. Explain the origin of the Missouri River.
9. Why did a large number of lakes develop in the arid part of the western United States during each major advance of the ice during the Pleistocene Ice Age?

Multiple-Choice Questions

1. If the front of an active glacier is observed to be stationary, it is correct to infer that the ice in the glacier is
   a. not moving at all.
   b. melting as fast as it moves.
   c. moving faster than it melts.
   d. moving more slowly than it melts.
   e. melting faster than it moves.
2. Glaciers are said to be *advancing* when
   a. the rate of melting at the glacier terminus exceeds the rate of flow.
   b. the rate of ice flow exceeds the rate of melting at the glacier terminus.
   c. the rate of ice flow is the same as the rate of melting at the glacier terminus.
   d. the thickness of the ice is reduced below 50 m.
   e. the glacier develops an outwash plain.
3. The removal of large blocks of bedrock by glacial ice is called
   a. abrasion.
   b. surging.
   c. glacial plucking.
   d. grooving.
   e. scouring.
4. Which is *not* associated with glaciation?
   a. change in sea level
   b. isostatic adjustment
   c. modification of drainage
   d. folding and faulting of bedrock
   e. deposition of stratified and unstratified sediment
5. Lateral moraines of a valley glacier are principally composed of rock debris
   a. deposited by glacial meltwater.
   b. freed from the enclosing ice by frost heaving.
   c. derived from the weathering of the valley walls.
   d. deposited by streams flowing along the ice margins.
   e. derived from the floor of the valley.
6. Moraines are made up of
   a. outwash.
   b. erratics.
   c. loess.
   d. till.
   e. varves.

7. Which of the following landforms is *not* associated with continental glaciers?
   a. eskers
   b. drumlins
   c. aretes
   d. kettles
   e. moraines
8. Glaciers erode mainly by
   a. wetting and drying.
   b. thermal expansion and contraction.
   c. plucking and abrasion.
   d. ablation and melting.
   e. solution and frost wedging.
9. A stream-lined hill composed of glacial drift, which indicates the direction of glacier movement, is termed a(n)
   a. esker.
   b. roche moutonnée
   c. drumlin.
   d. dome.
   e. erratic.
10. Which of the following glacial deposits is a heterogeneous mixture of unstratified fragments?
    a. kames
    b. outwash plain
    c. ground moraine
    d. esker
    e. glacial lake deposits

True/False Questions

1. *T F* The length of a glacier is an indication of the amount of ice that has moved through the system.
2. *T F* The weight of a continental glacier causes the Earth's crust to subside, so the land surface commonly slopes toward the glacier.
3. *T F* One of the largest continental glaciers was in Siberia.
4. *T F* Glacial outwash has all of the characteristics of stream deposits; the sediment typically is rounded, sorted, and stratified.
5. *T F* Evidence now shows that a number of periods of growth and retreat of continental glaciers occurred during the Ice Age.

Fill in the Missing Terms

1. The lifting out and removal of fragments of bedrock is called glacial _____.
2. The heads of glacial valleys terminate in large amphitheater-shaped or bowl-like landforms called _____.
3. Streams of meltwater flow in tunnels within and beneath a glacier; they carry a large bed load, which is deposited to form a long, sinuous ridge called an _____.
4. Deposits on the bottom of glacial lakes typically are stratified in a series of alternating light and dark layers called _____.
5. Before the Ice Age, much of North America was drained by rivers flowing in a _____ direction.

13

SHORELINE SYSTEMS

Water in oceans and lakes is in constant motion. It moves by wind-generated waves, tides, tsunamis (seismic sea waves), and a variety of density currents. As it moves, it constantly modifies the shores of the continents and islands of the world, reshaping coastlines with the ceaseless activity of waves and currents. Shoreline processes can change in intensity from day to day and from season to season, but they never stop.

The present shorelines of the world, however, are not the result of present-day processes alone. Nearly all coasts were profoundly affected by the rise in sea level caused by the melting of the Pleistocene glaciers, between 8,000 and 20,000 years ago. The rising sea flooded large parts of the low coastal areas, and shorelines moved inland over landscapes formed by continental processes. The configuration of a given coastline may, therefore, be largely the result of processes other than marine. It may originally have been shaped by stream erosion or deposition, glaciation, volcanism, Earth movements, or even the growth of organisms.

In this chapter we will consider these and other questions of coastal dynamics as part of the hydrologic system. Shorelines are especially important to our society because of the concentration of population on or near the coast. To live in harmony with this rapidly changing environment, we must understand its history and dynamics.

MAJOR CONCEPTS

1. Wind-generated waves provide most of the energy for shoreline processes.
2. Wave refraction concentrates energy on headlands and disperses it in bays.
3. Longshore drift, generated by waves advancing obliquely toward the shore, is one of the most important shoreline processes.
4. Erosion along a coast tends to develop sea cliffs by the undercutting action of waves and longshore currents. As a cliff recedes, a wave-cut platform develops, until equilibrium is established between wave energy and the configuration of the coast.
5. Sediment transported by waves and longshore current is deposited in areas of low energy to form beaches, spits, and barrier islands.
6. Erosion and deposition along a coast tend to develop a straight or gently curving shoreline that is in equilibrium with the energy expended.
7. Reefs grow in a special environment and form coasts that can evolve into atolls.
8. The worldwide rise in sea level associated with the melting of the Pleistocene glaciers drowned many coasts between 8,000 and 20,000 years ago.
9. Tides are produced by the gravitational attraction of the Moon (and to a lesser extent the sun) and exert a major local influence on shorelines.
10. Tsunamis are waves generated by faulting, volcanic eruptions, and submarine landslides that cause disturbance of the seafloor.

WAVES

What is the nature of the motion of water in wind-generated waves?
Why do breakers occur only along coasts and not in the open ocean?

Statement

Most types of ocean waves are generated by winds in storms far offshore. As waves move out from the storm area and approach the shore, they commonly are bent, or refracted, so that their crest line tends to become parallel to the shore. Wave refraction is an important shoreline process because it influences the distribution of energy along the shore as well as the direction in which coastal water and sediment move.

Discussion

As wind moves over the open ocean, the turbulent air distorts the surface of the water. Gusts of wind depress the surface where they move downward, and as they move upward, they cause a decrease in pressure that elevates the water surface. These changes in pressure produce an irregular, wavy surface in the ocean and transfer part of the wind's energy to the water. In a stormy area, waves tend to be choppy and irregular, and wave systems of different sizes and orientations may be superposed on each other. As the waves move out from their place of origin, however, the shorter waves move slowly and are left behind, while the wave patterns develop some measure of order.

As a wave approaches the shore, it collapses forward, or breaks, into a surf. The water then rushes forward to the shore and returns as backwash. It is this energy, the surge of the surf and subsequent backwash, that causes most erosion, transportation, and deposition along the shores of all the continents and islands of the world.

Wave motion can easily be observed by watching a floating object move forward as the crest of a wave approaches and then sinks back into the trough that follows. Viewed from the side, the object moves in a circular orbit with a diameter equal to the wave height (Figure 13.1). Beneath the surface this orbital motion dies out rapidly, becoming negligible at a depth equal to about one-half the wavelength. This level is known as the wave base (Figure 13.2). The motion of water in waves is therefore distinctly different from the motion in currents, in which water moves in a given direction and does not return to its original position.

The energy of a wave depends on its length and height. The greater the wave height, the greater the

Direction of wave advancement

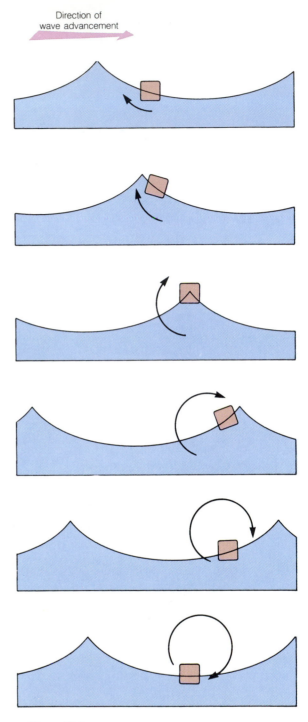

Figure 13.1
The motion of a water particle as a wave advances is indicated by the movement of a floating object. As the wave advances (from left to right), the object is lifted up to the crest and then returns to the trough. The wave form advances, but the water particles move in an orbit, returning to their original position.

size of the orbit in which the water moves. The total energy of a wave can be represented by a column of water in orbital motion.

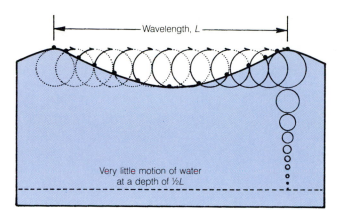

Figure 13.2
The orbital motion of water in a wave decreases with depth and dies out at a depth equal to about half the wavelength.

Breakers

Wave action produces little or no net forward motion of the water because the water moves in an orbital path as the waveform advances. As a wave approaches shallow water, however, some important changes occur (Figure 13.3). First, the wavelength decreases because the wave base encounters the ocean bottom, and the resulting friction gradually slows down the wave. Second, the wave height increases as the column of orbiting water encounters the seafloor. As the waveform becomes progressively higher and the velocity decreases, a critical point is reached at which the forward velocity of the orbit distorts the waveform. The wave crest then extends beyond the support range of the underlying column of water, and the wave collapses, or breaks. At this point, all of the water in the column moves forward, releasing its energy as a wall of moving, turbulent surf called a breaker.

After a breaker collapses, the swash (turbulent sheet of water) flows up the beach slope. The swash is a powerful surge, which causes a landward movement of sand and gravel on the beach. As the force of the swash is dissipated against the slope of the beach, the water flows down the beach slope as backwash, although some seeps into the permeable sand and gravel.

Wave Refraction

Wave refraction (the bending of water waves as they encounter different depths or bottom conditions) occurs because part of a wave in shallow water begins to "drag bottom" and slow down. The process is illustrated in Figure 13.4. The unrefracted wave is divided into three equal parts (AB, BC, and CD), each having an equal amount of energy. As the wave moves toward the shore, segment BC, in front of the headland, interacts with the shallow floor first and is slowed down. Meanwhile, the rest of the wave (segments AB and CD) moves forward at normal velocity. This difference in velocity causes the crest line of the wave to bend as it advances shoreward. The wave energy between points B and C is concentrated on a relatively short segment (B′C′) of the headland, whereas the equal amounts of energy between A and B and between C and D are distributed over much greater distances (A′B′ and C′D′). Breaking waves are, thus, powerful erosional agents on the headlands but relatively weak in bays, where they commonly deposit sediment to form beaches. Where major wave fronts are refracted around islands and headlands, the refraction patterns are obvious from the air (Figure 13.5).

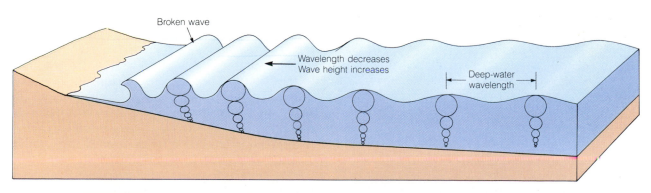

Figure 13.3
A wave approaching the shore undergoes several significant changes as the water in orbital motion encounters the seafloor. (1) The wavelength decreases due to frictional drag, and the waves become crowded together as they move closer to shore. (2) Wave height increases as the column of water, moving in an orbit, stacks up on the shallow seafloor. (3) The wave becomes asymmetrical because of increasing height and frictional drag on the seafloor, and ultimately breaks. The water then ceases to move in an orbit and rushes forward to the shore.

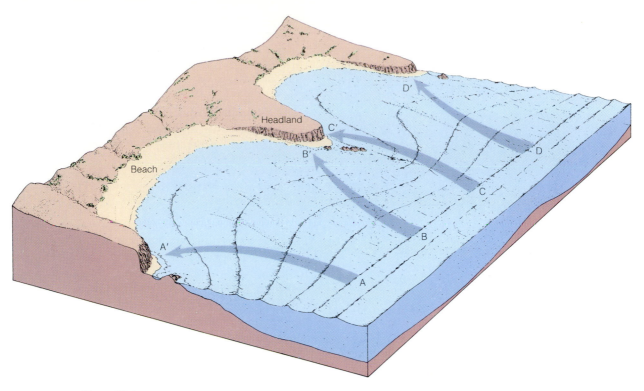

Figure 13.4
Wave refraction concentrates energy on headlands and disperses it across bays. Each segment of the unrefracted wave—AB, BC, and CD—has the same amount of energy. As the wave approaches shore, segment BC encounters the seafloor sooner than AB or CD and moves more slowly. This difference in the velocities of the three segments causes the wave to bend; the energy contained in segment BC is then concentrated on the headland, while the energy contained in AB and CD is dispersed along the beach.

Figure 13.5
Wave refraction around headlands and islands is clearly shown in aerial photographs taken along the coast of Oregon. The energy concentrated on the headlands has reduced some of them to offshore islands.

Longshore Drift

Longshore drift is the movement of sediment along a beach by the advance and backwash of waves as they approach the shore at an angle. It is one of the most important shoreline processes and is easily observed on almost any beach as the sand is moved to and fro with each advancing wave (see Figure 13.6). As a wave strikes the shore at an angle of less than 90°, water and sediment moved by the breaker are transported obliquely up the beach in the direction of the wave's advance. When the wave's energy is spent, the water and sediment return with the backwash directly down the beach, perpendicular to the shore. The next wave moves the material obliquely up the shore again, and the backwash returns it again directly down the beach slope. A single grain of sand is thus moved in an endless series of small steps, with a resulting net transport parallel to the shore.

Longshore drift results in the movement of an enormous volume of sediment. A beach can be thought of as a river of sand, moving by the action of beach drift. If the wave direction is constant, longshore drift occurs in only one direction. If waves approach the shore at different angles during different seasons, longshore drift is periodically reversed. Longshore

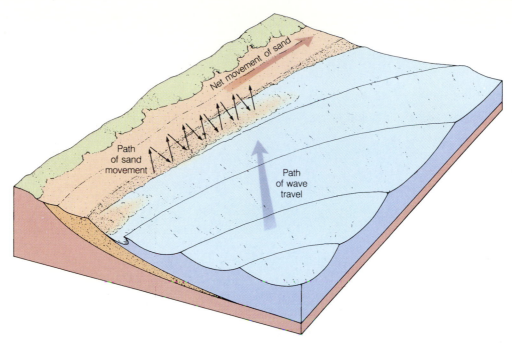

(A) As a wave breaks on the shore, sediment lifted by the surf is moved diagonally up the beach slope. The backwash then carries the particles back down the beach at a right angle to the shoreline. This action is repeated by each successive wave and transports the sediment along the coast in a zigzag pattern. Particles also are moved underwater in the breaker surf zone by this action.

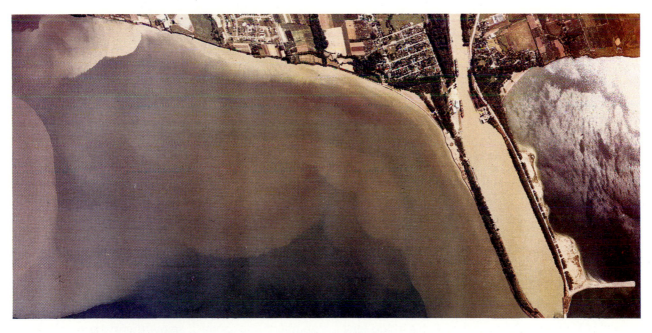

(B) Longshore drift can be seen in patterns of sediment on aerial photographs. The sediment is moved parallel to the shore in a series of waves.

Figure 13.6
Longshore drift occurs where waves strike the beach at an oblique angle.

currents can pile significant volumes of water on the beach, which return seaward through the breaker zone as a narrow rip current. These currents can be strong enough to be dangerous to swimmers.

Example
One of the best ways to appreciate the process of longshore drift is to consider how it has influenced human affairs. A good example occurred in the area of Santa

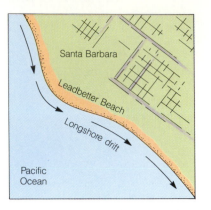

(A) The Santa Barbara coast had significant longshore drift before the breakwater was built.

(B) The initial breakwater prevented the generation of longshore currents in the protected area behind it, and therefore the harbor filled with sand.

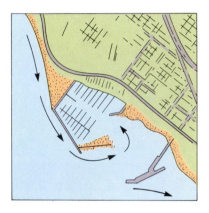

(C) After the breakwater was connected to the shore, longshore currents moved sand around the breakwater and filled the mouth of the harbor. Sand is now dredged from the harbor and pumped down the coast.

(D) Santa Barbara Harbor.

Figure 13.7
The effect of a breakwater on longshore drift in Santa Barbara, California, is documented by a series of maps on the coast from 1925 to 1938.

Barbara, along the southern coast of California, where data have been collected over a considerable period. Santa Barbara is a picturesque coastal town at the base of the Santa Ynez Mountains. It is an important educational, agricultural, and recreational area, and the people there wanted a harbor that could accommodate deep-water vessels. Studies by the United States Army Corps of Engineers indicated that the site was unfavorable because of the strong longshore currents, which carry large volumes of sand to the south (Figure 13.7A). Rivers draining the mountains of the coastal ranges supply new sediment to the coast at a rate of 592 m³ per day. Longshore drift continually moves the sand southward from beach to beach. The currents are so strong that boulders of 0.6 m in diameter can be transported. Ultimately, the sand transported by longshore drift is delivered to the head of a submarine canyon and then moves down the canyon to the deep-sea floor.

In spite of reports advising against the project, a breakwater 460 m long was built and a deep-water

harbor was constructed in 1925. This breakwater was not tied to the shore, and sand, moved by longshore drift, began to pour through the gap and fill the harbor, which was protected from wave refraction and longshore currents by the breakwater (Figure 13.7B). To stop the filling of the harbor, it was necessary to connect the breakwater to the shore. Sand then accumulated behind the breakwater, at its southern end. Soon a smooth, curving beach developed around the breakwater, and longshore drift carried sand around the breakwater and deposited it inside the harbor (Figure 13.7C). This produced two disastrous effects: First, the harbor became so choked with sand that it could accommodate only small vessels, and second, the beaches downcoast were deprived of their source of sand and began to erode. Within 12 years, more than $2 million worth of damage had been done to property down the coast from Santa Barbara, as the beach in some areas was cut back 75 m. To solve the problem, a dredge was installed in the Santa Barbara harbor to pump out the sand and return it to the longshore drift system on the downcurrent side of the harbor. Most of the beaches have been partly replenished, but expensive dredging must operate throughout the entire year.

EROSION ALONG COASTS

Why are most shorelines undergoing vigorous erosion? Can coastal erosion be stopped?

Statement

Most coastal regions of the world have been inundated by the geologically recent rise in sea level associated with the melting of the last glaciers. As a result,

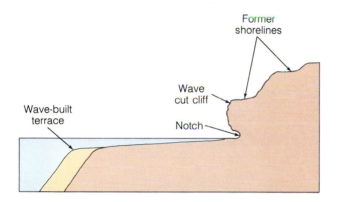

(A) Wave action operates like a horizontal saw cutting at the base of the cliff. The cliff is undermined and collapses. The debris is soon removed by wave action, and undercutting continues. As erosion continues, the cliff recedes farther, and a gently sloping wave-cut platform is left. Some sediment eroded from the shore can be deposited in deeper water to form a complementary wave-built terrace.

(B) A wave-cut platform along the Alaskan coast. Note the flat surface of the uplifted wave-cut platform and the new platform being eroded at sea level.

Figure 13.8
A profile of a wave-cut platform shows the features commonly produced by wave erosion.

many are subject to vigorous erosion. Erosion along coasts typically develops a wave-cut cliff and an associated wave-cut platform. Sea arches, caves, and sea stacks may develop from differential erosion along headlands.

Discussion

To understand the nature of wave erosion and the principal features of its forms, consider what happens along a profile at right angles to a shore (Figure 13.8). Where steeply sloping land descends beneath the water, waves act like a horizontal saw, cutting a notch into the bedrock at sea level (Figure 13.9). This undercutting produces an overhanging sea cliff, or wave-cut cliff, that ultimately collapses. The fallen debris is broken up and removed by the action of waves, and the process is repeated on the fresh surface of the new cliff face. As the sea cliff retreats, a wave-cut platform is produced at its base, the upper part of which commonly is visible near the shore at low tide. Sediment derived from the erosion of the cliff and transported by longshore drift may be deposited in deeper water to form a wave-built terrace. Stream valleys that formerly reached the coast at sea level are shortened and left as hanging valleys when the cliff recedes.

As the platform is enlarged, the waves break progressively farther from shore, losing much of their energy by friction as they travel across the shallow platform. Wave action on the cliff is consequently greatly reduced. Beaches can then develop at the base

Figure 13.9
Wave erosion along a coast of Mexico has produced this notch and overhanging cliff. Collapse appears imminent. The process will then be repeated, causing retreat of the sea cliff and development of a wave-cut platform.

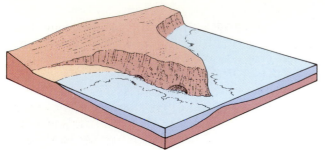

(A) Wave energy is concentrated on a headland as a result of wave refraction. Zones of weakness—joints, faults, and non-resistant beds—erode faster; sea caves develop in those areas.

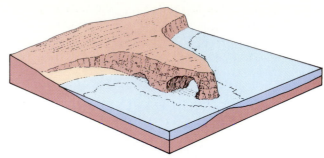

(B) Sea caves enlarge to form a sea arch.

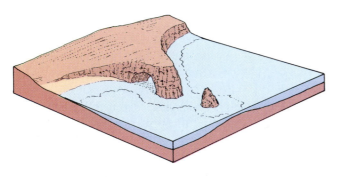

(C) Eventually the arch collapses, leaving a sea stack. A new arch can develop from the remaining headland.

Figure 13.10
The evolution of sea caves, sea arches, and sea stacks is associated with differential erosion of a headland.

of the cliff, and the cliff face is gradually worn down, mainly by weathering and mass movement. Wave-cut platforms effectively dissipate wave energy and thus limit the size to which they can grow. Some volcanic islands, however, appear to have been truncated completely by wave action and slope retreat, so that only a flat-topped platform is left near low tide.

Sea Caves, Sea Arches, and Sea Stacks

The rate at which a sea cliff erodes depends on the durability of the rock and the degree to which the coast is exposed to direct wave attack. Zones of weakness (such as outcrops with joint systems, fault planes, and beds of shale between harder sandstones) are zones of accelerated erosion. If a joint extends across a small peninsula or headland jetting out from the coast, wave action can hollow out an alcove, which may later enlarge to a sea cave. Because the headland commonly is subjected to erosion from two sides, caves excavated along a zone of weakness can join to form a sea arch (Figures 13.10 and 13.11). Eventually, the arch collapses, and an isolated pinnacle called a sea stack is left in front of the cliff. In the

(A) Sea arches and stacks along the coast of southern California, 1969.

(B) Same area as shown in (A), 1987. Erosion has eliminated the sea stack and caused collapse of the sea arch.

Figure 13.11
Sea stacks and a sea arch along the coast of California.

erosion of a shoreline, marine and terrestrial agents operate together to produce erosion above wave level. The seepage of groundwater, frost action, wind, and mass movement all combine with the undercutting action of waves to erode the coast.

The Evolution of a Wave-Cut Platform

The evolution of erosional features along a coast is shown in Figure 13.12. In the initial stage (Figure 13.12A), sea level rises over a stream-eroded land-scape, and an irregular shoreline is formed. Wave action develops a small notch, and abrasion of the platform begins. Continued wave erosion enlarges the platform and develops a high wave-cut cliff (Figure 13.12B). Minor features, such as sea caves, sea arches, and sea stacks, form by differential erosion in weak places in the bedrock. These are continually being formed and destroyed as the cliff recedes. In the advanced stage of development (Figure 13.12C), the platform is so enlarged that it absorbs most of the wave energy. Weathering, mass movement, and stream erosion subdue the cliff, and a beach develops as a result of the low energy level along the coast. The net result is a broad wave-cut platform.

Example

The reality of coastal erosion is made painfully clear by the passion of Americans to live and vacation on the seashore. Development projects unwittingly put more and more people and property on the shore, an area that by its very nature is dynamic and mobile. About 86 percent of California's coast is receding at an average rate of 15–75 cm per year. Parts of Monterey Bay lose as much as 2 to 3 m per year. Cape Shoalwater, Washington, about 70 miles west of Olympia, has been eroding at a rate of more than 300 m per year. Part of Chambus County, Texas, has lost 3 m of coast in nine months. In North Carolina erosion in one year has cut into beach-front property up to 25 m in places.

To combat these losses, sea walls and breakwaters have been erected, but these are temporary solutions at best. A sea wall or jetty (a long concrete or rock structure that juts out into water to block waves and currents) may protect threatened property near it, but it often hastens erosion in other areas. There appear to be no simple answers. Coasts are dynamic features, and by their very nature are mobile. In our battle with nature, retreat might be the ultimate solution.

(A) Initial stage: *Wave action begins to develop a notch at sea level, which evolves to form a wave-cut cliff.*

(B) Intermediate stage: *Continued wave erosion causes the cliff to recede, and a wave-cut platform develops. Sea stacks, sea arches, and sea caves result from differential erosion along zones of weakness.*

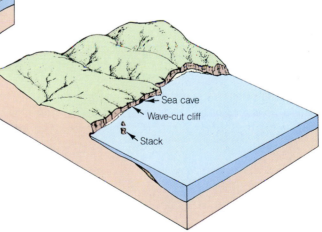

(C) Late stage: *The wave-cut platform grows so large that wave energy is dissipated across it. Erosion along the shore is greatly reduced. Beaches develop, and the sea cliff retreats through mass movement.*

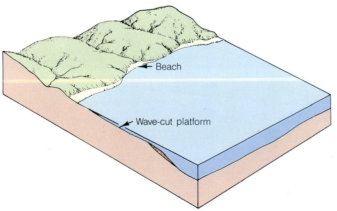

Figure 13.12
The evolution of a shoreline involves a series of stages in which the configuration of the coast is modified by both erosion and deposition until, finally, only a minimum of energy is expended on it. The most stable shoreline form has a smooth, straight coast and a wide platform.

DEPOSITION ALONG COASTS

What is the source of the sand on beaches?

Statement

A shoreline is a dynamic system that involves the input of sediment from various sources, transportation of the sediment, and the ultimate deposition of sediment. Much of the sediment is derived from the land and delivered to the sea by major rivers. The sediment is then transported by waves and longshore currents and deposited in quiet waters where wave energy is low. There, it commonly builds a variety of landforms, including beaches, spits, tombolos, and barrier islands. Shoreline material is in constant motion, and the configuration of the coast is continually modified by both erosion and deposition. Changes continue until the configuration of the coast is at equilibrium with the available wave energy. The final configuration of a shoreline is usually a smooth and straight or gently curving coastline.

Discussion

Figure 13.13 shows some of the important elements in a coastal system. The primary sources of sediment for beaches and associated depositional features are the rivers that drain the continents. Sediment from the rivers is transported along the shore by longshore drift and is deposited in areas of low wave energy such as protected bays. Erosion of headlands and sea cliffs is also a source of sediment. In tropical areas the greatest source of sand commonly is shell debris derived from wave erosion of near-shore coral reefs. Sediment can leave the system when coastal sand dunes migrate inland or where it is transported into deep areas of the ocean floor.

Beaches

A beach is a shore built of unconsolidated sediment. Sand is the most common material, but some beaches are composed of cobbles and boulders, and others of fine silt and clay. The physical characteristics of a beach (such as slope, composition, and shape) depend largely on wave energy, but the supply

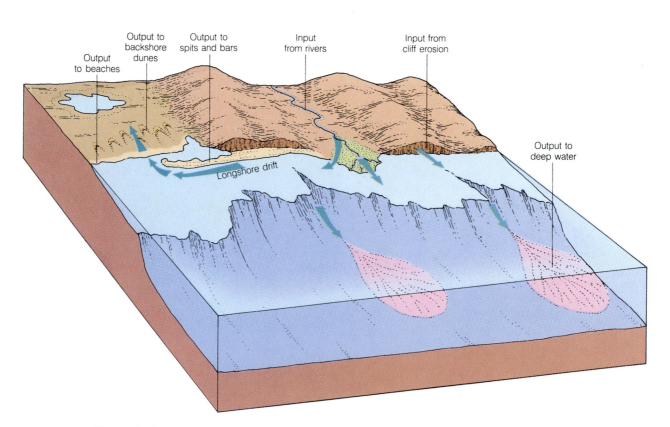

Figure 13.13
A shoreline is a dynamic system of moving sediment. Most of the sediment in a shoreline system is supplied by rivers bringing erosional debris from the continent and by the erosion of sea cliffs by wave action. This material is transported by longshore drift and can be deposited on growing beaches, spits, and bars. Some sediment, however, leaves the system either by transportation to deeper water or by the landward migration of coastal sand dunes.

and size of available sediment particles are also important. Beaches composed of fine-grained material generally are flatter than those composed of coarse sand and gravel.

Spits

In areas where a straight shoreline is indented by bays or estuaries, longshore drift can extend the beach from the mainland to form a spit. A spit can grow far out across the bay as material is deposited at its end (Figure 13.14). Eventually, it may extend completely across the front of the bay, forming a baymouth bar.

Tombolos

Beach deposits can also grow outward and connect the shore with an offshore island to form a tombolo. This feature commonly is produced by the island's effect on wave refraction and longshore drift (Figure 13.15). An island near the shore can cause wave refraction to such an extent that little or no wave energy strikes the shore behind it. Longshore drift, which moves sediment along the coast, is not generated in part of the shore behind the island. Sediment carried by longshore currents is therefore deposited and eventually forms a bar or beach (a tombolo) that connects the shore to the island. Longshore currents then move uninterrupted along the shore and around the island and tombolo.

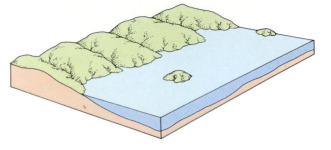

(A) *An offshore island acts as a breakwater to incoming waves and creates a wave shadow along the coast behind it.*

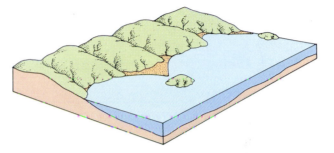

(B) *Sediment moved by long shore drift is trapped in the shadow zone.*

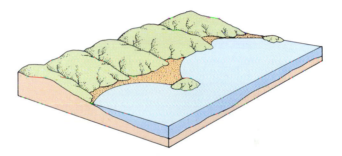

(C) *The zone of sediment deposit eventually grows until it connects with the island. Longshore drift will then move sediment along the shore and around the tombolo.*

(D) *An aerial photograph of a tombolo.*

Figure 13.15
A tombolo is a bar or beach that connects an island to the mainland. It forms because the island creates a wave shadow zone along the coast, in which longshore drift cannot occur.

Figure 13.14
Curved spits develop as longshore drift moves sediment parallel to the shoreline.

Barrier Islands

Barrier islands are long, low offshore islands of sediment trending parallel to the shore (Figure 13.16). Almost invariably, they form long shorelines that are adjacent to gently sloping coastal plains; they typically are separated from the mainland by a lagoon. Most barrier islands are cut by one or more tidal inlets. Most of the eastern seaboard, from Chesapeake Bay to Florida, has a continuous fringe of barrier islands, as does most of the Gulf Coast west of the Florida Peninsula.

Barrier islands may result from the growth of spits across irregularities in the shore, as illustrated in Figure 13.17. Certainly many barrier islands develop in this manner. Others may form by the shoreward migration of offshore bars.

Figure 13.16
A barrier island along the Atlantic coast has a smooth seaward face where wave action and longshore drift actively transport sediment. A tidal inlet forms a break in the island, and sediment transported through it is deposited as a tidal delta in the lagoon.

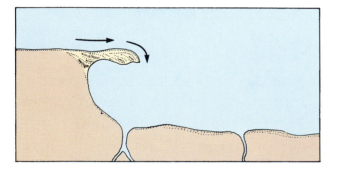

(A) Sediment moving along the shore is deposited as a spit in the deeper water near a bay.

(B) The spit grows parallel to the shore by longshore drift.

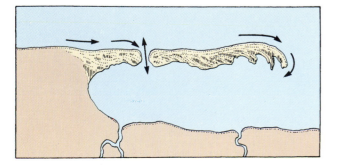

(C) Tidal inlets cut the spit, which is then long enough to be considered a barrier island.

Figure 13.17
A barrier island may form by migration of a spit.

EVOLUTION OF SHORELINES

What is the shape of a coastline that has reached equilibrium?

If Earth is so old, why are shorelines still changing?

Statement

Processes of erosion and deposition along a shore tend to develop a long and straight or gently curving coastline. Headlands are eroded back, and bays and estuaries are filled with sediment. The configuration of the shoreline evolves until wave energy is distributed equally along the coast and neither large-scale erosion nor deposition occurs. A shoreline with such a balance of forces in which the energy of the waves is just sufficient to transport the sediment that is supplied is called a shoreline of equilibrium. As is the case with equilibrium in river systems, a delicate balance is maintained between the landforms and the geologic processes operating on them.

Discussion

Very few present-day shorelines approach a state of equilibrium, however. They are all very young, having been formed during the last few thousand years by the worldwide rise in sea level associated with the melting of the Pleistocene glaciers. Most coastlines are therefore presently being vigorously modified by erosion and deposition and are just evolving toward equilibrium.

We can construct a simple conceptual model of a shoreline's evolution toward equilibrium and show the types of changes that would be expected to occur as the processes of erosion and deposition operate. One such model is shown in Figure 13.18. Diagram A

(A) A rise in sea level floods a landscape eroded by a river system and forms bays, headlands, and islands.

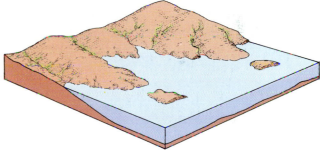

(B) Wave erosion cuts cliffs on the islands and peninsulas.

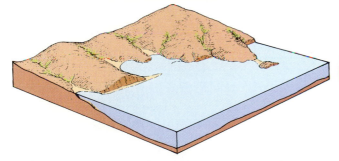

(C) Wave-cut cliffs recede and grow higher, and headlands are eroded back to a sea cliff. Sediment begins to accumulate, forming beaches and spits.

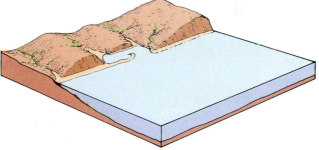

(D) Islands are completely eroded away, beaches and spits enlarge, and lagoons form in the bays.

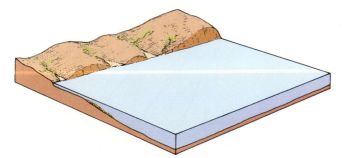

(E) A straight shoreline is produced by additional retreat of the cliffs and by sedimentation in bays and lagoons. The large wave-cut platform then limits further erosion by wave action.

Figure 13.18
The evolution of a shoreline of equilibrium from an embayed coastline involves changes due to both erosion and deposition. Eventually a smooth coastline is produced, and the forces acting on it are essentially at equilibrium; thus, neither erosion nor deposition occurs on a large scale.

shows an area originally shaped by stream erosion and subsequently partly drowned by rising sea level. River valleys are invaded by the sea to form irregular, branching bays, and some hilltops form peninsulas and islands. Next, as is shown in Figure 13.18B, marine erosion begins to attack the shore. The islands and headlands are eroded into sea cliffs. As erosion proceeds (Figure 13.18C), the islands and headlands are worn back, and the cliffs increase in height. A wave-cut platform enlarges, reducing wave energy, so that a beach forms at the base of the cliff. In a more advanced stage of development (Figure 13.18D), the islands are eroded away and bays become sealed off, partly by the growth of spits, forming lagoons. The shoreline then becomes straight and simple. In the final stages of marine development (Figure 13.18E), the shoreline is cut back beyond the limits of the bay. Sediment moves along the coast by longshore drift, but the wave-cut platform is so wide that it effectively eliminates further erosion of the cliff by wave action. The shoreline of equilibrium is straight and essentially in equilibrium with the energy acting on it; that is, neither larger-scale erosion nor deposition occurs. Additional modification of the cliffs results from weathering and stream erosion.

The development of a shoreline is interrupted in many areas by tectonic uplift, which abruptly elevates sea cliffs and wave-cut platforms above the level of the waves. When this happens, wave erosion begins at a new, lower level, and the elevated marine terraces, stranded high above sea level, are attacked and eventually obliterated by weathering and stream erosion (Figure 13.19).

REEFS

What types of shorelines are built by living organisms?

Statement

Reefs form a unique type of coastal feature because they are of organic origin. Modern reefs are built by a complex community of corals, algae, sponges, and other marine invertebrates. Most grow and thrive in the warm, shallow waters of semitropical and tropical regions. Only the upper part of a reef is organically active because sunlight is required for vigorous growth. The living animals and plants of the reef community build new structures on the remains of dead organisms and the accumulation of shell debris. Wave action, however, continually breaks up the reef. Large bodies of calcium carbonate shell debris, originally secreted by the organisms of the reef community, accumulate along the reef flanks, adding mass to the structure. Reefs can grow upward with rising sea level if the rate of rise is not excessive. They can also grow seaward over the flanks of reef debris.

Discussion

Reef Ecology

The marine life that forms a reef can flourish only under strict conditions of temperature, salinity, water clarity, and water depth. Most modern coral reefs occur in warm tropical waters between the limits of

Figure 13.19
A series of elevated beach terraces resulted from tectonic uplift along the southern coast of California and the offshore islands. This photograph of San Clemente Island was taken with the sun at low angle to emphasize the sequence of terraces.

Figure 13.20
A barrier reef in the Society Islands, French Polynesia, is typical of the intermediate stage in the evolution of an atoll. Note the outer margin of the reef, where the growth of organisms is most active. The shallow lagoon inside the reef, shown in light blue, is mostly calcarious sand formed by erosion of the reef. The remnant of volcanic island in the center is highly dissected by stream erosion, indicating that a long period of time has elapsed since the volcano was active.

25° south latitude and 30° north latitude (Figure 13.20). Colonial corals need sunlight, and they cannot live in water deeper than about 76 m. They grow most luxuriantly just a few meters below sea level. Dirty water inhibits rapid, healthy growth because it cuts off sunlight, and the suspended mud chokes the organisms that filter feed. Corals are therefore absent or stunted near the mouths of large, muddy rivers. They can survive only if the salinity of the water ranges from 27 to 40 parts per thousand; thus, a reef can be killed if a flood of fresh water from the land appreciably reduces the salinity. Coral reefs are remarkably flat on top, the upper surface being positioned at the level of the upper third of the tidal range. They usually are exposed at low tide but must be covered at high tide. In summary, corals thrive only in clear, warm, shallow oceans where wave action brings sufficient oxygen and food. The fact that reefs form in such restricted environments makes them especially important as indicators of past climatic, geographic, and tectonic conditions.

Types of Reefs

The most common types of reefs existing in the present oceans are fringing reefs, barrier reefs, and atolls (Figure 13.21). Fringing reefs, generally ranging from 0.5 to 1 km wide, are attached to such landmasses as the shores of volcanic islands. The corals grow seaward, toward their food supply. They are usually absent near deltas and the mouths of rivers, where the waters are muddy. Heavy sedimentation and high runoff also make some tropical coasts of continents unattractive to fringing reefs.

Barrier reefs are separated from the mainland by a lagoon, which can be more than 20 km wide. As seen from the air, the barrier reefs of islands in the South Pacific are marked by a zone of white breakers. At intervals, narrow gaps occur, through which excess shore and tidal water can exit. The finest example of this type is the Great Barrier Reef, which stretches for 800 km along the northern shore of Australia, from 30 to 160 km off the Queensland coast.

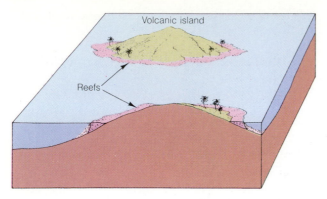

(A) A reef begins to grow along the coast of a newly formed volcanic island.

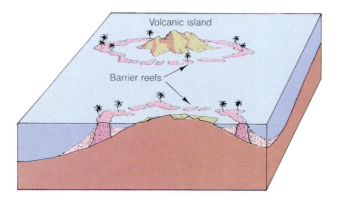

(B) As the island subsides, the reef grows upward and develops a barrier that separates the lagoon from open water.

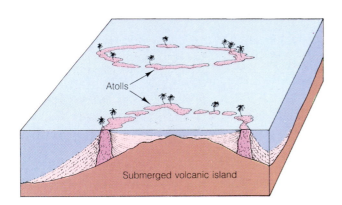

(C) Further subsidence completely submerges the island, but if subsidence is not too rapid, the reef continues to grow upward to form an atoll.

Figure 13.21
The evolution of an atoll from a fringing reef was first recognized by Charles Darwin. The theory assumes that continued slow subsidence of the ocean floor allows the reef to continue growing upward.

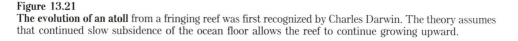

Atolls are roughly circular reefs that rise from deep water, enclosing a shallow lagoon in which there is no exposed central landmass. The outer margin of an atoll is naturally the site of most vigorous coral growth. It commonly forms an overhanging rim, from which pieces of coral rock break off, accumulating as submarine talus on the slopes below. A cross-sectional view of a typical atoll shows that the lagoon floor is shallow and is composed of calcareous sand and silt with rubble derived from erosion of the outer side (see the foreground of Figure 13.21).

Atolls are by far the most common type of coral reef. Over 330 are known, of which all but 10 lie within the Indo-Pacific tropical area. Drilling into the coral of atolls tends to confirm the theory that atolls form on submerged volcanic islands (see the following section, "The Origin of Atolls"). In one instance coral extends down as much as 1,400 m below sea level, where it rests on a basalt platform carved on an ancient volcanic island. Because coral cannot grow at that depth, it presumably grew upward as the volcanic island sank. A reef this thick probably accumulated over 40 or 50 million years.

Platform reefs grow in isolated oval patches in warm, shallow water on the continental shelf. They were apparently more abundant during past geologic periods of warmer climates. Most modern platform reefs seem to be randomly distributed, although some appear to be oriented in belts. The latter feature suggests that they were formed on submarine topographic highs, such as drowned shorelines.

The Origin of Atolls

In 1842 Charles Darwin first proposed a theory to explain the origin of atolls. As is indicated in Figure 13.21, the theory is based on the continued relative subsidence of a volcanic island. Darwin suggested that coral reefs are originally established as fringing reefs along the shores of new volcanic islands. As the island gradually subsides, the coral reef grows upward along its outer margins. The rate of upward growth essentially keeps pace with subsidence. With continued subsidence, the area of the island becomes smaller, and the reef becomes a barrier reef. Ultimately, the island is completely submerged, and the upward growth of the reef forms an atoll. Erosional debris from the reef fills the enclosed area of the atoll to form a shallow lagoon. The subsidence of the seafloor and the volcanic islands built upon it is commonly the result of the oceanic plate moving off from the high region of the oceanic ridge.

TYPES OF COASTS

Why are there so many different types of coastlines? What factors determine the configuration of a shoreline?

Statement

Nearly all coasts are complex, both in the types of landforms and in their geologic history. All are dependent upon the landforms that preceded them, all are subject to the effects of changes in sea level during the Ice Age, and all are influenced by the operation of present coastal processes. In spite of these complexities, important insight into the understanding of coasts can be gained by considering them in light of the processes responsible for their configuration. To

do this effectively, we must consider coastal features on two different scales:

1. Features on a global scale—those associated with tectonic plates
2. Features on a local scale—those mainly associated with processes of erosion and deposition, which modify the regional features

In tectonic terms, on a global scale, coasts can be classified as follows:

1. Convergence coasts
   a. Continental convergence
   b. Island arc convergence
2. Passive-margin coasts
3. Marginal sea coasts

On a smaller scale, coastal features are often associated with local geologic processes: erosion, deposition, structure volcanism, and so on. Two principal subdivisions are recognized:

1. Primary coasts shaped mainly by terrestrial processes (stream erosion, deposits, and so on)
2. Secondary coasts shaped by marine processes

Discussion

Classification Based on Plate Tectonics

Collision Coasts. (coasts found near convergent plate margins). The classification of the world's coastlines based on tectonics is shown in Figure 13.22. The collision coasts are all relatively straight and mountainous and are further characterized by sea cliffs, raised marine terraces, and narrow continental shelves. They typically contain zones of active volcanism and seismicity. The west coasts of North and South America are excellent examples of continental collision coasts. The Aleutian Islands and Japan are typical collision coasts of island arcs.

Passive-Margin Coasts. Passive-margin coasts are formed by rifting. As the oceans spread and the continents move apart, the coasts are tectonically passive. In the initial stages of rifting, the coasts may be topographically high and marked by steep escarpments. With time the coast subsides as it moves off the uparched, spreading center. Sediment derived from erosion of the mountain belt on the active margin is transported to the passive margin, where it is deposited and helps to construct a broad continental shelf. Typically, passive-margin coasts are low and have been modified by marine deposits.

The east coasts of the Americas are typical passive-margin coasts. They have low relief and broad coastal plains and are bordered by a wide continental shelf.

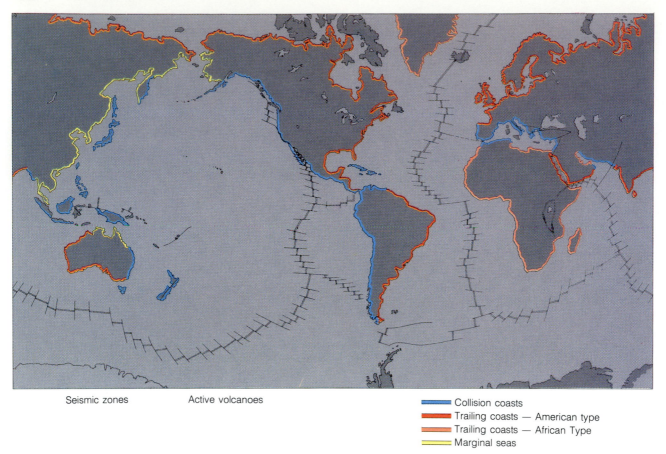

Seismic zones   Active volcanoes

—— Collision coasts
—— Trailing coasts — American type
—— Trailing coasts — African Type
—— Marginal seas

Figure 13.22
The tectonic classification of coasts is based upon the tectonic setting of the continental margins. Passive margins characterize the eastern coasts of North and South America, Africa, and Australia. Mountainous coasts are typical of converging plate margins.

The coasts of the Red Sea and Baja, California, are also passive-margin coasts, but they were formed by recent uplift and rifting and, as a result, have high cliffs and narrow continental slopes. The passive-margin coasts of Africa, India, and Greenland are distinct in that both the east and west coasts face spreading centers, and the coasts have relatively high relief.

Marginal Sea Coasts. Marginal sea coasts are protected from the open ocean by island arcs and are frequently modified by large rivers and their deltas; Vietnam and southern China are examples.

Classification Based on Geologic Processes

On a smaller scale (lengths of 100 km), coasts are highly diverse and are commonly controlled by local erosion and deposition or by the growth of reefs. These effects are superposed on the larger, regional tectonic features. The particular climatic region of Earth may be a major controlling factor in that it controls glacial systems, the locations of major deltas, and the growth and destruction of reefs.

Primary Coasts

The configuration of primary coasts is largely the result of geologic processes operating on the land, such as streams, glaciers, volcanism, and Earth movements. These produce highly irregular coastlines characterized by bays, estuaries, fiords, headlands, peninsulas, and offshore islands. The landforms can be either erosional or depositional but are only slightly modified by marine processes. Some of the more common types are illustrated in Figure 13.23.

Secondary Coasts

Secondary coasts are shaped by marine erosion and deposition. They are characterized by wave-cut cliffs, beaches, barrier bars, spits, and (in some cases) sediment deposited through the action of biological agents, such as marsh grass, mangroves, and coral reefs. Marine erosion and deposition smooth out and straighten shorelines and establish a balance between the energy of the waves and the configuration of the shore. The most common types of secondary coasts are illustrated in Figure 13.24.

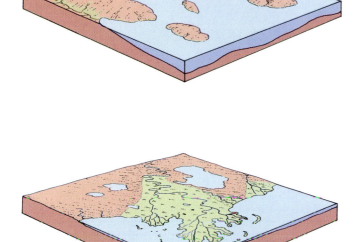

(A) Stream Erosion Coasts: *If an area eroded by running water is subsequently flooded by the rise of sea level, the landscape becomes partly drowned. Stream valleys become bays or estuaries, and hills become islands. The bays extend up the tributary valley system, forming a coastline with a dendritic pattern. Chesapeake Bay is a well-known example.*

(B) Stream Deposition—Deltaic Coasts: *At the mouths of major rivers, fluvial deposition builds deltas out into the ocean, which dominate the configuration of the coast. Deltas can assume a variety of shapes and are locally modified by marine erosion and deposition. The Louisiana coast is an excellent example because its configuration is entirely the result of the growth of the Mississippi Delta.*

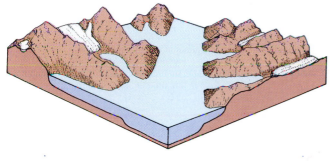

(C) Glacial Erosion Coasts: *Drowned glacial valleys usually are called fiords. They form some of the most rugged and scenic shorelines in the world. Fiords are characterized by long, troughlike bays that cut into mountainous coasts, extending inland as much as 100 km. The fiords of Norway and Canada are excellent examples. In polar areas glaciers still remain at the heads of many fiords. The walls of fiords are steep and straight. Hanging valleys with spectacular waterfalls are common.*

Figure 13.23
Primary coasts are those in which the configuration of the shoreline is produced by nonmarine processes. Rivers and glaciers are the most important processes forming this type of coast.

TIDES

What are the major effects of the rise and fall of tides?

Statement

Tides are produced by a combination of external forces: the principal ones are the gravitational attraction of the Moon and the centrifugal force of the Earth-Moon system. Many variables affect the height of tides, however, including the Coriolis effect (the deflection of moving bodies due to a force arising from Earth's rotation), the position of the sun in relation to the Moon and Earth, and the configuration of coastlines and the ocean floor. Tides affect coasts in two major ways:

1. By initiating a rise and fall of the water level
2. By generating currents

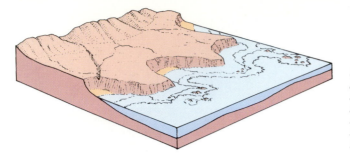

(A) Wave Erosion Coasts: *Wave erosion begins to modify primary coasts as soon as the landscape produced by other agents is submerged. Wave energy is concentrated on the headlands, and a wave-cut platform develops slightly below sea level. Ultimately a straight cliff is formed, with hanging stream valleys and a large wave-cut platform. The cliffs of Dover, England, are a prime example.*

(B) Marine Deposition Coasts: *Where abundant sediment is supplied by streams or ocean currents, marine deposits determine the characteristics of the coast. Barrier islands and beaches are the dominant features. The shoreline is modified as waves break across the barriers and transport sand inland. The barriers also increase in length and width as sand is added. The lagoons behind the barriers receive sediment and fresh water from streams; thus, they are often capable of supporting dense marsh vegetation. Gradually a lagoon fills with stream sediment, with sand from the barrier bar (which enters through tidal deltas), and with plant debris from swamps. The barrier coasts of the southern Atlantic and Gulf Coast states are excellent examples.*

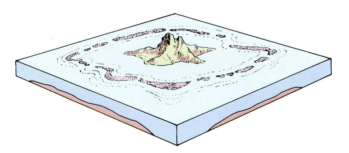

(C) Coasts Built by Organisms: *Coral reefs develop a type of coast that is prominent in the islands of the southwestern Pacific. The reefs are built up to the surface by corals and algae, and they can ultimately evolve into an atoll. Another type of organic coast prevalent in the tropics is formed by the intertwined root systems of the mangrove trees that grow in the water, particularly in shallow bays.*

Figure 13.24
Secondary coasts are those in which the configuration of the shoreline is produced by marine processes, including erosion, deposition, and the growth of organisms.

Discussion

The diagram in Figure 13.25 illustrates on a highly exaggerated scale the principal forces that produce tides. The gravitational force exerted by the Moon tends to pull the oceans facing the Moon into a bulge. Another tidal bulge, on the side of Earth opposite the Moon, is caused by centrifugal force. Earth and the Moon rotate around a common center of mass, which lies approximately 4,500 km from the center of Earth on a line directed toward the Moon. The eccentric motion of Earth as it revolves around the center of mass of the Earth-Moon system creates a large centrifugal force, which forms the second tidal bulge.

The Earth rotates beneath the bulges, so that the tides rise and fall twice every 24 hours.

The major effect of the rise and fall of tides is the transportation of sediment along the coast and over the adjacent shallow seafloor. Extremely high tides are produced in shallow seas where the rising water is funneled into bays and estuaries. For example, in the Bay of Fundy, between New Brunswick and Nova Scotia, the tide range (the difference in height between high tide and low tide) is as much as 21 m. Where fine-grained sediment is plentiful and the tide range is great, the configuration of the coast is greatly influenced by tides and tidal currents.

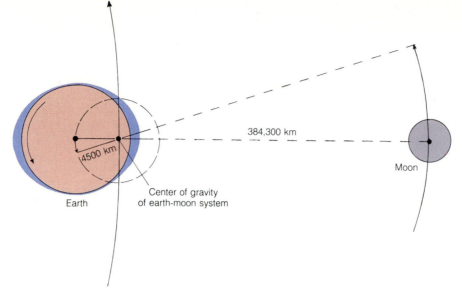

Figure 13.25
Ocean tides are caused by the gravitational attraction of the Moon and the centrifugal force of Earth-Moon system. On the side of Earth facing the Moon, the gravitational attraction is greater, forming a tidal bulge in the ocean's water. On the other side of Earth, the centrifugal force is greater, causing another tidal bulge.

TSUNAMIS

Why is it that a tsunami (tidal wave) is so small it is imperceptible in the open ocean but yet may be more than 30 m high when it reaches a coastline?

Statement

Movement of the ocean floor by earthquakes, volcanic eruptions, or submarine landslides frequently produces an unusual wave called a seismic sea wave, or a tsunami. Tsunamis have long wavelengths and travel across the open ocean at high speeds. As they approach shore, the wavelength decreases and the wave height increases, so tsunamis can be formidable agents of destruction along the shoreline.

Discussion

Tsunamis differ from wind-produced ocean waves in that energy is transferred to the water from a seafloor disturbance, so the entire ocean depth of water participates in the wave motion. The wave front travels from its point of origin at high speeds, ranging from 500 to 800 km per hour, and can traverse an entire ocean. The wave height is only from 30 to 60 cm, and the wavelength ranges from 55 to 200 km. In an open ocean, therefore, a tsunami can pass unnoticed. As the wave approaches the shore, however, important changes take place. The energy distributed in the deep column of water becomes concentrated in an increasingly shorter column, resulting in a rapid increase in wave height. Waves that were less than 60 cm high in the deep ocean can build rapidly to heights exceeding 15 m in many cases and well over 30 m in rare instances. They exert an enormous force against the shore and can inflict serious damage and great loss of life.

Example

A number of tsunamis have been well documented by seismic stations and coastal observers. For example, the tsunami that hit Hawaii on April 1, 1946, originated in the Aleutian Trench off the island of Unimak. The waves moving across the open ocean were imperceptible to ships in their path because the wave height was only 30 cm. Moving at an average speed of 760 km per hour, they reached the Hawaiian Islands—3,200 km away—in fewer than five hours. Because the wavelength was 150 km, the wave crests arrived about 12 minutes apart. As the waves approached the island, their height increased to at least 17 m, which produced an extremely destructive surf. It swept inland and completely demolished houses, trees, and almost everything else in its path.

SUMMARY

1. Shorelines are dynamic systems involving the forces of waves and currents. Wind-generated waves provide most of the energy of erosion, transportation, and deposition of sediment, but tides and tsunamis can be locally important.
2. Waves approaching a shore are bent, or refracted, so that energy is concentrated on headlands and dispersed in bays.
3. Longshore drift, one of the most important shoreline processes, is generated as waves strike the shore at an angle. Water and sediment move obliquely up the beach face but return with the backwash directly down the beach, perpendicular to the shoreline. This results in a net transport of water and sediment parallel to the shore.
4. Erosion along coasts results from the abrasive action of sand and gravel moved by the waves and currents and, to a lesser extent, from solution and hydraulic action. The undercutting action of waves and currents typically produces sea cliffs. As a sea cliff recedes, a wave-cut platform develops. Minor erosional forms associated with the development of sea cliffs include sea caves, sea arches, and sea stacks.
5. Sediment transported by waves and longshore drift is deposited in areas of low energy such as protected bays. It produces a variety of landforms, including beaches, spits, tombolos, and barrier islands.
6. Erosion and deposition processes tend to develop a straight or gently curving coastline where neither large-scale erosion nor large-scale deposition occurs. A shoreline with such a balance of forces is called a shoreline of equilibrium.
7. Reefs are a special type of coastal feature because the active process is organic. They commonly evolve through a series of types, from fringing reefs to barrier reefs to atolls.
8. The worldwide rise in sea level associated with the melting of the Pleistocene glaciers drowned many coasts. Coasts are classified on the basis of the process that has been most significant in developing their configurations.

KEY TERMS

atoll (p. 230)
backwash (p. 217)
barrier island (p. 226)
barrier reef (p. 229)
breaker (p. 217)
coral (p. 228)
fiord (p. 233)
fringing reef (p. 229)

headland (p. 217)
longshore current (p. 218)
longshore drift (p. 218)
primary coast (p. 232)
sea arch (p. 222)
sea cave (p. 222)
sea cliff (p. 222)
sea stack (p. 222)

secondary coast (p. 232)
shoreline of
  equilibrium (p. 223)
spit (p. 225)
swash (p. 217)
tide (p. 233)
tombolo (p. 225)
tsunami (p. 235)

wave-built terrace (p. 221)
wave-cut cliff (p. 221)
wave-cut platform
  (terrace) (p. 221)
wave height (p. 216)
wavelength (p. 216)
wave refraction (p. 217)

REVIEW QUESTIONS

1. Describe the motion of water in a wind-generated wave.
2. Explain how wave refraction alters the form of a coastline.
3. Explain the origin of long shore drift.
4. Describe the stages in the evolution of a sea cliff and wave-cut platform.
5. Name the major depositional landforms along a coast and explain the origin of each.
6. How are marine terraces formed?
7. What conditions are necessary for the formation of a coral reef?
8. Explain the origin of atolls.
9. Describe six common types of shoreline.
10. Explain how ocean tides are generated.
11. Explain the origin of tsunamis.

Multiple-Choice Questions

1. The movement of a water particle in a wave in the open ocean is dominantly
   a. circular.
   b. horizontal.
   c. vertical.
   d. elliptical.
   e. none of the above.

2. Ordinary ocean waves derive their energy from
   a. winds.
   b. surface currents.
   c. tides.
   d. refraction.
   e. density currents.

3. On an irregular shoreline, the energy of a wave is
   a. dissipated at the line where the wave first "feels bottom."
   b. evenly spread over the entire shoreline.
   c. usually completely dissipated before reaching the shore.
   d. concentrated mostly on the headlands by wave refraction.
   e. concentrated along the beaches by wave refraction.

4. Longshore currents are a result of
   a. tides.
   b. winds moving parallel to shore.
   c. waves striking the shore obliquely.
   d. differences in temperature along the shore.
   e. trade winds.

5. The predominant geologic process responsible for most of the depositional features along a shoreline is
   a. density currents.
   b. longshore currents.
   c. surface currents.
   d. tidal currents.
   e. rip currents.

6. Which feature is *not* a direct product of shoreline erosion?
   a. baymouth bar
   b. stack
   c. sea arch
   d. sea cave
   e. sea cliff

7. The single major source of beach sediments in most regions is
   a. offshore bars.
   b. rivers.
   c. wind-blown sand from the beach.
   d. the continental shelf.
   e. wave erosion of the coast.

8. Secondary coasts are characterized by
   a. bays and estuaries.
   b. headlands and fiords.
   c. barrier bars and spits.
   d. offshore islands.
   e. deltas.

9. In general, we can expect that on an irregular coastline
   a. headlands will be eroded and bays will be the site of deposition.
   b. bays will be cut back deeper and headlands will be built out.
   c. erosion and deposition will be uniform throughout.
   d. breakers will not exist.
   e. normal processes of erosion and deposition will not occur.

10. Shoreline processes tend to develop a coastline such that a
    a. maximum of energy is expended.
    b. minimum of energy is expended.
    c. minimum of energy is created.
    d. maximum of energy is created.
    e. none of the above.

True/False Questions

1. *T F* Wave action in the open sea produces little or no net forward motion of the water.
2. *T F* The final configuration in a shoreline at equilibrium is usually a smooth and straight or gently curving coastline.
3. *T F* Many present-day shorelines are almost in a state of equilibrium.
4. *T F* The configuration of many coasts is not the result of marine processes at all, but is the result of geologic processes operating in the area before sea level rose.
5. *T F* Primary coasts are shaped by marine erosion and deposition.

Fill in the Missing Terms

1. The most important types of ocean wave are generated by _____.
2. Along a coast, wave refraction concentrates energy on _____.
3. A sea arch will commonly evolve into a sea _____.
4. _____ are the primary sources of sediment for beaches and associated depositional features.
5. _____ are roughly circular reefs that rise from deep water and enclose a shallow lagoon in which no central landmass is exposed.

14

EOLIAN SYSTEMS

In regions of low precipitation and high evaporation, wind may be the principal geologic agent by which sand and dust are transported, and a variety of sand dunes may locally dominate the landscape. Foremost among these are the great "seas of sand" in large deserts, but coastal dunes are also common along many of the world's shorelines, and mountain ranges may block the flow of moisture-laden air, causing a "rain shadow" that forms desert conditions. Also, blankets of windblown dust, called loess, cover millions of square kilometers of the middle-latitude continents, further attesting to the work of the wind in developing certain landscapes.

Wind is probably the least effective agent of erosion, although many curious erosional landforms are mistakenly attributed to it. Even in the desert, most erosional landforms are the products of running water, and the greatest effects of wind are shifting sand dunes. The transportation and deposition of loose, fine-grained sediment eroded by other geologic agents is the most significant effect of wind.

In this chapter we will examine the transportation and deposition of sediment by wind and the resulting distinctive landforms they create.

MAJOR CONCEPTS

1. Wind is not a major agent in eroding the landscape, but it can transport loose, unconsolidated fragments of sand and dust.
2. Wind transports sand by saltation and surface creep. Dust is transported in suspension, and it can remain high in the atmosphere for long periods of time.
3. Sand dunes migrate as sand grains are blown up and over the windward side of the dune and accumulate on the lee slope. The internal structure of a dune consists of thin layers of sand inclined in a downwind direction.
4. Various types of dunes form, depending on the wind velocity, the sand supply, the constancy of wind direction, and the characteristics of the surface over which the sand migrates.
5. Windblown dust (loess) forms blanket deposits, which can mask the older landscape beneath them. The source of loess is desert dust or the fine rock debris deposited by glaciers.

WIND AS A GEOLOGIC AGENT

What is the principal cause of wind?
What controls the origin and evolution of deserts?

Statement

Wind is an effective local geologic agent, capable of lifting and transporting loose sand and dust, but its ability to erode solid rock is limited. The main effects of wind as a geologic agent are the transportation and deposition of sand and dust in arid regions.

Discussion

Geologists once thought that wind, like running water and glaciers, had great erosional power—power to abrade and wear down Earth's surface. It has become increasingly apparent, however, that few major topographical features are formed by wind erosion. Abrasion by wind-transported sand aids in eroding and shaping some rock surfaces, but only on a small scale. Even in the desert, where water is not an obvious geologic agent, wind is not the major agent in eroding the regional landscape. Most erosional landforms in deserts were produced by weathering and by running water in times of wetter climate. They are, in a sense, "fossil" landscapes, formed by processes that are no longer active. Even minor alcoves and niches, or "wind caves," and certain topographic features called pedestal rocks, which are often thought to be caused by wind erosion, are actually produced by differential weathering, not by wind.

Although it is relatively insignificant as an erosional agent, wind is effective in transporting loose, unconsolidated sand, silt, and dust. It is responsible for the formation of the great "seas of sand" in the Sahara, the Arabian, and other deserts, as well as the blankets of windblown dust covering millions of square kilometers in China, the central United States, and parts of Europe. It is estimated that windblown dust covers one-tenth of the land surface. This fact is important because soils from these deposits constitute some of Earth's richest farmland.

Prevailing wind patterns are determined by variations of solar radiation with latitude, the Coriolis effect (deflection of the moving atmosphere due to Earth's rotation), the configuration of continents and oceans, and the location of mountain ranges (Figure 14.1). The world's great deserts, such as the Sahara and the deserts of Asia, are mostly located in low-latitude belts (Figure 14.2). The reason for this geographic control of deserts is found in the circulation patterns of the atmosphere. Equatorial air, heated by solar radiation, rises and moves to the north or south. As it rises it cools and descends again in a zone about 30 or 35° north and south of the equator. As the air rises to higher altitudes and cools, it releases moisture, which falls as tropical rains in the equatorial regions. The air is then much drier as it continues to move northward and southward. The dry air descends to the surface near latitude 30 to 35° north and south of the equator. It is capable of holding more moisture, so it rarely releases any as precipitation. As a result, evaporation of the surface moisture, rather than precipitation, occurs in the low latitudes where the air descends. When the air reaches the equator, it again rises and cools, releasing this moisture as rain. Other deserts lie downwind, behind higher mountain ranges, which intercept the moisture-laden air. As the air is forced to rise over a mountain range, it cools and precipitates its moisture. There, too, the dry descending air is heated by compression. A good example of an arid region in the rain shadow of a high mountain range is in Nevada and Utah, which lie downwind from the Sierra Nevada.

Wind action is most significant in desert areas, where the effects of running water are at a minimum, but is also important in other areas. Many coasts for example are modified by winds that pick up loose sand on the beach and transport it inland. Also, wind action is a major process in polar regions.

WIND EROSION

Why is the erosive power of wind less effective than that of running water?
What are the major features produced by wind erosion?

Statement

Wind erosion acts in two ways:

1. By deflation, which is the lifting and removal of loose sand and dust particles from Earth's surface
2. By abrasion, which is the sandblast action of windblown sand

Discussion

Deflation

In areas where weak, unconsolidated sediment is exposed at the surface, wind may blow away much of the dust and fine sand and create large depressions called deflation basins, or blowouts. Deflation basins commonly develop where calcium carbonate cement in sandstone formations is dissolved by groundwater, leaving loose sand grains, which are picked up and transported by the wind. Perhaps the best example of wind erosion in the United States is in the Great Plains, especially the High Plains of Colorado, Kan-

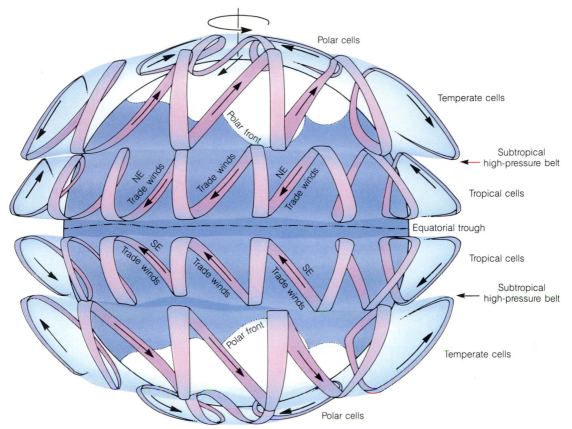

Figure 14.1

Atmospheric circulation and prevailing wind patterns are generated by solar radiation and by Earth's rotation. Air heated in the equatorial regions rises in convection columns and moves toward higher latitudes. There it is cooled, compressed, and forced to descend, forming subtropical high-pressure belts. The air then moves toward the equator as trade winds. In the Northern Hemisphere, this air is deflected by the Earth's rotation to flow southwestward; in the Southern Hemisphere, it is deflected to flow northwestward. Cold polar air tends to wedge itself toward the lower latitudes and forms polar fronts.

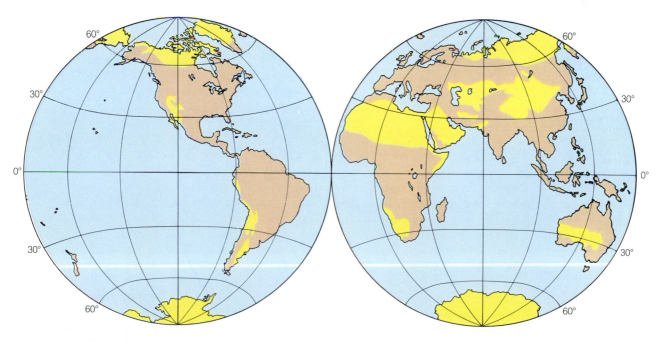

Figure 14.2

The major desert areas of the world—the Sahara, the Arabian, the Kalahari, and the deserts of Australia— all lie between 10 and 30° north or south of the equator. These areas are under almost constant high atmospheric pressure and are characterized by subsiding dry air and low humidity. Desert and near-desert areas cover nearly one-third of the land surface.

Figure 14.3
Deflation basins in the Great Plains are produced where solution activity in the layers of horizontal bedrock dissolves the cement that binds the sand grains. The loose sand is removed by the wind, and a basin is formed. Water trapped in the basin dissolves more cement and the basin is enlarged.

Figure 14.4
Wind selectively transports sand and fine sediment, leaving the coarser gravels to form a lag deposit called desert pavement. The protective cover of lag gravel limits future deflation.

sas, and Texas. In this area innumerable deflation basins, ranging from small dimples 1 ft deep and 10 ft in diameter to larger basins 50–60 ft deep and more than a mile across, are scattered across the landscape (Figure 14.3). Large deflation basins, covering areas of several hundred square kilometers, are associated with the great desert areas of the world, particularly in North Africa in the vicinity of the Nile delta.

Generally, wind can move only sand and dust-sized particles, so that deflation leaves concentrations of coarser material called lag deposits, or desert pavements (Figure 14.4). These striking desert features of erosion stand out in contrast to deposits in dune fields and playa lakes. Deflation occurs only where unconsolidated material is exposed at the surface. It does not occur where there are thick covers of vegetation or layers of gravel. The process is therefore limited to areas such as deserts, beaches, and barren fields.

Abrasion

Wind abrasion is similar to the artificial sandblasting used to clean building stone; the grooves and polished surfaces produced by wind abrasion can be seen on the surface of the bedrock in most desert regions (Figure 14.5A). Some pebbles, called ventifacts (literally meaning "wind-made"), are shaped and polished by the wind (Figures 14.5B and 14.6). Such pebbles are commonly distinguished by two or more flat faces that meet at sharp ridges and are generally well polished. Other ventifacts have a variety of shapes. Some have surface irregularities and grooves aligned with the wind direction (Figure 14.5A).

Larger landforms produced by wind abrasion are less common, but in some desert regions distinctive linear ridges, called yardangs, are produced by wind erosion. These features were first discovered in the Taklimankan Desert of China. The name is derived from the Turkistani word *yar*, meaning "ridge" or "bank." Typical yardangs have the form of an inverted boat hull (Figure 14.7) and commonly occur in cluster, that are oriented parallel to the prevailing wind that formed them (Figure 14.8).

(A) Pitted and polished surfaces

(B) Faceted pebbles and cobbles called ventifacts

Figure 14.5
Features produced by wind abrasion are apparent on the bedrock surface in most deserts of the world.

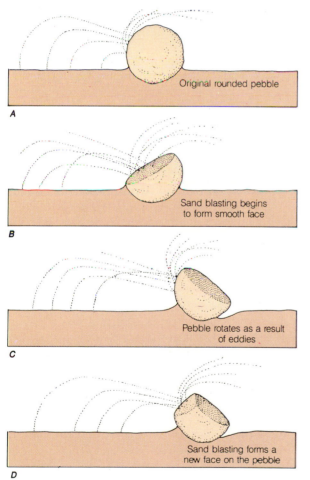

A — Original rounded pebble

B — Sand blasting begins to form smooth face

C — Pebble rotates as a result of eddies

D — Sand blasting forms a new face on the pebble

Figure 14.6
Ventifacts are pebbles shaped and polished by wind action. They commonly have two or more well-polished sides, which are formed as the sandblasting action of the wind reshapes one side of the pebble and then another. In diagram B a facet (smooth face) is formed by sandblasting on the side facing the prevailing wind. Removal of the sand by deflation causes the pebble to rotate so that other faceted and polished surfaces are produced on its other side (D).

Figure 14.7
The erosion of rock outcrops forms elongate ridges called yardangs.

Figure 14.8
Yardangs in the Lut Basin, southeastern Iran, are carved out of fine-grained, horizontally bedded, silty clay.

TRANSPORTATION OF SEDIMENT BY WIND

What is the difference in the way wind and water transport sediment?

Statement

Field observations and wind-tunnel experiments indicate that windblown sand grains move by skipping or bouncing into the air (a process called saltation) and by rolling or sliding along the surface (called surface creep). Fine silt and dust are carried in suspension over great distances and settle back to the ground only after the turbulent wind ceases.

Discussion

Movement of Sand

Although both wind and water transport sand by saltation, the mechanics of motion involved are different because the viscosity of water is so much greater than that of air. The energy that initially lifts sand grains into the air comes from collision with other grains (Figure 14.9). If the wind reaches a critical velocity, the grains begin to roll or slide. As one grain moves along, it strikes others. The impact can cause one or more of the colliding grains to bounce into the air, where they are driven forward by the stronger wind above the ground surface. Gravity soon pulls them back, and the grains strike the ground at an angle generally ranging from 10 to 16°. If the sand is moving over solid rock, the grains bounce back into the air. If the surface is loose sand, the impact of a falling grain can knock other grains into the air. This sets up a chain reaction, which eventually sets in motion the entire sand surface.

Some sand grains, too large to be ejected into the air, move by surface creep (rolling and sliding). These large grains are moved by the impact of saltating grains, not directly by the wind. Approximately one-fifth to one-fourth of the sand moved by a sandstorm travels by rolling and sliding. Particles with a diameter greater than 1 cm are rarely moved by wind.

Movement of Dust

Dust storms are a major process in deserts. They are capable of transporting thousands of tons of sediment hundreds of kilometers; a major dynamic feature of the planet, they subtly, but constantly, change its surface.

Throughout human history, dust storms have been a major cause of soil erosion. References to dust storms were recorded in 1150 B.C. in China and in biblical times in the Middle East. However, because of their regional extent and violent nature, it has been difficult to study them in a meaningful way. Our understanding of dust storms, however, has been greatly enhanced with the advent of satellite photography, which reveals intricate patterns of dust movement and associated dune fields and has clarified the behavior and mechanics of dust storms.

Dust storms are initiated by the downdraft of cool air from a cumulonimbus cloud. Such a cloud develops to the point that rain begins to fall from it, and the rain cools the air as it falls. Because the cool air is denser than the surrounding air, it descends in a downdraft. As the heavy, cooled air reaches the

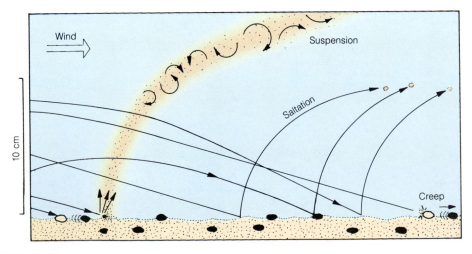

Figure 14.9
The transportation of sediment by wind is accomplished by surface creep, saltation, and suspension. Coarse grains move by impact from other grains and slide or roll (surface creep). Medium grains move by skipping or bouncing (saltation). Fine silt and dust move in suspension.

ground, it is deflected forward and moves in a large, tongue-shaped pattern. It flows across the ground as a density current—that is, a body of moving air that is heavier than the surrounding air because it is cooler. As the dense air moves across the dry surface, it sweeps up dust and sand by the churning action of its turbulent flow. Dust storms of this type are called *haboobs,* from the Arabic word for violent wind (Figure 14.10).

Great dust storms sometimes reach elevations of 2,500 m and advance at speeds of up to 200 m per second. It has been estimated that 500×10^6 tons of windblown dust are carried from the deserts each year. (This is only slightly less than the amount of sediment deposited each year by the Mississippi River.) Some is deposited downwind from the desert, such as in China (Figure 14.2), but because of the prevailing wind pattern in the Sahara, Australia, and South America, large quantities of windblown dust are carried out to sea. Some of the larger dust storms in the Sahara have even carried dust across the Atlantic to the eastern coast of South America.

MIGRATION OF SAND DUNES

How can the wind move an entire sand dune?

Statement

Wind commonly deposits sand in the form of dunes (mounds or ridges), which generally migrate downwind. In many respects sand dunes are similar to ripple marks (formed either by air or water) and to the large sand waves or sandbars found in many streams and in shallow marine water.

Discussion

Many dunes originate where an obstacle—such as a large rock, a clump of vegetation, or a fence post—creates a zone of quieter air behind it (Figure 14.11). As sand is blown up or around the obstruction and into the protected area (the wind shadow), its velocity

Figure 14.10
Dust storm in the Blue Nile area (Sudan, Africa) results when cool air descends and moves laterally over the surface as a density current. As the dense, cool air moves across the surface, it sweeps up dust and sand by its turbulent flow, creating a dust storm or "Haboob."

Figure 14.11
Sand dunes commonly originate in wind shadows. Any obstacle that diverts the wind, such as a bush or a fence post, creates eddies and reduces wind velocity. Wind-blown sand is deposited in protected areas, and eventually enough sand accumulates in the wind shadow area to form a dune. The dune itself then acts as a barrier, making its own wind shadow, and thus causes additional accumulation of sand.

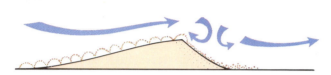

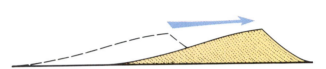

Figure 14.12
A sand dune migrates as sand grains move up the slope of the dune and accumulate in a protected area on the downwind face. The dune slowly moves grain by grain. As the grains accumulate on the downwind slope, they produce a series of layers (cross beds) inclined in a downwind direction.

Figure 14.13
Cross bedding in the Navajo Sandstone in Zion National Park, Utah, is excellent evidence that the rock formed in an ancient desert. The inclination of the strata shows that the wind blew from north to south (from left to right, in the photograph) during most of the time this formation was being deposited.

is reduced and it settles to the ground. Once a small dune is formed, it acts as a barrier itself, disrupting the flow of air and causing continued deposition downwind. Dunes range in size from a few meters high to as much as 200 m high and 1 km wide. The Empty Quarters of Saudi Arabia contain excellent examples (see Figure 14.15).

The movement of sand in a typical dune is diagramed in Figure 14.12. Dunes are asymmetrical, with a gently inclined windward slope and a steeper downwind slope, called the lee slope, or slip face. The steep slip face of the dune indicates the direction of the prevailing wind. Dunes migrate grain by grain as wind transports sand by saltation and surface creep up the windward slope. The wind continues upward past the crest of the dune, creating diverging air currents just over the lee slope. Beyond the crest, the sand drops out of the wind stream and accumulates on the slip face. As more sand is transported from the windward slope and accumulates on the lee slope, the dune migrates downwind. The internal structure of a migrating dune consists of crossbedding formed as the saltating grains accumulate on the inclined downwind slope of the dune (Figures 14.12 and 14.13). Layers of grains formed on the lee slope are therefore inclined in a downwind direction. Geologists map the directions of ancient eolian systems by measuring the directions in which the crossstrata of windblown sandstone are inclined (Figure 14.13).

TYPES OF SAND DUNES

What determines the shape of the various types of sand dunes?

Statement

Sand dunes can assume a variety of fascinating shapes and patterns, depending on such factors as sand supply, wind velocity, variability of wind direction, and characteristics of the surface over which the sand moves. The most common varieties are transverse dunes, barchan dunes, longitudinal dunes, star dunes, and parabolic dunes.

Discussion

Transverse Dunes

Transverse dunes typically develop where there is a large supply of sand and a constant wind direction (Figure 14.14A). These dunes cover large areas and develop a wavelike form, with sinuous ridges and troughs perpendicular to the prevailing wind. Transverse dunes commonly form in desert regions where exposed ancient sandstone formations provide an ample supply of sand. They usually cover large areas known as sand seas, so called because the wavelike dunes produce a surface resembling a stormy sea.

Barchan Dunes

Where the supply of sand is limited and winds of moderate velocity blow in a constant direction, crescent-shaped barchan dunes tend to develop (Figure 14.14B). Typically they are small, isolated dunes from 1 to 50 m high. The tips (or horns) of a barchan point downwind, and sand grains are swept around them as well as up and over the crest. With a constant wind direction, beautiful symmetrical crescents form. With shifts in wind direction, however, one horn can become larger than the other. Although barchans typically are isolated dunes, they may be arranged in a chainlike fashion extending downwind from the source of sand.

Longitudinal Dunes

Longitudinal dunes, also called seif dunes (from the Arabic for "sword"), are long, parallel ridges of sand, elongate in a direction parallel to the vector, which results from two slightly different wind directions (Figure 14.14C). They develop where strong prevailing winds converge and blow in a constant direction over an area having a limited supply of sand. Many longitudinal dunes are less than 4 m high, but they can extend downwind for several kilometers. In larger desert areas, they can grow to 100 m high and 120 km long, and they are usually spaced from 0.5 to 3 km apart. Longitudinal dunes occupy a vast area of central Australia called the Sand Ridge Desert. They are especially well developed in some desert regions of North Africa and the Arabian Peninsula.

Star Dunes

A star dune is a mound of sand having a high central point from which three or four arms, or ridges, radiate (Figure 14.14D). This type of dune is typical of parts of North Africa and Saudi Arabia. The internal structure of these dunes suggests that they were formed by winds blowing in three or more directions.

Parabolic (Blowout) Dunes

Parabolic dunes typically develop along coastlines where vegetation partly covers a ridge of windblown sand transported landward from the beach (Figure 14.14E). Where vegetation is absent, small deflation basins are produced by strong onshore winds. These blowout depressions grow larger as more sand is exposed and removed. Usually the sand piles up on the lee slope of the shallow deflation hollow, forming a crescent-shaped ridge. In a map view, a parabolic

(A) **Transverse dunes** develop where the wind direction is constant and the sand supply is large.

(B) **Barchan dunes** develop where the wind direction is constant but the sand supply is limited.

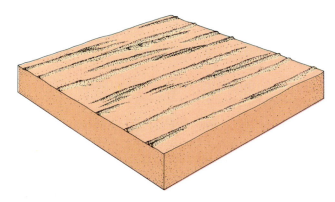

(C) **Longitudinal dunes** are formed by converging winds in an area with a limited sand supply.

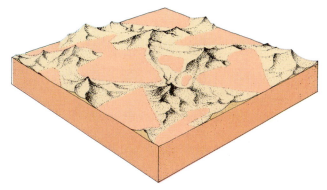

(D) **Star dunes** develop where the wind direction is variable.

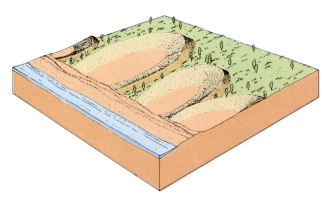

(E) **Parabolic dunes** are formed by strong onshore winds.

Figure 14.14
Each of the major types of sand dunes represents a unique balance between sand supply, wind velocity, and variability in wind direction.

dune is similar to a barchan, but the tips of the parabolic dune point upwind. Because of their form, parabolic dunes are also called hairpin dunes.

Sand Seas

Although Earth is commonly referred to as the water planet, several continents have vast areas where precipitation is rare and the surface is covered with windblown sand. These areas are referred to as sand seas, or ergs (North Africa). It has been calcu-lated that 99.8 percent of all windblown sand is in the great sand seas of the world. The largest are in Africa, Asia, and Australia (Figure 14.15). In Africa about 800,000 km^2 (or one-ninth of the entire area of the Sahara) is covered by stable or active sand dunes. One-third of Saudi Arabia, approximately 1 million km^2, is covered by dunes, and in the vast "empty quarter" (the Rub-al-Khah Desert), dunes may be over 200 m high. The Australian sand seas are mainly in the western and central portions of the continent.

Figure 14.15
Sand seas in North Africa have been mapped with the use of satellite photography. The main sand seas in Algeria, Tunisia, and western Libya are located in structural and topographic basins, separated by plateaus and low mountain ranges. A variety of dune types is formed as a result of variations in wind velocity and direction, supply of sand, and the nature of the surface over which the sand moves.

DESERTIFICATION

Why do deserts expand?

Statement

The great deserts of the world were formed by natural processes over long periods of time, as continents migrated into dry climates, which were produced in low-latitude, high-pressure zones. During development, a desert may expand and shrink in response to cyclic climatic fluctuations. The margins of deserts, therefore, have always been transitional or gradational to the adjacent, more humid environments.

Discussion

The desert fringes have very delicately balanced ecosystems. Sparse vegetation serves to inhibit wind erosion, but when it is destroyed, the desert expands. Along the desert margins, human activity is commonly superimposed upon the natural processes that cause expansion and contraction of the deserts. Grazing livestock, the pounding of soil by hooves, and even collecting firewood can reduce the plant cover and stress the ecosystem beyond its tolerance. This results in a degradation of the land and, in many cases, expansion of the desert. This process, in which productive land is converted to desert, is referred to as desertification. When the general climatic trend is toward increasing aridity, desertification can occur with remarkable speed.

Desertification does not occur in a broad, even swath that can easily be mapped along the desert fringe. Deserts advance erratically, forming patches on their borders, and areas far from the desert may quickly degrade into barren rock and sand.

Desertification presents an enormous problem for human existence. About one-third of Earth's land is arid or semiarid, but only about one-half of this area is so dry that it cannot support human life. More than 600 million people live in the dry areas, and about 80 million live on land that is nearly useless because of desertification. The most severe problems are in Africa and Asia.

LOESS

What geologic features are produced by windblown dust?

Statement

Loess is a deposit of windblown silt (dust) that accumulates slowly and ultimately blankets large areas, often masking preexisting landforms. It may cover as much as one-tenth of the world's land surface and is particularly widespread in semiarid regions along the margins of great deserts (Figure 14.2). The equatorial tropics are free from loess because it is washed away by the heavy rainfall as soon as it is deposited. Areas formerly covered by continental glaciers are also loess-free because until only a few thousand years ago, they were covered with ice.

Discussion

Loess is a distinctive sedimentary deposit. Its particles are very similar, if not identical, to the dust in the air at the present time. Small, angular fragments of quartz predominate, with lesser amounts of feldspar, mica, and calcite. Normally loess deposits lack stratification and erode into vertical cliffs (Figure 14.16). Large loess deposits are derived from (1) desert regions and (2) glacial deposits.

Desert Loess

In northern China extensive loess deposits consist primarily of disintegrated rock material brought by the prevailing westerly winds from the Gobi Desert, in central Mongolia. The yellow-colored loess, which reaches a thickness of more than 60 m, blankets a large area. It is easily eroded and transported in suspension by running water and is responsible for the characteristic color of the Hwang Ho (the Yellow River) and the Hwang Hai (the Yellow Sea). In the nearly vertical walls of loess deposits, a great number of Chinese have excavated caves in which they live (Figure 14.17).

Glacial Loess

Much of the loess of North America and Europe appears to have originated from rock debris pulverized by glaciers and deposited as outwash. This sediment, commonly called rock flour, was then picked up by the wind, transported, and deposited as a blanket of loess. In the central United States, most loess occurs in the bluffs and uplands of the Mississippi valley, the major drainage system for the meltwaters of the glaciers. Near rivers where floodplain muds provide a ready source of dust, the loess is 30 m or more thick, but it thins out away from the river channels. The greatest accumulations are east of the floodplains.

In many areas loess lies directly on glacial or glaciofluvial deposits at the glacier margins. As the glaciers melted, the outwash of fine rock flour was apparently deposited so rapidly that a protective plant cover could not be established—a situation typical of many outwash plains. In Europe loess deposits also appear to have been derived from the adjacent glaciated areas.

Figure 14.16
A deposit of loess, composed of fine, closely packed silt particles, commonly erodes into nearly vertical cliffs. Most of the loess in the United States is found in the Great Plains and in the Mississippi Valley region. This area is in the lower Mississippi Valley.

Figure 14.17
Loess deposits in China are widespread and have been excavated into an elaborate system of chambers for dwellings.

SUMMARY

1. The great deserts of the world form in low-latitude regions in a zone roughly 30 degrees north and south of the equator.
2. Wind erosion forms ventifacts, deflation basins, and yardangs.
3. Wind transports sand by saltation and surface creep. Silt and dust-sized particles are carried in suspension.
4. Windblown sand commonly accumulates in dunes that migrate downwind as sand is transported up the windward slope and accumulates in the relatively quiet areas on the lee slope.
5. A variety of dune types results from variations in sand supply, wind direction, and velocity and from characteristics of the desert or shoreline surface. The most significant include (a) transverse dunes, (b) barchan dunes, (c) longitudinal dunes, (d) star dunes, and (e) parabolic dunes.
6. The margins of deserts can be disturbed by human activity and slight climatic fluctuations, causing desertification.
7. Loess accumulates as a blanket deposit that can completely cover the preexisting surface. The dust is derived either from the rock flour near glacial margins or from desert regions.

KEY TERMS

barchan dune (p. 247)
blowout (p. 247)
Coriolis effect (p. 240)
deflation (p. 240)
deflation basin (p. 242)
desertification (p. 250)
desert pavement (p. 242)

dune (p. 245)
lag deposit (p. 242)
lee slope (p. 247)
loess (p. 250)
longitudinal (seif) dune (p. 247)

parabolic (blowout) dune (p. 247)
saltation (p. 244)
sand sea (erg) (p. 248)
slip face (p. 245)
star dune (p. 247)

surface creep (p. 244)
transverse dune (p. 247)
ventifact (p. 242)
wind shadow (p. 245)
yardang (p. 242)

REVIEW QUESTIONS

1. Describe the processes involved in wind erosion.
2. What controls the distribution of the major desert regions of Earth?
3. What landforms are produced by wind erosion?
4. Explain the origin of desert pavements.
5. Draw a simple diagram showing how sand is transported by the wind.
6. Why is the wind an effective agent in sorting sand and dust?
7. Describe how a sand dune forms and how it migrates.
8. List the five major types of dunes and state the conditions under which each type forms (wind direction and velocity, sand supply, and the characteristics of the surface over which the sand moves).
9. What is the origin of loess?
10. What are the major areas of loess deposits in the world today?

Multiple-Choice Questions

1. Most of the world's major deserts are located in the
   a. Southern Hemisphere.
   b. coastal areas.
   c. subtropical high-pressure belts.
   d. tundra regions.
   e. rain shadows.
2. In general the dominant erosional landforms in desert regions are developed by
   a. wind.
   b. groundwater.
   c. running water.
   d. chemical weathering.
   e. sand blasting.
3. A lag deposit, or desert pavement, is a concentration of
   a. sand left behind as dust is removed by the wind.
   b. sand cemented by calcium carbonate.
   c. dust or loess remaining after a windstorm.
   d. gravel-sized particles left behind by the wind.
   e. slow-moving sand.
4. A cobble that has been shaped, polished, and abraded by wind action is called a
   a. yardang.
   b. seif.
   c. barchan.
   d. deflation rock.
   e. ventifact.
5. The steepest slope of a sand dune is toward the southwest if the wind blows steadily from the
   a. northeast.
   b. northwest.
   c. southeast.
   d. southwest.
   e. none of these because the Coriolis effect makes it unpredictable.
6. The migration of sand dunes results from
   a. gravity sliding.
   b. erosion of sand on the slip face and deposition on the windward side.
   c. changes in the wind velocity.
   d. erosion of sand on the windward side and deposition on the slip face.
   e. none of the above.

7. Barchans are
   a. crescent-shaped sand dunes.
   b. desert sandstorms.
   c. star-shaped sand dunes.
   d. wind-eroded depressions.
   e. wind-eroded cobbles.
8. Loess is deposited
   a. as a comparatively uniform blanket over hundreds of square miles.
   b. by the wind.
   c. as dust.
   d. all of these.
   e. a and c only.
9. Sand dunes on the shores of oceans and lakes are likely to be
   a. barchan.
   b. longitudinal.
   c. transverse.
   d. parabolic (blowout).
   e. star.
10. The source of the loess that covers portions of the central United States is
    a. sand from western deserts.
    b. sediment from the Dust Bowl of the 1930s.
    c. carbonate sediment from karst regions to the southeast.
    d. sediment from glacial outwash plains.
    e. dust blown south from the Canadian Shield.

True/False Questions

1. *T F* Wind is one of the most effective agents of erosion.
2. *T F* The steep slip face of a dune indicates the direction of the prevailing wind.
3. *T F* The tips (or horns) of a barchan dune point upwind.
4. *T F* The tips of a parabolic dune point downwind.
5. *T F* The tropics and those areas formerly covered by continental glaciers commonly contain thick deposits of loess.

Fill in the Missing Terms

1. Where the supply of sand is limited and winds of moderate velocity blow in a constant direction, crescent-shaped dunes called _____ dunes tend to develop.
2. _____ dunes, also called seif dunes, are long parallel ridges of sand that elongate in a direction parallel to the vector resulting from two slightly different wind directions.
3. _____ are pebbles shaped and polished by wind action.
4. Wind selectively transports sand and dust and leaves the coarse gravels as _____ deposits.
5. Large depressions formed by wind erosion are called _____ _____.

15

PLATE TECTONICS

The theory of plate tectonics is universally accepted by geologists throughout the world. Yet only a few years ago, the continents and the ocean basins were considered to be permanent features that had existed since the beginning of Earth's history. We know that the lithosphere is in motion and that the continents have drifted with the moving plates thousands of kilometers across Earth's surface. The movement of lithospheric plates causes earthquakes, mountain building, metamorphism, and igneous activity. What is the nature of the mobile plates? What causes them to move? How do we measure rates and directions of plate movement? Have the plates always been in motion? Will they continue to move in the future? Why have we discarded the old concept of permanent continents and ocean basins?

In this chapter we will briefly review the development of the theory of plate tectonics and the evidence upon which it is based. Then we will consider the nature of the lithospheric plates, their boundaries, and their relative motion, which produces the tectonic features of Earth.

MAJOR CONCEPTS

1. The theory of continental drift was proposed in the early 1900s and was supported by a variety of geologic evidence. Without a knowledge of the nature of the oceanic crust, however, a complete theory of Earth dynamics could not have been developed.
2. A major breakthrough occurred in the early 1960s, when the topography of the ocean floors was mapped and magnetic and seismic characteristics were determined. This led to the development of the plate tectonic theory.
3. Most tectonic activity occurs along plate boundaries.
4. Divergent plate boundaries are zones where the plates split and spread apart. The major geologic processes are (a) rifting and (b) the generation of basaltic volcanism and new oceanic lithosphere.
5. Convergent plate boundaries are zones where plates collide. Important geologic processes include (a) subduction, (b) generation of granitic magma, (c) mountain building, and (d) metamorphism.
6. Transform fault boundaries are zones where plates slide horizontally past each other.
7. Plate motion can be described in terms of a pole of rotation. The direction of relative motion of plates is indicated by (a) the trend of oceanic ridge and associated fracture zones, (b) seismic data, (c) magnetic stripes on the seafloor, and (d) the ages of chains of volcanic islands and seamounts.
8. The major forces acting on plates are (a) slab-pull, (b) ridge-push, (c) basal drag, and (d) friction along transform faults and in subduction zones.
9. Thermal convection in the mantle is believed to be the fundamental process responsible for plate motion.

CONTINENTAL DRIFT

What evidence indicates that continents split and drift apart?
Why did most geologists reject the idea?

Statement

Continental drift was first considered a serious theory in 1915, when Alfred Wegener, a German meteorologist, published his book *The Origin of the Continents and Oceans*. The theory was based on a variety of geologic evidence including:

1. the geographic fit of the continents.
2. fossils found in Brazil and Africa.
3. rock types and structural fit of the continents.
4. patterns of glaciation.
5. paleoclimatic records.

Most geologists and geophysicists rejected the theory because no one could explain the forces that could cause continents to move through oceanic crust.

Discussion

The theory of plate tectonics wrought a sweeping change in our understanding of Earth and the forces that shape it. Some scientists consider this conceptual change as profound as those that occurred when Darwin reorganized biology in the nineteenth century, or when Copernicus, in the sixteenth century, determined that Earth is not the center of the universe. Yet, the concept of continental drift is an old idea.

In 1915 Wegener based his theory of continental drift not only on the shape of the continents, but also on geologic evidence, such as similarities in the fossils found in Brazil and Africa (Figure 15.1). He drew a series of maps showing three stages in the drift process, beginning with an original large landmass, which he called Pangaea (meaning "all lands"). Wegener believed that the continents, composed of light granitic rock, somehow plowed through the denser basalts of the ocean floor, driven by forces related to the rotation of the Earth.

Most geologists and geophysicists rejected Wegener's theory, although many scientific observations supporting it were known at the time. A few noted scholars, however, seriously considered the theory. Arthur Holmes, in England, developed it in his textbook *Principles of Physical Geology* (1944). Alexander L. du Toit, from South Africa, compared the landforms and fossils of Africa and South America and further expounded the theory in his book *Our Wandering Continents* (1937).

The early arguments for the theory of continental drift were supported by some important and imposing evidence, most of which resulted from regional geologic studies.

Paleontological Evidence

The striking similarity of certain fossils found on the continents on both sides of the Atlantic is difficult to explain unless the continents were once connected. The fossil record indicates that a new species appears at one point and disperses outward from there. Floating and swimming organisms could migrate in the ocean from the shore of one continent to another, but the Atlantic Ocean would present an insurmountable obstacle for the migration of land-dwelling animals, such as reptiles and insects, and certain land plants. Consider the following examples (Figure 15.1).

Fossils of Glossopteris, a fernlike plant, have been found in rocks of the same age from South America, South Africa, Australia, and India and within 480 km of the South Pole, in Antarctica. Mature seeds of this plant were several millimeters in diameter, too large to have been dispersed across the ocean by winds. The simultaneous presence of Glossopteris on all of the southern continents, therefore, is strong supporting evidence that the continents were once connected. The distribution of Paleozoic and Mesozoic reptiles provides similar evidence, as fossils of several species have been found in the now separated southern continents. An example is a mammal-like reptile belonging to the genus Lystrosaurus. This creature was strictly a land dweller. Its fossils are found in abundance in South Africa, South America, and Asia, and in 1969 a United States expedition discovered them in Antarctica. This genus thus inhabited all of the southern continents. Clearly, these reptiles could not have swum thousands of kilometers across the Atlantic and Antarctic oceans, so some previous connection of the continents must be postulated. A former land bridge, similar to present-day Central America, would explain the presence of Lystrosaurus in distant parts of the world, but surveys of the ocean floor show no evidence of such a submerged land bridge.

Evidence from Structure and Rock Type

A number of geologic features end abruptly at the coast of one continent and reappear on the facing continent across the Atlantic (Figure 15.2). Folded mountain ranges at the Cape of Good Hope, at the southern tip of Africa, trend from east to west and terminate sharply at the coast. An equivalent structure, of the same age and style of deformation, appears near Buenos Aires, Argentina.

The folded Appalachian Mountains are another example. The deformed structures of the mountain belt extend northeastward across the eastern United States and through Newfoundland and terminate abruptly at the ocean. They reappear at the coast of Ireland and Brittany.

Other examples could be cited, but the important point is that the continents on both sides of the Atlantic fit together, not only in outline, but in rock type

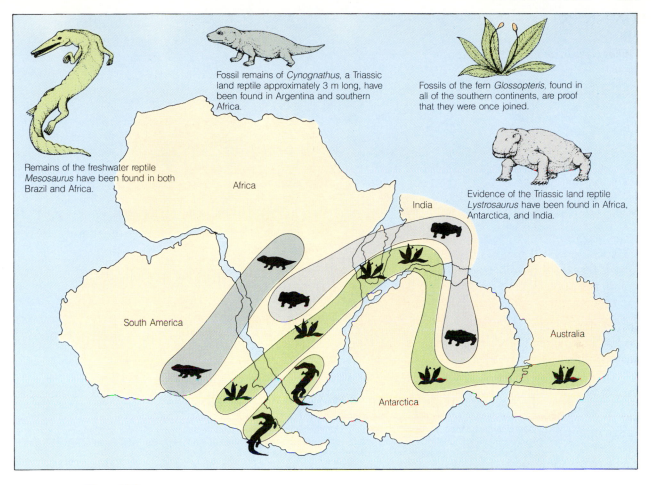

Figure 15.1

Paleontologic evidence of continental drift can be appreciated by considering the distribution of some fossil plants and animals found in South America, Africa, Madagascar, India, Antarctica, and Australia. Mesosaurus, a Permian freshwater reptile, is found in both Brazil and South Africa. Glossopteris, a fossil fern, is found in all of the southern continents in the zone shown on the map. Lystrosaurus, a Triassic land reptile, is found in South Africa, South America, India, and Antarctica. Cynognathus, an older Triassic reptile, is found in Argentina and South Africa.

and structure. They are related much like matching pieces of a torn newspaper. The jagged edges fit, and the printed lines (structure and rock types) join together in a coherent unit. One important point needs emphasis. The geologic similarities on opposite sides of the Atlantic are found only in rocks older than the Cretaceous period, which began about 137 million years ago. The continents are believed to have split and begun drifting apart in Jurassic time, about 200 million years ago.

Evidence from Glaciation

During the latter part of the Paleozoic Era (about 300 million years ago), glaciers covered large portions of the continents in the Southern Hemisphere. The deposits left by these ancient glaciers can be readily recognized, and striations and grooves on the underlying rock show the direction in which the ice

moved (Figure 15.3A). Except for Antarctica, all of the continents in the Southern Hemisphere now lie close to the equator. In contrast, the continents in the Northern Hemisphere show no trace of glaciation during this time. In fact, fossil plants indicate a tropical climate in that area. This evidence is difficult to explain in the context of fixed continents because the climatic belts are determined by latitude.

Even more difficult to explain is the direction in which the glaciers moved. Regional mapping of striations and grooves indicates that, in South America, India, and Australia, the ice moved inland from the oceans. Such movement would be impossible unless there had been a landmass where the oceans now exist.

If the continents were grouped together as Wegener proposed, the glaciated areas would have comprised a neat package near the South Pole (Figure 15.3B), and Paleozoic glaciation could be explained nicely. The

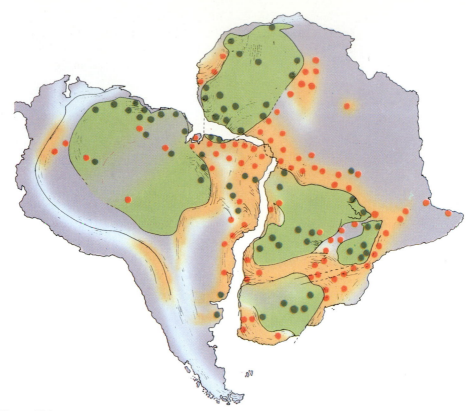

Figure 15.2
South America and Africa fit together, not only in outline, but in rock types and geologic structure. The green areas represent the shields of metamorphic and igneous rocks, formed at least two billion years ago. The gray areas represent younger rock, much of which has been deformed by mountain building. Structural trends such as fold axes are shown by dashed lines. Most of the deformation occurred from 450 million to 650 million years ago. Several fragments of the African Shield are stranded along the coast of Brazil. Green dots represent rocks that are more than two billion years old. Orange dots represent younger Precambrian rocks (0.5–2.0 billion years old).

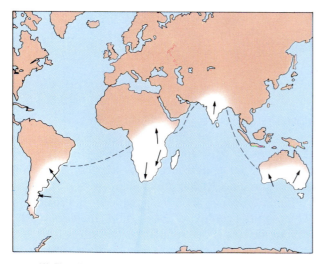

(A) The distribution of late Paleozoic glacial deposits is restricted to the Southern Hemisphere (except for India). Arrows indicate that the direction of ice movement was from the sea toward the land, an impossible situation because glaciers flow from centers of accumulation on the continents outward toward the sea. These areas are now close to the tropics. The present-day cold latitudes in the Northern Hemisphere show no evidence of glaciation during the late Paleozoic.

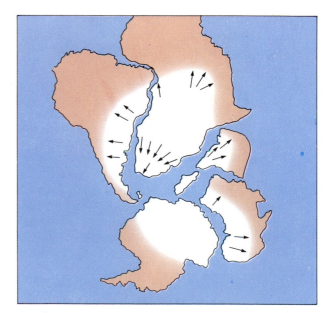

(B) If the continents were restored to their former positions according to Wegener's theory of continental drift, and if the former South Pole were located approximately where South Africa and Antarctica meet, the location of late Paleozoic glacial deposits and the directions in which the ice flowed would be explained nicely.

Figure 15.3
Distribution and flow direction of late Paleozoic glaciers provide further evidence of continental drift.

pattern of glaciation was considered strong evidence of continental drift, and many geologists who worked in the Southern Hemisphere became ardent supporters of the theory because they could see the evidence with their own eyes.

Evidence from other Paleoclimatic Records

Other evidence of striking climatic changes tends to support the drift theory. Great coal deposits in Antarctica show that abundant plant life once flourished on that continent, now mostly covered with ice.

On the other continents, salt deposits, formations of windblown sandstone, and coral reefs provide additional clues that permit us to reconstruct the climatic zones of the past. The paleoclimatic patterns are baffling with the continents in their present positions, but if they are grouped together in their predrift positions, the patterns are easily explained (Figure 15.4).

The evidence for the theory of continental drift was considered and debated for years. Wegener was criticized for failing to explain what forces would permit continents of granite to plow through oceans of rock. The idea of a moving lithosphere was yet to come. In the absence of a reasonable mechanism for drift, there was little further development of the theory until after World War II. An explosion of knowledge then provided renewed support for the drift hypothesis and also led to the discovery of a possible mechanism.

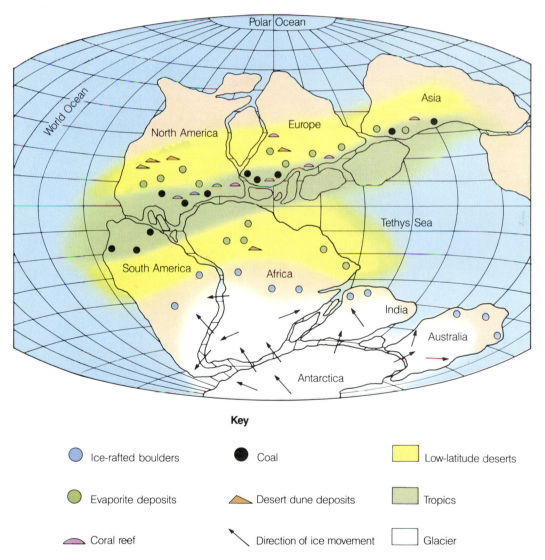

Figure 15.4
Paleoclimatic evidence for continental drift includes deposits of coal, desert sandstone, rock salt, windblown sand, gypsum, and glacial deposits. Each indicates a specific climatic condition at the time of its formation. The distribution of these deposits is best explained if we assume that the continents were once grouped as shown in this diagram.

DEVELOPMENT OF THE THEORY OF PLATE TECTONICS

The theory of plate tectonics is a simple, straightforward explanation of how Earth works. Why wasn't it discovered earlier?

Statement

Although the theory of continental drift was supported by some convincing evidence, the data upon which it was originally based came only from the continents because, prior to the 1950s, there was no effective means of studying the ocean floor. Before 1950, therefore, geologists faced an almost total absence of data about the geology of three-fourths of the Earth's surface; then, in the 1950s and 1960s, a burst of new data and new ideas resulted from research efforts in three areas:

1. The topography and geology of the ocean floor
2. Rock magnetism
3. Studies of earthquakes

Discussion

The Geology of the Ocean Floor

In the 1950s and 1960s, newly developed echo-sounding devices enabled marine geologists and geophysicists to map the topography of the ocean floor in considerable detail (see Chapter 18). When the results of these studies were compiled, they revealed that the ocean basins are divided by a great ridge approximately 84,000 km long and about 1,500 km wide. Moreover, at the crest of the ridge is a central valley, from 1 to 3 km deep. This feature appears to be a rift valley, which is splitting apart under tension. No one could imagine why the ridge was there, but no one could dispute that it was the longest mountain range on the planet, and along its crest was the longest valley. Other evidence shows that ocean basins are relatively young. Seismic studies have established that the oceanic crust (composed largely of basalt) has a completely different composition from the continental crust and is much thinner. Furthermore, the oceanic crust is not deformed into folded mountain structures and apparently is not subjected to strong compressional forces.

In 1960 H. H. Hess, a noted geologist from Princeton University, proposed a theory of seafloor spreading that took into account the new data from echo soundings and also suggested a possible mechanism for continental drift. Hess postulated that the ocean floors are spreading apart, propelled by thermal convection currents in the mantle, and are moving away from the oceanic ridge. According to his theory, this continuous spreading produces fractures in the rift valley, and magma from the mantle is injected into these fractures to become new oceanic crust. The convection currents in the mantle carry the continents away from the ocean ridge and toward deep-sea trenches. There the oceanic crust descends into the mantle with the descending convection current and is reabsorbed. In this way the entire ocean floor is completely regenerated in 200 or 300 million years.

Hess thus elaborated on the theory of continental drift in the light of fresh knowledge and redefined it in the scheme of seafloor spreading. A test of his ideas, using new studies in paleomagnetism, was soon to follow.

Evidence from Sediment on the Ocean Floor. To many geologists some of the most convincing evidence for the plate tectonic theory comes from recent drilling in the sediment on the ocean floor. Deep-sea drilling confirms the conclusions drawn from paleomagnetic studies by providing samples of the fossils that first accumulated on different portions of the ocean floor. As is predicted by the plate tectonics theory, the youngest sediment resting on the basalt of the ocean floor is found near the oceanic ridge, where new crust is being created. Away from the ridge, the sediments that lie directly above the basalt become progressively older, with the oldest sediment nearest the continental borders (Figure 15.5).

Measurements of rates of sedimentation in the open ocean show that about 3 mm of red clay accumulate every 1,000 years. If the present ocean basins were old enough to have existed since Cambrian time (600 million years ago), for example, the sediments would be 1.5 km thick (Figure 15.5A); but the average thickness of deep-ocean sediments measured to date is only 300 m, suggesting that the ocean basins are younger geologic features (Figure 15.5B). In fact, the oldest sediments yet found on any ocean floor are only about 200 million years old. In contrast, the metamorphic rocks of the continental shields are as much as 3.8 billion years old.

The theory of plate tectonics is now firmly established and accepted as the fundamental theory of the Earth's dynamics. It was first used to explain the meaning of features on the ocean floor. Now the emphasis has switched to the continents, and most previous geologic observations of the continents are being reexamined in its light.

Rock Magnetism

The study of rock magnetism developed during the 1950s with the perfection of new, highly sensitive magnetometers. Certain rocks, such as basalt, are fairly rich in iron and become weakly magnetized by Earth's magnetic field as they cool. In a sense the mineral grains in the rock become "fossil" magnets,

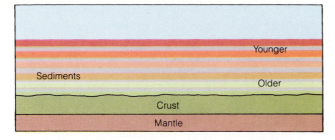

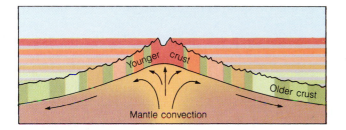

(A) With no seafloor spreading, the entire ocean floor would be covered with a thick sequence of oceanic sediment, with alternating polarity preserving a record of Earth's magnetism since the Precambrian.

(B) With seafloor spreading, the blanket of oceanic sediment *thins progressively toward the crest of the oceanic ridge and is almost nonexistent in the rift valley. The edge of each layer of magnetized sediment lies upon basaltic crust, which was generated at the spreading center during the same time interval the sediment was deposited.*

Figure 15.5
The thickness of sediment and magnetic reversals on the oceanic ridge confirm the theory of seafloor spreading.

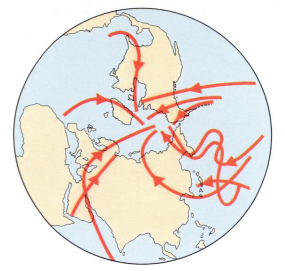

(A) The magnetic properties of rocks in North America suggest that the north magnetic pole has apparently migrated in a sinuous path over the last several hundred million years. Evidence from other continents shows similar migration, but along different paths. How could different continents show different paths of polar migration? The paleomagnetic evidence implies that, if the continents had remained fixed, different continents would have had different magnetic poles at the same time, but that would be impossible.

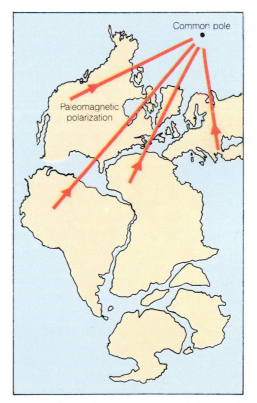

(B) The question can be answered if the pole has remained fixed while the continents have drifted. If, for example, Europe and North America were previously joined, the paleomagnetic field preserved in their rocks would indicate a single pole location until they drifted apart. The sequence of rocks on each continent would then show the pole taking a different path of migration to its present position.

Figure 15.6
Changes in the position of the magnetic poles during the geologic past are indicated by paleomagnetic studies of rocks.

oriented with respect to Earth's magnetic field at the time when the rock was formed, and thus preserving a record of paleomagnetism. Similarly, the iron in grains of red sandstone becomes oriented in Earth's magnetic field as the sediment is deposited, so red sandstone also can indicate the orientation of the paleomagnetic fields. These rocks therefore retain an imprint of Earth's magnetic field at the time of their formation.

The magnetic field of Earth resembles that of a simple bar magnet with its axis inclined 11° from Earth's geographic axis. The mantle and core of Earth are far too hot to retain a permanent magnetic field. Earth's magnetism, therefore, must be generated electromagnetically. The electromagnetic, or dynamo, theory postulates that the outer core of liquid iron slowly rotates with respect to the surrounding mantle. Such motion would generate electrical currents, which would establish a magnetic field.

Apparent Polar Wandering.   Studies of paleomagnetism in European rocks of widely different ages demonstrate that Earth's north magnetic pole apparently has steadily changed its position with time. As is shown in Figure 15.6, the pole appears to have migrated slowly northward and westward to its present position. The change in position was systematic, not random. A similar migration of the magnetic pole was found from paleomagnetic work in North America, and, although the path of migration was systematically different, it paralleled that of the European shift. These observations could be explained nicely by drifting continents, so students of paleomagnetism became leading proponents of the theory of continental drift. Soon results collected from the southern continents were reported. Again, a systematic change in the position of the magnetic pole through time was documented—but with different paths for different continents.

It is impossible that there were numerous magnetic poles migrating systematically and eventually merging. The most logical explanation is that there has always been only one magnetic pole, which has remained fixed, while the continents moved with respect to it. The results of paleomagnetic studies make sense if the continents were once arranged as shown in Figure 15.5 and then drifted to their present positions. This discovery brought renewed interest in the theory of continental drift and lent support to the conclusion that the Atlantic Ocean opened relatively recently.

Patterns of Magnetic Reversals on the Seafloor.
Recent studies of the magnetic properties of numerous samples of basalt from many parts of the world demonstrate that Earth's magnetic field has been reversed many times over the last 70 or 80 million years. Epochs of normal polarity (that is, periods when the magnetic field was oriented as it is today and the magnetic poles were thus close to their present location), lasting from one to three million years, have been followed by similar periods during which the north magnetic pole and the south magnetic pole were reversed. At least nine magnetic reversals have occurred in the last 4.5 million years. The present period of normal polarity began about 700,000 years ago. It was preceded by a period of reversed polarity, which began 2.5 million years ago and lasted about 2 million years. That period of generally reversed polarity contained two short periods of normal polarity.

The major intervals of alternating polarity (about one million years apart) are termed polarity epochs, and the intervals of shorter duration are termed polarity events. The pattern of alternating polarities has been clearly defined, and evidence of the occurrence of polarity epochs has been found in widely separated parts of Earth. From the sequence of magnetic anomalies and their radiometric ages, a reliable chronology of magnetic reversals has been established for the last four million years (Figure 15.7). In addition, extrapolation as far back as 76 million years reveals a sequence of at least 171 reversals.

In 1963 Fred Vine and D. H. Matthews saw a way to test the idea of seafloor spreading put forth by Hess. If seafloor spreading has occurred, they suggested, it should be recorded in the magnetism of the basalts in the oceanic crust. (The same idea was developed independently by L. W. Morely.) If Earth's magnetic field reversed intermittently, new basalt forming at the crest of the oceanic ridge would be magnetized according to the polarity at the time it cooled. As the ocean floor spreads, a symmetrical series of magnetic stripes, with alternating normal and reversed polarities, would be preserved in the crust along either side of the oceanic ridge. Subsequent investigations have conclusively proved this theory, proposed by Vine and Matthews and by Morley.

To understand better the origin of these magnetic patterns, consider how the seafloor could have evolved during the last few million years. Figure 15.8A shows the seafloor as it is considered to have been about 2.75 million years ago, during the Gauss normal polarity epoch (named for the German mathematician Karl Friedrich Gauss). Basalt, injected into the fractures of the oceanic ridge, formed dikes or was extruded over the seafloor as submarine flows. As it solidified, it became magnetized in the direction of the existing (normal) magnetic field; thus, basalt extruded along the oceanic ridge formed a zone of new crust with normal magnetic polarity. As the seafloor spread, this zone of crust split and migrated away from the ridge but remained parallel to it. About 2.5 million years ago, Earth's magnetic polarity was reversed. New crust generated at the oceanic ridge was magnetized in the opposite direction (Figure 15.8B), producing a zone of crust with reverse polarity. When the polarity changed to normal again, the newest crustal material was magnetized in the normal direction. In this way the sequence of polarity reversals became imprinted as magnetic stripes on the oceanic crust.

Note that the patterns of magnetic stripes on the ocean floor on either side of the ridge match the patterns found in a sequence of recent basalts on the continents (Figures 15.8A and 15.8B; see also Figure 15.7); that is, the crest of the ridge shows normal polarity and is flanked on either side by a broad stripe of rocks with reversed polarity (formed during a reversed epoch) containing two narrow bands of rocks with normal polarity (formed during normal epochs). Then follows a stripe with normal polarity containing one narrow band with reversed polarity, and so on. In brief, the patterns of magnetic reversals away from the rest of the ridge are the same as those found in a vertical sequence of rocks on the continents, from youngest to oldest. These data provide compelling evidence that the seafloor is spreading and that continents drift.

An important aspect of these reversal patterns is that they enable us to determine rates of plate movement. Magnetic reversals in rock sequences on the continents have been radiometrically dated. These

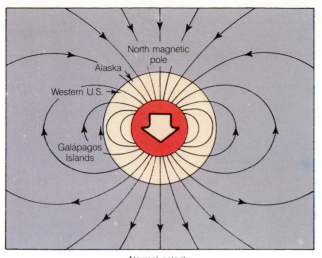

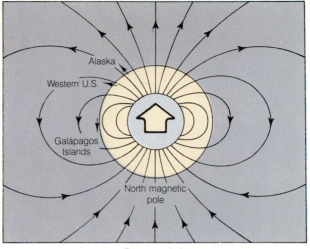

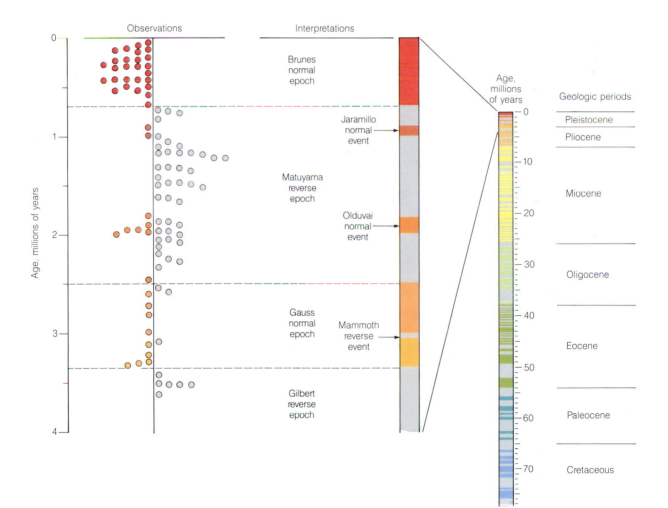

Figure 15.7
Reversals of lines of force in the Earth's magnetic field are documented by paleomagnetic studies of numerous rock samples from throughout the world. Periods of time with normal polarity are shown in color. Periods of reverse polarity, where the lines of force were oriented in the opposite direction, are shown in gray. Diagram C shows the patterns of changing polarity with time. The pattern of change during a period of one or two million years is distinctive, and it can be used to help establish the age of rock sequence.

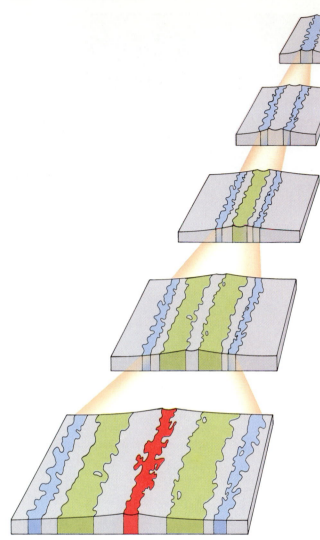

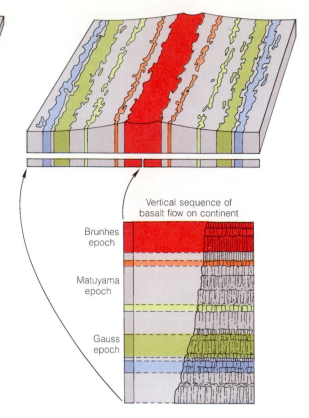

(B) Patterns of magnetic reversals appear in a vertical sequence of basalts on the continents. The youngest (upper) continental rocks correlate with the youngest oceanic crust (at the center of the oceanic ridge).

(A) As magma cools and solidifies along the ridge in dikes and flows, it becomes magnetized in the direction of the magnetic field existing at that time (normal polarity). As seafloor spreading continues, the magnetized crust formed during earlier periods separates into two blocks. Each block is transported laterally away from the ridge, as though on a conveyor belt. New crust, formed at the ridge, becomes magnetized in the opposite direction. Note that the pattern of magnetic reversals away from the ridge is the same as the pattern found in a sequence of basalt flows on the continents.

Figure 15.8
Specific patterns of magnetism are preserved in the newly formed crust, which is generated at the oceanic ridge as the lithosphere moves laterally. The patterns of magnetic reversals away from the ridge are identical to the patterns of magnetic reversals in a vertical sequence of rocks on the continents.

studies show that the present normal polarity has existed for the last 700,000 years and was preceded by the pattern shown in Figure 15.7. Because the same pattern exists in the ocean crust, we can assign provisional ages to the magnetic anomalies on the ocean floor.

Magnetic surveys have now determined patterns of magnetic reversal for much of the ocean floor, and from these patterns, the age of various segments of the seafloor has been established (Figure 15.9). These studies show that most of the deep-sea floor was formed during Cenozoic time (during the last 65 million years). It now seems probable that very little or none of the present ocean basin was formed before the Jurassic period, roughly 200 million years ago. From

the pattern of magnetic reversals, the rate of seafloor spreading appears to range from 1 to 17 cm per year.

Iceland, because it is essentially a large exposure of the mid-Atlantic Ridge, offers a unique opportunity to study the physical mechanism of seafloor spreading. Geologic studies there show that the island is being pulled apart by the spreading crust beneath. The tension causes faults and fissures parallel to the axis of the ridge. Volcanic eruptions occur through these fissures, and swarms of parallel dikes are injected into them with each increment of crustal extension. The aggregate width of these dikes is about 400 km, which corresponds to the total amount of crustal extension since the beginning of Tertiary time, about 65 million years ago. A geologic map of Iceland

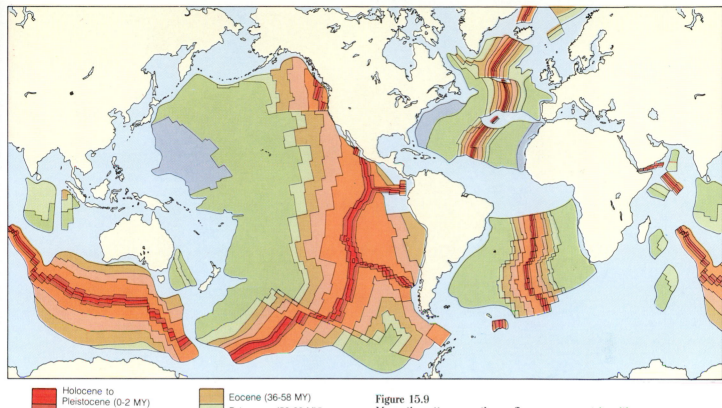

Holocene to Pleistocene (0-2 MY)	Eocene (36-58 MY)
Pliocene (2-5 MY)	Paleocene (58-66 MY)
Miocene (5-23 MY)	Cretaceous (66-144 MY)
Oligocene (23-38 MY)	Jurassic (144-208 MY)

Figure 15.9
Magnetic patterns on the seafloor are symmetric with respect to the oceanic ridge. The youngest crust is along the crest of the ridge. Away from the ridge, the crust is progressively older.

(Figure 15.10) shows that the oldest rocks are at the extreme eastern and western ends of the island. Rocks become progressively younger toward the center of the island, where most of the present-day fissures and volcanism occur.

Evidence from Earthquakes

The establishment of a worldwide network of sensitive new seismic stations in 1966 to monitor nuclear testing has enabled seismologists to amass an amazing amount of data concerning earth dynamics. Earthquakes can be pinpointed with great precision, and the depth and direction of movement that caused the quakes can be established. This new information indicates a remarkable relationship between earthquake activity and plate boundaries (see Figure 16.5). One of the major zones of earthquake activity surrounds the Pacific Ocean and parallels the line of intense volcanic activity. Comparisons of the earthquake map (Figure 16.5) with a map showing plate boundaries (Figure 15.11), will immediately show that the Pacific belt marks the outer margins of the Pacific, Nazca, and Cocos plates and that most of this boundary is a subduction zone.

A second belt of intense earthquake activity runs down the crest of the midoceanic ridge and coincides

with the spreading plate boundaries. Clearly, plate movement is causing the major earthquakes of the world. Further details of earthquake activity and its relationship to plate tectonics are presented in Chapter 16.

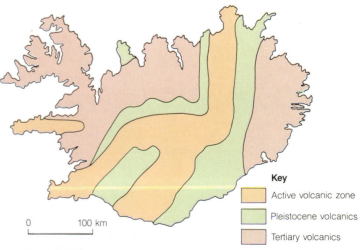

Key
Active volcanic zone
Pleistocene volcanics
Tertiary volcanics

0   100 km

Figure 15.10
A geologic map of Iceland shows that the oldest rocks are along the eastern and western margins, and the youngest are near the center of the island. The pattern is identical to that of rocks on either side of the oceanic ridge.

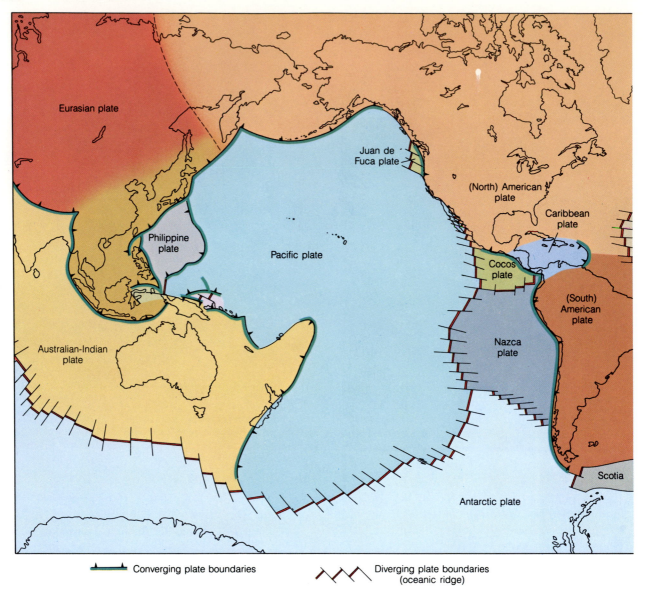

Converging plate boundaries

Diverging plate boundaries (oceanic ridge)

PLATE GEOGRAPHY

How does the geography of tectonic plates differ from classical physical geography?
Why is it more important geologically?

Statement

The shorelines of the continents are major geographic features but have little significance from the standpoint of Earth's tectonics. Plate boundaries are the most significant structural elements of the planet, and to understand plate tectonics, you must learn a new geography: the geography of plate boundaries. This should not be difficult because plate boundaries generally are marked by major topographic features. Understanding requires only that you focus your attention on the structural features of Earth, rather than on the boundaries between land and ocean.

Discussion

The new geography of tectonic plates is illustrated in Figure 15.11. The outer rigid layer of Earth—the lithosphere—is divided into a mosaic of seven major plates and a number of smaller subplates. The major plates are outlined by oceanic ridges, trenches, and young mountain systems. These include the Pacific, Eurasian, North American, South American, African,

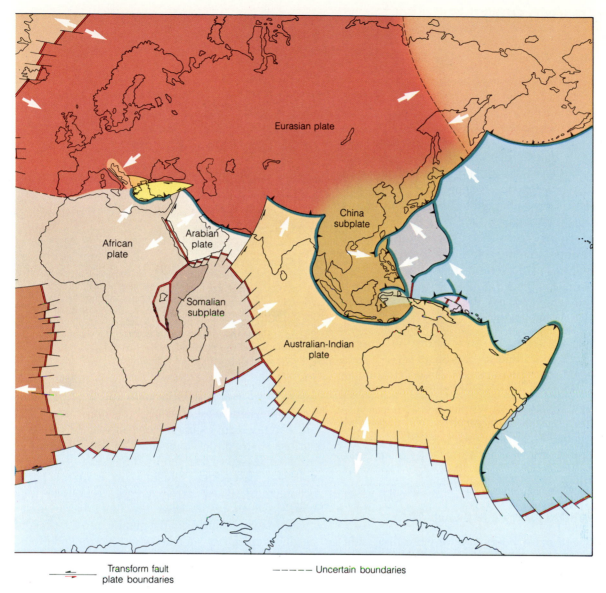

Transform fault plate boundaries

----- Uncertain boundaries

Figure 15.11
The major tectonic plates are delineated by the major tectonic features of the globe: (1) the oceanic ridge, (2) deep-sea trenches, and (3) young mountain belts. Plate boundaries are outlined by earthquake belts and volcanic activity. Most plates (such as the North American, African, and Australian) contain both continental and oceanic crust. The Pacific, Cocos, and Nazca plates contain predominantly oceanic crust.

Australian, and Antarctic plates. The largest is the Pacific plate, which is composed almost entirely of oceanic crust and covers about one-fifth of the Earth's surface. The other large plates contain both continental crust and oceanic crust. No major plate is composed entirely of continental crust. Smaller plates include the China, Philippine, Arabian, Juan de Fuca, Cocos, Caribbean, and Scotia plates, plus a number of others that have not yet been precisely defined. The smaller plates appear to form near convergent boundaries of major plates, where collisions between continents occur, or between a continent and an island arc.

The smaller plates are thus characterized by rapid and complex movement.

Individual plates are not permanent features. They are in constant motion and continually change in size and shape. Plates that do not contain continental crust can be completely consumed in a subduction zone. Plate margins are not fixed. A plate can change its shape by splitting along new lines, by welding itself to another plate, or by the addition of new oceanic crust along its passive margin. The movement and modification of a plate margin can change the size and shape of the entire plate.

PLATE BOUNDARIES

What major geologic processes occur at the three types of plate boundaries?

Statement

Each tectonic plate is rigid and moves as a single mechanical unit; that is, if one part moves, the entire plate moves. It can be warped or flexed slightly as it moves, but relatively little change occurs in the middle of a plate. Nearly all major tectonic activity occurs along the plate boundaries, and thus geologists and students of geology focus their attention on the plate margins, the ones that are active as well as the ancient plate boundaries preserved on the continents.

Three kinds of plate boundaries are recognized. These define three fundamental kinds of deformation and geologic activity (Figure 15.12):

1. Divergent plate boundaries are zones of tension, where plates split and spread apart.
2. Convergent plate boundaries (also called subduction zones) are zones where plates collide and one plate moves down into the mantle.
3. Transform fault boundaries are zones of shearing, where plates slide past each other without diverging or converging.

Discussion

Processes at Divergent Plate Boundaries

A divergent plate boundary, or spreading axis, forms where a plate splits and is pulled apart. Where a zone of spreading extends into a continent, rifting occurs, and the continent splits (Figure 15.13). The separate continental fragments drift apart with the diverging plates, so that a new and continually enlarging ocean basin is formed at the site of the initial rift zone. Divergent plate boundaries are thus characterized by tensional stresses that tend to pull the crust apart. This produces block faulting, fractures, and open fissures along the margins of the separating plates. Basaltic magma derived from the partial melting of the mantle is injected into the fissures or extruded as fissure eruptions. The magma then cools and becomes part of the moving plates. Divergent plate boundaries are some of the most active volcanic areas on Earth; however, they are generally characterized by unspectacular, quiet fissure eruptions, most of which are concealed beneath the sea. The importance of volcanism along this zone is underlined by the fact that more than half of the Earth's surface has been created by volcanic activity along divergent plate boundaries during the last 200 million years.

Except for a few rift zones in Africa and western North America, essentially all present divergent plate

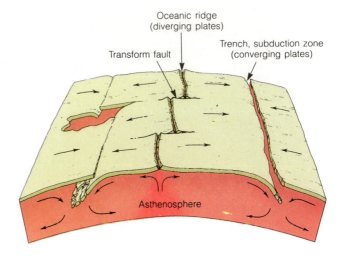

Figure 15.12
Types of plate margins are depicted in this idealized diagram. Constructive margins (divergent plate boundaries) occur along the oceanic ridge, where plates move apart. Destructive margins (convergent plate boundaries) occur along the deep trenches. Margins with no change in seafloor area during displacement occur along transform faults.

margins are submerged beneath the sea, and these regional characteristics cannot be observed. There are several clear examples of continental rifting in various stages of development. The initial stage (Figure 15.13A) is represented by the system of great rift valleys in East Africa. The long, linear valleys, partly occupied by lakes, are huge, downdropped fault blocks, which result from the crust being pulled apart. Magma rising from the mantle into the rift zone produces volcanoes such as Mount Kenya and Mount Kilimanjaro. The Red Sea illustrates a more advanced stage of rifting (Figure 15.13B). The Arabian Peninsula has been completely separated from Africa, and a new linear ocean basin is just beginning to develop. The Atlantic Ocean represents a still more advanced stage of continental drift and seafloor spreading (Figures 15.13C and D). The American continents have become separated from Africa and Europe by thousands of kilometers. The mid-Atlantic ridge is the boundary between the diverging plates, with the American plates moving westward, in relation to Africa and Europe.

Processes at Convergent Plate Boundaries

Convergent plate boundaries, or subduction zones, are areas of complicated geologic processes, including igneous activity, crustal deformation, and mountain building. The specific processes that are active along a convergent plate boundary depend on the types of crust involved in the collision of the converging plates.

If both plates at a convergent boundary contain oceanic crust, one is thrust under the margin of the

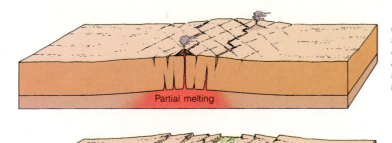

(A) Continental rifting begins when the crust is uparched and stretched, so that block faulting occurs. Continental sediment accumulates in the depressions of the downfaulted blocks, and basaltic magma is injected into the rift system. Flood basalt can be extruded over large areas of the rift zone during this phase.

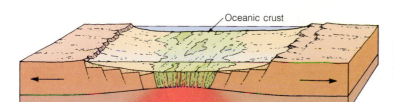

(B) Rifting continues and the continents separate enough for a narrow arm of the ocean to invade the rift zone. The injection of basaltic magma continues and begins to develop new oceanic crust.

(C) As the continents separate, new oceanic crust and new lithosphere are formed in the rift zone, and the ocean basin becomes wider. Remnants of continental sediment can be preserved in the downdropped blocks of the new continental margins.

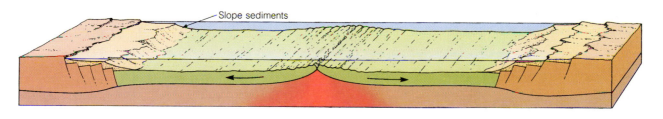

(D) As spreading continues, the ocean basin grows larger. The continents move off from the uparched spreading zone, and parts of the continental crust can be covered by the ocean.

Figure 15.13
Stages of continental rifting are shown in this series of diagrams. The major geologic processes at divergent plate boundaries are tensional stress, block faulting, and basaltic volcanism.

other in a process called subduction. The subducting plate descends into the asthenosphere, where it is heated and ultimately absorbed in the mantle. If one plate contains a continent, the lighter continental crust resists subduction and tends to override the ocean plate. If both converging plates contain continental crust, neither can subside into the mantle, although one can override the other for a short distance. Both continental masses are instead compressed, and the continents are ultimately "fused" or "welded" together into a single continental block, with a mountain range marking the line of suture.

The zone of convergence between two plates is a zone of deformation, mountain building, and metamorphism. If the overriding plate contains continental crust, compression deforms the margins into a folded mountain belt, and the deep roots of the mountains are metamorphosed.

The major processes and geologic phenomena that characterize convergent plate margins are shown in Figure 15.14. A subduction zone usually is marked by a deep-sea trench, and the movement of the descending plate generates an inclined zone of seismic activity. Magma is generated by partial melting of the descending plate and is usually andesitic or granitic in composition.

Some of the magma is extruded at the surface as lava and forms an island arc or a chain of volcanoes in the mountain belt of the overriding plate. Usually most of the magma intrudes into the deformed mountain belt to produce batholiths. Both extrusion and intrusion add new material to the continental plate, and thus continents grow by accretion. This is an important mechanism in the differentiation of Earth, whereby less dense material, enriched in elements such as Si, Al, K, and Na, is concentrated in the outer layers of the planet.

Examples of the three types of convergent plate boundaries include the west coast of South America,

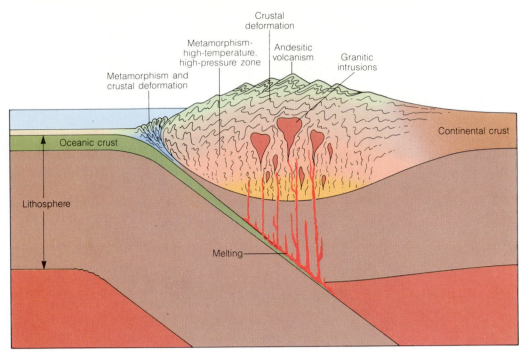

Crustal
deformation

Metamorphism-
high-temperature,
high-pressure zone

Andesitic
volcanism

Granitic
intrusions

Metamorphism and
crustal deformation

Continental crust

Oceanic crust

Lithosphere

Melting

Figure 15.14
The major geologic processes at convergent plate boundaries include the deformation of continental
margins into folded mountain belts, metamorphism due to high temperatures and high pressures in the
mountain roots, and partial melting of the descending plate, which produces granitic intrusions and
andesitic volcanism on the overriding plate.

the island arc of Japan, and the collision of the Australian plate with Asia that has formed the Himalaya Mountains.

Chapter 2 emphasized that continents are composed of low-density granitic material that rides passively on the moving plates. Except for the small amount of continentally derived sediment subducted back into the mantle, this material is so buoyant that it cannot sink into the denser mantle material below. It is simply not dense enough. The continents, therefore, move with the plates, sometimes colliding and sometimes splitting. They are much older than the ocean basins. Because continents are not consumed back into the mantle, they preserve records of plate movements in the early history of Earth—records in the form of ancient faults, old mountain belts, granitic batholiths, and sediment deposited along ancient continental margins.

Processes at Transform Fault Boundaries

The third type of plate boundary occurs where adjacent plates slide horizontally past each other, neither creating nor destroying the lithosphere. These boundaries occur along a special type of fault called a transform fault, which is simply a strike-slip fault between plates (that is, movement along it is horizontal and parallel to the fault). The term *transform* is used because the kind of motion between plates is changed—transformed—at the ends of the active

part of the fault. For example, the diverging motion between plates at an oceanic ridge can be transformed along the fault to the converging motion between plates at a subduction zone (Figure 15.15).

Transform faults connect convergent and divergent plate boundaries in various combinations (Figure 15.15). Where segments of the oceanic ridge have been fractured and displaced, a transform fault connects the two divergent plate boundaries and creates a major topographic feature called a fracture zone (see Figure 15.16). Fracture zones, however, are not what they might at first seem to be. The apparent offset of the oceanic ridge may suggest a simple strike-slip fault with displacements of thousands of kilometers, but one must keep in mind the relative motion of the plates at the spreading axis. Relative motion between the plates and seismic activity occurs only in the area between the offset segments of the ridge (Figure 15.16). This zone is the only place where the fault forms a boundary between the plates. Beyond this zone, the plates on either side of the fracture are moving in the same direction and at the same rate and can be considered to be linked. Note that the oceanic ridge is not being displaced by motion along the transform fault. It was displaced previously and may represent an old line of weakness in the rifted continental crust that preceded the development of oceanic crust. Moreover, major volcanic activity usually is not associated with transform faults. As the plates slide past each other, their boundaries are frac-

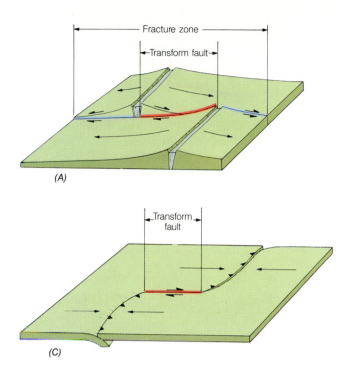

(A)

(B)

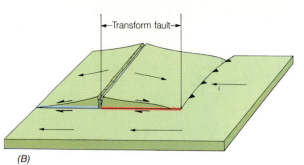

(C)

Figure 15.15
Transform faults can connect convergent and divergent plate boundaries in various combinations. Diagram A shows the common ridge-ridge transform fault. Note that relative motion occurs only along the boundary of the plates between the two segments of the ridge. Diagram B shows a ridge-trench transform fault, and Diagram C shows a trench-trench transform fault. In all cases the trend of a transform fault is parallel to the direction of relative motion between plates. This characteristic is helpful in determining the direction of plate motion.

tured and broken. This fracturing produces parallel ridges and troughs along the fault zone.

Transform faults can also join ridges to trenches and trenches to trenches. In all cases transform faults are parallel to the direction of relative plate motion, so that little divergence or convergence occurs along this type of boundary. The plates slide along the fracture system, and their movement produces only fracturing and seismic activity.

The direction of movement of the major plates in relation to their neighbors can be determined in several ways. As we have seen, the trends of the oceanic ridge and the associated fracture zones are related to plate motion. Indications of movement are also drawn from seismic data (see page 287), from relative ages of different regions of the seafloor, and from the ages of chains of volcanic islands and seamounts (see page 309). From these data, geologists have determined the motion of the present tectonic plates. This motion can be summarized using Figures 15.11 and 15.17.

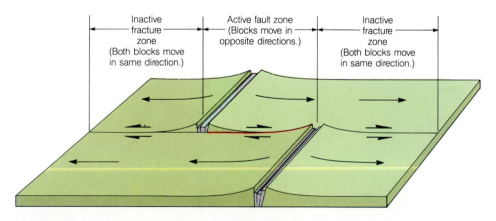

Figure 15.16
The relative movement of plates at a ridge-ridge transform fault changes along the trend of the fracture zone. The plates are moving away from the ridge, but the relative motion between the plates along the transform fault depends on the position of the spreading center. Along an active fault, between two segments of the ridge, plates on opposite sides of the fault move in opposite directions. Beyond the spreading centers, however, the plates move in the same direction on both sides of the fault, with no relative motion along the fault plane.

RATES OF PLATE MOTION

How fast is London moving away from New York?
How fast is New York moving away from Kansas City?
How do we know?

Statement

The velocity of movement of one plate relative to another can be measured most easily by dating the magnetic reversals on the seafloor, using the time scale of magnetic stratigraphy (page 263). The timing system is the oscillation of Earth's magnetic field. Earth's magnetic field has switched back and forth between "normal" polarity (as exists today) and reverse polarity, in which the north and south poles are reversed. The switching back and forth between these two states has been at irregular intervals, some as short as 20,000 years, others longer than ten million years. The pattern of this "irregularly ticking clock" is known and calibrated by radiometric dates. Thus the pattern of magnetic reversals in the rocks on the seafloor can be used to establish magnetic time lines (isochrons), which can be used like tree rings to date the age of the rocks on the seafloor.

Discussion

The pattern of magnetic reversals has been worked out for most of the seafloor and shows an incredible record of plate motion (Figure 15.9). As seen in Figure 15.9, magnetic time lines are symmetric around the ridge that formed them. The distance from the ridge axis to any isochron indicates the amount of seafloor created during that interval of time. Thus, the wider the band, the faster the rate of plate movement.

It is apparent from Figure 15.17 that the plates are moving at significantly different rates. The Pacific, Nazca, Cocos, and Indian plates are moving faster than the American, African, Eurasian, and Antarctic plates.

You will note from the data shown in Figure 15.17 that the fastest-moving plates are those in which a large part of the margins is being subducted, and that the slower-moving plates are those that lack subducting boundaries or that have large continental blocks imbedded in them. This has been interpreted by some geologists as evidence that the tectonic plates are part of the convection system of Earth and that plate motion is, to a large extent, a result of a cold, dense plate sinking into the mantle.

In addition to measuring plate velocities by magnetic reversals, current plate motion can be measured directly, using satellites and lasers. A narrow beam of light is emitted from an Earthbound laser and bounced off an orbiting satellite whose position is known very accurately. The light is collected at the surface of Earth again, and the elapsed time is determined. This allows the location of the laser to be determined to within a centimeter. If the locations of several such stations on different plates are repetitively determined, the relative motion between plates

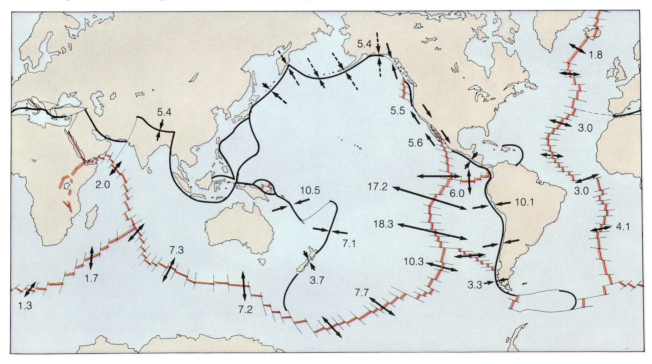

Figure 15.17
Relative velocities and directions of plate movement show how the major plates are presently interacting. The length of the arrows is proportional to the velocity of plate movement, and the numbers represent velocity in centimeters per year.

can be accurately measured. The velocities measured in this fashion correlate well with those measured by geological means.

MANTLE CONVECTION

What are the three major models explaining the driving mechanism of plate tectonics?
What do they all have in common?

Statement

Most scientists agree that some type of convection in the mantle is the fundamental process responsible for plate motion, but there are two major schools of thought about the nature of the convection and where it occurs:

1. Thermal convection cells within the mantle carry the plates like a conveyor belt. The plates move passively.
2. The plates themselves are an active part of the convecting process. They are not passive passengers.

Discussion

One of the first models to explain the driving mechanism of plate tectonics suggested that thermal convection cells within the mantle carried the plates and that the plates played little or no active part in the convection (Figure 15.18A). (You may want to refer again to Figure 3.10, page 39 for a discussion on thermal convection.) The rising limbs of the convecting cells in the mantle would therefore determine the position of the oceanic ridges. The convecting mantle would cause the lithosphere to split, and the moving mantle would carry the lithosphere laterally toward the subduction zone. The descending cell would mark the location of the trench and would drag the lithosphere down into the mantle. Movements in the asthenosphere were thought to be strongly coupled to the lithosphere. In other words, convection cells in the mantle supposedly cause ridges, trenches, and their effects. The distance between plate boundaries was thought to be caused by the size of the convection cell.

Another model of the thermal convection theory considers the plates themselves to be active participants in the convection process, not passive passengers on a churning mantle. In this model the lithosphere is considered to be the cold upper layer of the convection cell. Because of its greater density, the lithosphere tends to sink. Subduction occurs not because the plate is pulled down by the descending mantle, but because the plate is the dense sinking limb of the cell. If this is true, the plate may be driven by slab-pull (the gravitational pull of the plates them-

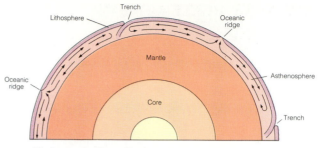

(A) Convection is confined to the upper asthenosphere.

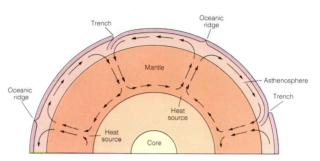

(B) Convection involves the entire mantle.

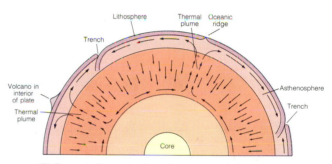

(C) Thermal plumes rise from the core-mantle boundary.

Figure 15.18
Three suggested models of convection in the mantle show how flow in the asthenosphere might move tectonic plates.

selves). Mantle convection may be driven by the plates and not vice versa. Thus, the plate is moved by forces largely independent of the mantle beneath.

Several examples demonstrate this process. If a large cauldron of molten metal is allowed to cool, a skin of solid metal forms on the surface and, because it is colder and more dense than the liquid, it founders and sinks into the molten liquid, thereby stirring the melt. The same process has been observed on a larger scale in lava lakes formed in the pit craters of Hawaiian volcanoes (Figure 15.19). As molten lava cools, a solid crust forms over the lake, but the crust splits into sheets and moves about. Because they are cooler and denser than the underlying liquid, large slabs of the crust break up, split, and sink. Molten lava rises from convection below and creates a zone of new cooling crust. Many features of plate tectonics are exhibited in Figure 15.19. Spreading ridges, trans-

Figure 15.19
Convection in a lava lake in Hawaii simulates convection and plate motion. As fresh molten lava rises by convection, slightly older, chilled lava (darker) is shoved aside to sink at some other (compressive) zone in the crater (out of view). Note the transform fault near the middle of the view and the differential rate of spreading revealed by different widths of recently chilled lava on either side of the spreading line. The front edge of the view is about 50 m wide.

form faults, and subduction zones are all observed. From this perspective it is more accurate to think of the plates and the mantle as forming a single, though complex, system, each affecting the other. In fact, however, convective flow in the mantle may have a radically different aspect than the motion of lithospheric plates seen at the surface of Earth.

The size and shape of the convection cells within the mantle are also a matter of considerable debate. The principle competing models are (1) layered mantle convection and (2) whole-mantle convection. There is evidence for and against each. The basic problem is that we do not understand enough about the Earth's interior. Our current theories are based on what is happening at the surface and what geophysical and geochemical studies tell us about the deep interior.

In the first model, two separate convecting layers of the mantle are envisioned. The upper layer is confined largely to the asthenosphere and lithosphere (Figure 15.18A). Slabs of lithosphere are known to penetrate to depths of 700 km before becoming undetectable by their seismic activity, so that a significant part of the upper mantle must be involved in convection. Below 700 km, the mantle is thought to convect independently of the upper mantle, and probably at a very slow rate. The principal evidence for this model is the geochemical distinctiveness of midoceanic-ridge basalts, which come from the upper mantle. The lower mantle has not been processed by plate tectonics and is rarely sampled by magmas that reach the surface. There is thought to be little transfer of material across the boundary between the upper and lower mantle, but heat readily conducts across it.

The second model considers convection to involve the entire mantle. Heat in the whole-mantle convection is supplied from the outer core. The major difference in these models is the size of the convection cells (Figure 15.18B).

Another variety of mantle convection involves the rise of jetlike plumes of low-density material from the core-mantle boundary region. These plumes, or hot spots, are considered in the following section.

Until we are able to probe the interior of Earth more effectively and discover ways of mapping the fine details of its internal structures, we will have more questions than answers concerning convection systems in the mantle.

MANTLE PLUMES

What features are explained by mantle plumes that are not explained by moving plates?

Statement

Although most tectonic activity is located along plate margins, a number of volcanic centers exist far away from plate boundaries. These areas have been referred to as hotspots and are believed to be surface manifestations of mantle plumes. Mantle plumes are long, narrow columns of hot mantle material that originate at depths greater than 700 km. The location of mantle plumes moves very, very slowly and can be considered, for all practical purposes, to be stationary and independent of plate motion above them. Thus, they supplement our knowledge of how plate tectonics works and they provide information about the deep mantle.

Discussion

The idea of hotspots was first developed from observations made in 1963 concerning the geology of the Hawaiian Islands. It was well known that the Hawaiian Islands grow progressively older in a line that begins at the southernmost island of Hawaii and runs northwestward.

The chain of active and recently extinct volcanoes is located atop a broad rise in the center of the Pacific plate. The island of Hawaii is an active shield volcano. Maui, an island about 50 miles to the northwest, is barely extinct. Farther to the west, the chain of islands becomes progressively older (Figure 15.20). Other linear chains of volcanic islands and seamounts in the Pacific, Atlantic, and Indian oceans show similar trends, in which an active or young volcano is at one end of the chain and the series of volcanoes becomes progressively older toward the other end.

Why is there such an orderly and obviously meaningful pattern of volcanic activity in the interior of plates? We are not sure, but the simple idea of mantle plumes has generally become accepted as part of the plate tectonic theory.

Mantle plumes are believed to be isolated, long, slender columns of hot rock that originate deep inside the Earth's mantle and rise slowly toward the surface, lifting the crust and forming volcanoes. They are thought to have diameters of 100–240 km and to rise at rates of perhaps 2 m per year. Some scientists think they originate at depths of at least 700 km, and perhaps at points as deep as the core-mantle boundary. Their position appears to be relatively stationary, so that the lithospheric plates drift over them. The plumes are thus independent of the major tectonic elements of the crust, which are produced by plate movement. They rise up under continents and oceans alike, in the center of plates and along the oceanic ridges. The extra heat they bring to the lithosphere commonly produces domes up to 1,000 km in diameter, with uplift ranging from 1 to 2 km.

It should be emphasized that this "plume model" is just that: a model. Plumes, of course, have not been observed directly, but the indirect evidence for them is substantial. Geochemical studies show that the basalt erupted from hotspots is different from the basalt that forms from the upper mantle at spreading cen-

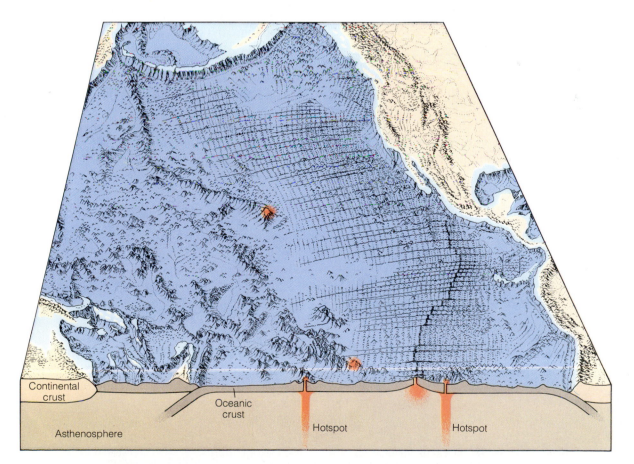

Figure 15.20
The motion of a plate over a fixed mantle plume produces a chain of volcanoes. The plumes originate deep in the mantle, and the string of volcanoes produced reveals the path of the moving plates.

ters. This suggests that the lavas are derived from well below the asthenosphere and is an important argument in favor of a layered mantle. In addition, satellite measurements of Earth's gravitational field have shown that hotspots are located on sites of anomalously high gravity and areas of excess mass. The excess mass can be attributed to the bulges produced by the upwelling plumes.

Like the theory of plate tectonics itself, the idea of mantle plumes is a simple, but powerful concept. It explains much of the geologic activity in the central parts of plates that never seemed to fit into any scheme of global tectonics. Further research will undoubtedly show that other geologic phenomena can be attributed to rising mantle plumes and their relation to moving plates.

SUMMARY

1. The theory of drifting continents was proposed in the early 1900s. It was developed by Alfred Wegener in his book *The Origin of Continents and Oceans*.

2. The theory was supported by various types of geologic evidence, including (a) matching geologic features on the margins of continents, (b) paleontologic findings, (c) paleozoic glaciation, and (d) other paleoclimatic evidence. Without an understanding of the oceanic crust, however, a complete theory of Earth's dynamics could not be developed.

3. During the 1960s new information about the topography and paleomagnetism of the ocean floor led to the development of the plate tectonic theory. Since then the theory has been supported by a wide variety of geologic and geophysical data from both the continents and oceans.

4. Paleomagnetic studies show that the continents have changed their position in relation to the magnetic poles and that each continent has followed a different path with respect to the poles throughout geologic time. These findings indicate that the continents have moved with respect to each other. Paleomagnetic reversals in the rocks of the seafloor occur in symmetrically matching sets on both sides of the oceanic ridge, a fact explained by seafloor spreading.

5. The theory of plate tectonics explains the Earth's crustal dynamics by the movement of rigid lithospheric plates. New lithosphere is created at oceanic ridges, where plates split and spread apart and basaltic magma wells up in the rift. Oceanic lithosphere is consumed where a plate descends into the mantle at a subduction zone.

6. The major structural features of Earth are formed along plate boundaries. At divergent plate boundaries, the lithosphere is under tension, and the major geologic processes are (a) rifting and block faulting, (b) the generation of basaltic magma and the formation of new oceanic lithosphere, and (c) the rifting of continents. At convergent plate boundaries, the important geologic processes include (a) subduction, (b) generation of andesitic to granitic magma, as well as (c) metamorphism and mountain building. Transform fault boundaries are zones where plates slide passively past each other.

7. Convection of the mantle with the active participation of oceanic lithosphere is probably the fundamental driving mechanism responsible for plate motion, but details about the type of convection heat remain unknown.

8. Rates of plate motion can be measured by dating patterns of magnetic reversals on the seafloor. Plates move at significantly different rates, ranging from less than 2 cm per year to more than 10 cm per year.

9. Mantle plumes are isolated, long, slender columns of hot rock that originate deep inside Earth's mantle. Rock within these plumes rises slowly toward the surface. The plumes are relatively fixed features, independent of the overlying moving plates. Where they approach the surface, they create hotspots of high heat flow, volcanic activity, and broad crustal upwarps. They explain many geologic features in the central parts of plates.

KEY TERMS

continental drift (p. 256)
convection cell (p. 273)
convergent plate
   boundary (p. 268)
divergent plate
   boundary (p. 268)

fracture zone (p. 270)
hotspots (p. 275)
magnetic anomaly (p. 262)
magnetic reversal (p. 262)

mantle plumes (p. 274)
normal polarity (p. 262)
Pangaea (p. 256)
paleomagnetism (p. 260)

polarity epoch (p. 262)
polarity event (p. 262)
reversed polarity (p. 262)
transform fault (p. 270)

REVIEW QUESTIONS

1. Briefly explain the theory of plate tectonics.
2. Describe the types of plate boundaries and give an example of each.
3. List the major processes that occur along each type of plate boundary.
4. Sketch a simple map of a part of the oceanic ridge and draw arrows to show the relative motion along ridge-to-ridge transform faults.
5. Explain the origin of the following features in the context of plate tectonics: (a) the Ural Mountains, (b) the Alps, (c) Iceland, (d) Hawaii, (e) the San Andreas Fault, (f) the Andes Mountains, and (g) volcanoes in Italy.
6. Draw a cross section showing a tectonic plate with divergent and convergent boundaries and label the major forces acting on the plate.
7. How fast are the plates moving? How do we determine rates of plate motion?
8. What are mantle plumes? Why do we believe they exist?

Multiple-Choice Questions

1. Paleoclimatic evidence of continental drift includes deposits of
   a. desert sandstone.
   b. rock salt.
   c. glacial deposits.
   d. gypsum.
   e. all of the above.
2. The youngest rocks of the oceanic crust are found
   a. in the abyssal depths of the oceans.
   b. along the oceanic ridge system.
   c. near continental margins.
   d. in the deep trenches.
   e. along fracture zones.
3. In addition to continental North America, the North American plate includes the
   a. western half of the north Atlantic Ocean crust.
   b. eastern half of the Pacific Ocean crust.
   c. Cocos plate.
   d. Nazca plate.
   e. east Pacific rise.
4. Which of the following represents the earliest stages of continental rifting?
   a. the Atlantic Ocean
   b. the valleys of East Africa
   c. the Red Sea
   d. the Aleutian Trench
   e. the North Sea
5. The zone of soft plastic rock beneath the lithosphere is called the
   a. stratosphere.
   b. thermosphere.
   c. asthenosphere.
   d. mantle.
   e. magnetosphere.
6. Which of the following indicates direction of plate motion?
   a. the trend of the oceanic ridge system
   b. the trend of fracture zones
   c. the trend of magnetic reversal
   d. all of the above
   e. only a and c

7. The movement of lithospheric plates is believed to be driven primarily by
   a. solar energy.
   b. gravity.
   c. isostasy.
   d. heat in the mantle.
   e. geomagnetism.
8. The Andes Mountains are believed to be the result of
   a. a divergence of lithospheric plates.
   b. magma welling up from the Peru-Chile trench.
   c. a volcanic arc.
   d. a convergence of lithospheric plates.
   e. processes at transform plate boundaries.
9. Which of the following is *not* a major process at divergent plate boundaries?
   a. normal faulting
   b. tensional stress
   c. eruption of andesitic magma
   d. shallow-focus earthquakes
   e. fissure eruptions of basalt
10. Which does *not* indicate the direction of plate movement?
    a. the trends of the oceanic ridge
    b. the ages of chains of volcanic islands and seamounts
    c. the trends of fracture zones
    d. the age of regions of the seafloor
    e. the trends of folds in seafloor sediment

True/False Questions

1. *T F* The continents on both sides of the Atlantic fit together, not only in outline, but in rock type and structure.
2. *T F* An important aspect of magnetic reversal patterns is that they enable us to determine the rate of plate movement.
3. *T F* Most of the deep-sea floor was formed before Cenozoic time.
4. *T F* Individual tectonic plates are permanent features. They are in constant motion but do not change in size or shape.
5. *T F* Volcanic activity is commonly associated with transform faults.

Fill in the Missing Terms

1. _____ plate boundaries are zones of tension where plates split and spread apart.
2. _____ plate boundaries are zones of compression where plates collide and one plate moves down into the mantle.
3. _____ fault boundaries are zones of shearing where plates slide past each other without diverging or converging.
4. A _____ zone is usually marked by a deep-sea trench, and the movement of the descending plate generates an inclined zone of seismic activity.
5. Volcanic activity is not associated with _____ plate boundaries.

16

EARTH'S SEISMICITY

Earthquakes, perhaps more than any other phenomenon, demonstrate that Earth continues to be a dynamic planet, changing each day by internal tectonic forces. Most earthquakes occur along plate boundaries. As the plates move, these boundaries (spreading centers, subduction zones, and transform faults) will be the sites of the most intense earthquake activity on Earth.

Earthquakes occur during sudden movements along faults. During long periods of slow deformation, elastic strain builds up between the rock bodies on opposite sides of a fault. Slip along the fault is prevented by friction until a threshold of strain is exceeded. Then the rocks snap past each other along the fault to release some of the stored energy.

Every year more than 150,000 earthquakes are recorded by the worldwide network of seismic stations and are analyzed with the aid of computers at the earthquake data center in Boulder, Colorado. With this network, the exact location, depth, and magnitude of all detectable earthquakes are plotted on regional maps. Information about the direction of fault movement associated with the shock is also determined. As a result, we can literally monitor the details of present plate motion. But that is not all. Seismic waves also provide our most effective probe of the Earth's interior, and they constitute the main method of collecting the data upon which we base our present concepts of the Earth's internal structure.

MAJOR CONCEPTS

1. Seismic waves are vibrations in the Earth caused by the rupture and sudden movement of rock.
2. Three types of seismic waves are produced by an earthquake shock: (a) P waves, (b) S waves, and (c) surface waves.
3. The primary effect of an earthquake is ground motion. Secondary effects include (a) landslides, (b) tsunamis, and (c) regional or local uplift or subsidence.
4. Most earthquakes occur along plate boundaries. Divergent plate boundaries and transform fault boundaries produce shallow-focus earthquakes. Convergent plate boundaries produce a zone of shallow-focus, intermediate-focus, and deep-focus earthquakes.
5. Infrequent shallow-focus earthquakes occur in the interior of the plates, away from the plate boundaries.
6. The velocities at which P waves and S waves travel through Earth indicate that Earth has a solid inner core, a liquid outer core, and a thick mantle with a soft asthenosphere and a rigid lithosphere.

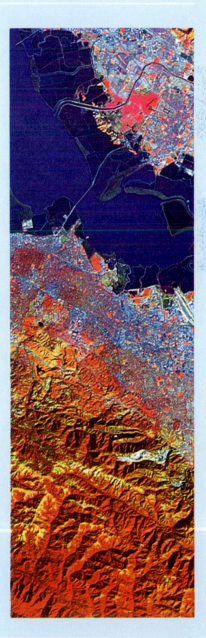

CHARACTERISTICS OF EARTHQUAKES

What causes earthquakes?

Statement

Earthquakes are vibrations of the Earth, caused by the rupture and sudden movement of rocks that have been strained beyond their elastic limits and by volcanic eruptions. If a strained rock breaks, it then snaps into a new position and, in the process of rebounding, generates vibrations called seismic waves. Three types of seismic waves are generated by an earthquake shock:

1. Primary waves (P waves): particles moving back and forth in the direction in which the wave travels
2. Secondary waves (S waves): particles moving back and forth at right angles to the direction in which the wave travels
3. Surface waves: waves that travel only in the Earth's outer layer and are similar to waves in water

Surface waves cause the most damage during an earthquake. Together with secondary effects from associated landslides, tsunamis, and fires, they result in the loss of approximately 10,000 lives and $100 million each year.

Discussion

Elastic-Rebound Theory

The origin of an earthquake can be illustrated by a simple experiment. Bend a stick until it snaps. Energy is stored in the elastic bending and is released when the stick breaks, causing the fractured ends to vibrate and send out sound waves. Detailed studies of active faults show that this model, known as the elastic-rebound theory, applies to all major earthquakes. Precision surveys across the San Andreas Fault in California show that railroads, fence lines, and streets are slowly deformed or bent out of shape. As strain builds up, they are broken and displaced when movement occurs along the fault releasing the elastic strain. The San Andreas Fault is the boundary between the Pacific and North American plates. Its movement is horizontal, with the Pacific plate moving toward the northwest. On a regional basis, the plates move quite steadily at a rate of roughly 5 cm per year. Along the fault, movement can be smooth or it can occur in a series of jerks because sections of the fault can be "locked" together until enough energy accumulates to cause the rocks to move past each other. The elastic-rebound theory explains earthquakes as the result of either rupture and sudden movement along a fault or recurrent movement along existing fractures.

The point within the Earth where the initial slippage generates earthquake energy is called the focus. The point on the Earth's surface directly above the focus is called the epicenter.

Types of Seismic Waves

Three major types of seismic waves are generated by an earthquake shock (Figure 16.1). Each type

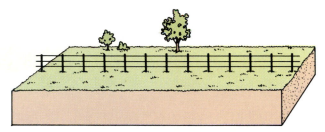

(A) Prior to seismic disturbance: *A straight fence line provides a good reference marker for future movement.*

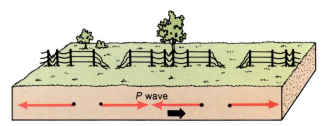

(B) Motion produced by a P wave: *Particles are compressed and then expanded in the line of wave progression. P waves can travel through any Earth material.*

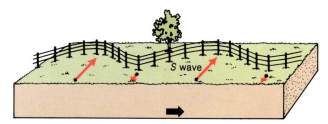

(C) Motion produced by an S wave: *Particles move back and forth at right angles to the line of wave progression. S waves travel only through solids.*

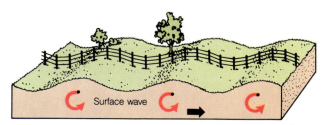

(D) Motion produced by a surface wave: *Particles move in a circular path at the surface. The motion diminishes with depth, like that produced by surface waves in the ocean (see Figure 13.2).*

Figure 16.1
Motion produced by the various types of seismic waves can be illustrated by the distortions they produce in a straight fence line.

travels through the Earth at a different speed, and each therefore arrives at a seismograph hundreds of kilometers away at a different time. The first waves to arrive are called primary waves (P waves). These are a kind of longitudinal wave, identical in character to sound waves passing through a liquid or gas. The particles involved in these waves move forward and backward in the direction of wave travel, causing relatively small displacements. The next waves to arrive are called secondary waves (S waves). In these, particles oscillate back and forth at right angles to the direction of wave travel. S waves cause strong movements to be recorded on a seismograph. The last waves to arrive are surface waves, which travel relatively slowly over the Earth's surface. Particles involved in surface waves move in an orbit similar to that of particles in water waves.

Occurrence of Earthquakes

The location of an earthquake focus is important in the study of plate tectonics because it indicates the depth at which the rupture and movement of rock bodies occur. Although movement of material within the Earth occurs throughout the mantle and core, it is only within the outer 700 km that the rocks are brittle enough to break and fracture and cause vibration or earthquakes. Earthquakes are thus concentrated in the upper 700 km.

Within the 700 km range, earthquakes can be grouped according to depth of focus. Shallow-focus earthquakes occur from the surface to a depth of 70 km. They occur in all seismic zones and are the largest percentage of earthquakes. Intermediate-focus earthquakes occur between 70 and 300 km below the surface, and deep-focus earthquakes between 300 and 700 km. Both intermediate-focus and deep-focus earthquakes are limited in number and distribution. Generally they are confined to convergent plate margins. The maximum energy released by an earthquake tends to become progressively smaller as the depth of focus increases. Also, seismic energy from a source deeper than 70 km is largely dissipated by the time it reaches the surface. Most large earthquakes therefore have a shallow focus, originating in the crust. The location of the focus of an earthquake is calculated from the time that elapses between the arrivals of the three major types of seismic waves.

The method of locating the epicenter of an earthquake is relatively simple and can be easily understood by reference to Figure 16.2. The P wave, traveling faster than the S wave, is the first to be recorded at the seismic station. The time interval between the arrival of the P wave and the arrival of the S wave is a function of the station's distance from the epicenter. By tabulating the travel times of P and S waves from earthquakes of known sources, seismologists have constructed time-distance graphs that can be used to

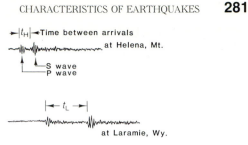

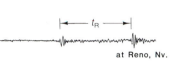

(A) The greater the distance between the seismic event and a recording seismograph station, the more time it takes for the first wave to arrive. Also, the greater the distance, the longer the interval between the arrival of the P waves and the arrival of the S waves.

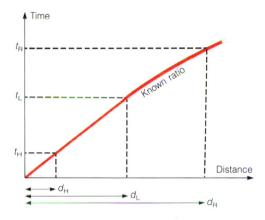

(B) The time between the arrival of the P waves and the arrival of the S waves is correlated with the distance between the seismic event and the recording station. For example, time at Helena, tH, yields distance from Helena, dH.

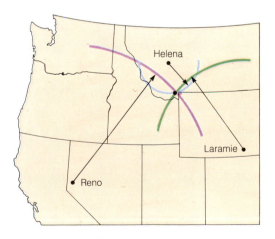

(C) The direction of the event from any single station is not known, but by simply plotting the intersection of three arcs that have radii the respective distances from the three stations, a common point is found. That point lies at the epicenter of the seismic event.

Figure 16.2
Locating the epicenter of an earthquake is accomplished by comparing the arrival times of P waves and S waves at three seismic stations.

determine the distance to the epicenter of a new quake. The seismic records indicate the distance, but not the direction, to the epicenter. Records from at least three stations are therefore necessary to determine the precise location of the epicenter.

Magnitude of Earthquakes

The magnitude of an earthquake is a measure of the amount of energy released. Earthquake magnitudes are based on direct measurements of the size (amplitude) of seismic waves, made with recording instruments, rather than on subjective observations of destruction. The total energy released by an earthquake can be calculated from the amplitude of the waves and the distance from the epicenter. Seismologists express magnitudes of earthquakes using the Richter scale, which arbitrarily assigns 0 to the lower limits of detection. Each step on the scale represents an increase in amplitude by a factor of 10 (Table 16.1). The vibrations of an earthquake with a magnitude of 2 are therefore 10 times greater in amplitude than those of an earthquake with a magnitude of 1, and the vibrations of an earthquake with a magnitude of 8 are one million times greater in amplitude than those of an earthquake with a magnitude of 2.

The largest earthquake ever recorded had a magnitude of approximately 8.8 on the Richter scale. Significantly larger earthquakes are not likely to occur because rocks are not strong enough to accumulate more energy.

Examples

The primary effect of earthquakes is the violent ground motion accompanying the movement of rock bodies along a fracture. This motion can collapse and destroy buildings, dams, tunnels, and other rigid structures (Figure 16.3). Secondary effects include landslides, tsunamis, and regional or local submergence of the land. The following are a few examples of well-documented earthquakes in historical times.

The earthquake that shook Peru on May 31, 1970, had a magnitude of 7.8. Claiming the lives of 50,000 people, it was the deadliest earthquake in Latin American history. Much of the damage to towns and villages was caused by the collapse of adobe buildings, which are easily destroyed by ground motion. Eighty percent of the adobe houses in an area of 65,000 km^2 were destroyed. Vibrations caused most of the destruction to buildings, but the massive landslides triggered by the quake in the Andes were a second major cause of fatalities. A huge debris flow buried 90 percent of the resort town of Yungay, with a population of 20,000, in a matter of a few minutes. Similar earthquake-triggered debris flows killed 41,000 in Ecuador and Peru in 1797 and, in 1939, killed 40,000 in Chile.

Table 16.1 Richter Scale of Earthquake Magnitude

Approximate Number	Magnitude per Year
1	700,000
2	300,000
3	300,000
4	50,000
5	6,000
6	800
7	120
8	20
>8	1 every few years

Another example of an earthquake at convergent plate margins occurred in the Andes, in Chile, in 1960, and caused extensive damage from ground motion, landslides, and flooding. It also produced a spectacular tsunami, which devastated seaports with a series of waves up to 7 m high. This tsunami crossed the Pacific at approximately 1,000 km per hour and built up to 11 m high at Hilo, Hawaii. Twenty-two hours after the quake, it reached Japan, causing $70 million in property damage.

The earthquake that devastated southern Alaska late in the afternoon of March 27, 1964, was one of the largest tectonic events of modern times. The magnitude was between 8.30 and 8.75, and its duration ranged from three to four minutes at the epicenter (by comparison, the San Francisco earthquake of 1906 lasted about one minute). The Alaskan earthquake was important to geologists because it had such marked effects at Earth's surface and because it was well documented. Despite its magnitude and severe effects, this quake caused far less property damage and loss of life than other national disasters (114 lives were lost and property worth $311 million was damaged) because, fortunately, much of the affected area was uninhabited. The crustal deformation associated with the Alaskan earthquake was the most extensive ever documented. The level of the land was changed in a zone 1,000 km long and 500 km wide (an area of 500,000 km^2 was elevated or depressed). Submarine and terrestrial landslides triggered by the earthquake caused spectacular damage to communities, and the shaking spontaneously liquified deltaic materials along the coast, causing slumping of the waterfronts at Valdez and Seward.

The Tangshan quake, which shook China on July 28, 1976, was probably the second most devastating earthquake in history. (The most destructive also occurred in China, in 1556.) The enormous shock registered 8.2 on the Richter scale. Although Chinese authorities have withheld many details about this disaster, scientists from Mexico and other countries were recently allowed to investigate the affected area. They say that Tangshan looked like Hiroshima after

(A) Earthquake of June 16, 1964, Niigata, Japan: *Liquification of the subsoil produced tilted apartment houses. Poor foundations were partly responsible for this type of damage. About one-third of the city subsided as much as 2 m as a result of sand compaction.*

(B) Earthquake of September 19, 1985, Mexico City, Mexico: *A 15-story reinforced concrete building collapsed.*

(C) Earthquake of February 9, 1971, San Fernando, California: *Southern Pacific railroad tracks near Los Angeles, California, were displaced laterally.*

Figure 16.3
The effects of historic earthquakes are dramatically displayed in the types of damage rendered to buildings and other structures.

the explosion of the atomic bomb. The devastation of the city of one million was complete. The Tangshan quake killed 750,000 people.

In addition to the descriptions of awesome damage at Tangshan, the reports of the phenomenon itself have been interesting. Just before the earth began to shudder, residents were awakened by a brilliant, incandescent light, which lit the early morning sky for hundreds of kilometers around. The glow reportedly was predominantly red and white. When the quake struck, many people were catapulted into the air, some as high as 2 m, by what was described as a violent, hammerlike subterranean blow. Trees and crops were blown over as if by a colossal steamroller, and leaves were burned to a crisp. In the quake's aftermath, many bushes were scorched on one side only.

EARTHQUAKE PREDICTION

Why is earthquake prediction so difficult?

Statement

Effective earthquake prediction might help minimize loss of life, yet pinpointing earthquakes, which seemed so close at hand during the 1960s, is proving to be an elusive goal.

Discussion

Chinese scientists claim to have been successful in predicting about fifteen earthquakes in recent years.

They rely heavily on the centuries-old idea that animals sense various underground changes prior to an earthquake and behave abnormally. Most of the bizarre behavior is simply increased restlessness. Cattle, sheep, and horses refuse to enter their corrals. Rats leave their hideouts and march fearlessly through houses. Shrimp crawl on dry land. Ants pick up their eggs and migrate en masse. Fish jump above the surface of the water, and rabbits hop aimlessly about.

Chinese scientists successfully predicted the Haicheng earthquake in February 1975 by means of unusual animal behavior. The most intriguingly bizarre behavior occurred in mid-December, when snakes came out of hibernation and froze to death on the icy ground, and groups of rats appeared and scurried about in the cold winter weather. These events were followed by a swarm of small earthquakes at the end of December. During January, Chinese scientists received thousands of reports of unusual animal behavior, especially in larger animals, in the area that proved to be the quake epicenter.

One hypothesis offered to explain abnormal animal behavior before an earthquake is that certain animals are sensitive to small variations in the Earth's magnetic field or to sounds produced by microfractures prior to the larger event. Animal sensors that detect light, sound, odor, touch, and temperature are well known, and they may have the ability to detect subtle changes in other physical phenomena.

Much of the work on earthquake prediction in the United States has been based on the dilatancy theory. Laboratory and field studies in recent years have indicated that a rock subjected to stress swells just before it ruptures. This dilation is caused by the opening and extension of numerous tiny cracks, and it begins at levels of stress that are about half as great as those needed to break the rock. As a rock dilates, changes occur in certain physical characteristics, such as electrical resistance, seismic wave velocities, and magnetic properties. Geologists therefore attempt to monitor uplift and tilting of the ground, electrical resistivity, the number of seismic events, and groundwater pressure.

For a prediction to benefit the populace, it must specify the time, the location, and the magnitude of the coming quake. Such accuracy is proving to be difficult to achieve. The problem is not so much that the dilatancy theory is wrong, but rather that it is inadequate. Different kinds of earthquakes apparently have different kinds of precursors. Instead of attempting to predict the time, place, and magnitude of an expected earthquake, geologists are now concentrating on the more modest goal of forecasting which areas of the world may be most susceptible to significant quakes. A major contribution to forecasting has been the compilation of a map showing the seismic potential of the world's major tectonic plate boundaries (Figure 16.4). This map essentially shows locations along the plate boundaries of the Pacific where major quakes are most likely to occur in the near future. Along the plate margins are several gaps in seismic activity, where stress may be building up to a critical level. The most susceptible areas are those where major tremors have occurred in the past but have not occurred within the last one hundred years. These include such heavily populated areas as southern California, central Japan, central Chile, Taiwan, and the west coast of Sumatra. These areas appear likely to experience a major earthquake (magnitude of 7.0 or greater) in the next few decades. One such seismic activity gap existed along the western coast of Mexico until a major quake struck the area on November 29, 1979.

If a reliable earthquake prediction system could give from one to ten years' advance warning of a "killer" quake, what would be the appropriate social response? Would the usual flow of mortgage money be terminated? New earthquake insurance would certainly become unavailable. What about fire insurance, business expansion, unemployment, tax revenues, and demands on local government? The fear that false alarms would lead to an adverse social response has led some to call for the withholding of predictions until prediction techniques are perfected to absolute certainty.

EARTHQUAKES AND PLATE TECTONICS

How is the distribution of earthquakes related to plate tectonics?

Statement

A worldwide network of sensitive seismic stations has been established to monitor nuclear testing. As a by-product, this network has enabled seismologists to compile an amazing amount of information concerning earthquakes and plate tectonics. With this network of seismic stations, the location and magnitude of the thousands of earthquakes occurring each year have been established together with other information, such as the direction of displacement along faults where earthquakes have occurred. The result is a new and important insight into the details of current plate motion. This is what has been found (see Figure 16.5):

1. The distribution of earthquakes delineates plate boundaries.
2. Shallow-focus earthquakes coincide with the crest of the oceanic ridge and with transform faults between ridge segments.

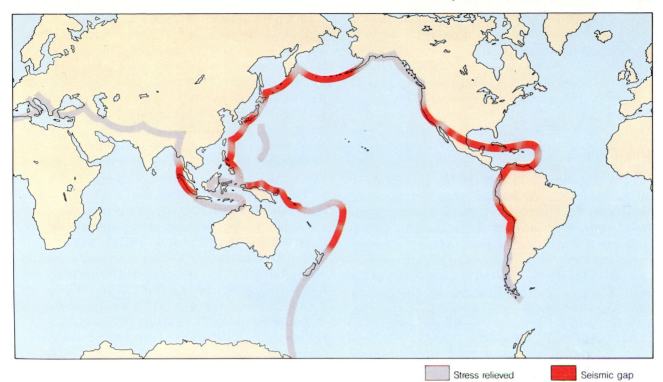

Stress relieved   Seismic gap

Figure 16.4
Seismic gaps are important in earthquake forecasting. Areas along plate margins that are not seismically active are believed to be building up stress and may be sites of significant future seismic activity. In the gray areas earthquakes have relieved strain within the last forty years, but in the red shaded areas, no large quakes have occurred and strain is still building.

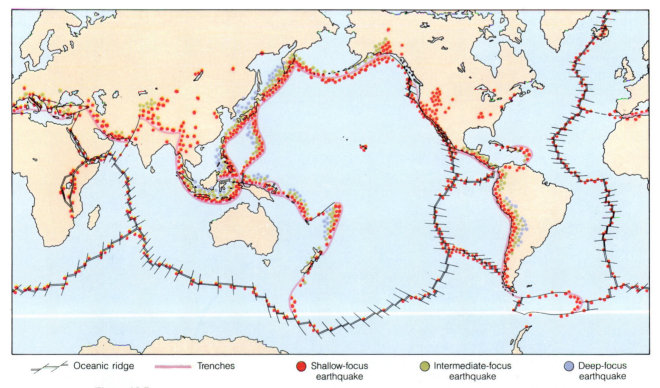

Oceanic ridge   Trenches   Shallow-focus earthquake   Intermediate-focus earthquake   Deep-focus earthquake

Figure 16.5
Earth's seismicity is clearly related to plate margins. This map shows the location of tens of thousands of earthquakes that occurred during a five-year period. Shallow-focus earthquakes occur at both divergent and convergent plate margins; intermediate-focus and deep-focus earthquakes are restricted to the subduction zones of converging plates.

3. Earthquakes at convergent plate margins occur in a zone inclined downward beneath the adjacent continent or island arc.
4. Fault motion associated with earthquakes along plate margins shows the present direction of plate motion.

Discussion

Global Patterns of Earth's Seismicity

Tens of thousands of earthquakes have been recorded since the establishment of a worldwide network of seismic observation stations. Their locations and depths are summarized in the seismicity map in Figure 16.5. From the standpoint of the Earth's dynamics, this map is an extremely significant compilation because it shows where and how the crust of the Earth is moving at the present time. This new insight into Earth's seismicity confirms, in a most remarkable way, the theory of plate tectonics because it shows the present-day movement of plates.

Seismicity at Divergent Plate Boundaries

The global patterns of Earth's seismicity show a narrow belt of shallow-focus earthquakes, coinciding almost precisely with the crest of the oceanic ridge and marking the boundaries between divergent plates. This zone is remarkably narrow in comparison to the zone of seismicity that follows the trends of young mountain belts and island arcs. The shallow earthquakes along divergent plate boundaries are less than 70 km deep and typically are small in magnitude. Although the zone looks like a nearly continuous line on regional maps, two types of seismic boundaries can be distinguished on the basis of fault motion. These are (1) spreading centers and (2) transform faults (Figure 16.6). Earthquakes associated with the crest of the oceanic ridge occur within or near the rift valley. They appear to be associated with normal faulting and intrusions of basaltic magmas. Locally earthquakes occur in swarms. Detailed studies indicate that earthquakes associated with the ridge crest are produced by vertical faulting, a process that appears to be responsible for the ridge topography.

Shallow-focus earthquakes also follow the transform faults that connect offset segments of the ridge crest, but they generally are not associated with volcanic activity. Studies of fault motion indicate horizontal displacement in a direction away from the ridge crest. Moreover, as is predicted by plate tectonic theory, earthquakes are restricted to the active transform fault zone—the area between ridge axes—and do not occur in inactive fracture zones.

Seismicity at Convergent Plate Boundaries

The most widespread and intense earthquake activity on Earth occurs along subduction zones at convergent plate boundaries. This belt of seismic activity is immediately apparent from the world seismicity map (Figure 16.5), which shows a strong concentration of shallow, intermediate, and deep earthquakes coinciding with the subduction zones of the Pacific Ocean. The three-dimensional distribution of earthquakes in this belt defines a seismic zone that is inclined at moderate to steep angles from the trenches and extends down under the adjacent island arcs or continental borders. This distribution is well illus-

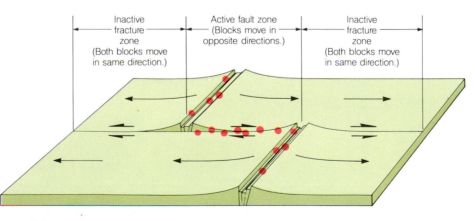

Figure 16.6
The distribution of earthquakes along divergent plate boundaries shows that seismicity along spreading centers results from normal faulting. Seismicity along transform faults results from strike-slip movement. No significant seismic activity occurs along the inactive fracture zone.

trated in the Tonga Trench in the South Pacific. The zone of seismicity there forms an inclined, nearly planar surface that plunges into the mantle to a depth of more than 600 km (see Figure 16.7).

Studies of fault motions from seismic waves generated in this zone indicate that the type of faulting varies with depth. Near the walls of the trench, normal faulting is typical, resulting from tensional stresses generated by the initial bending of the plate. In the zone of the shallow earthquakes, thrust faulting dominates as the descending lithosphere slides beneath the upper plate. At intermediate depths, extension or compression can occur, depending on the specific characteristics of the subduction zone. Extension and normal faulting result when a descending slab, which is denser than the surrounding mantle, sinks under its own weight. Compression results if the mantle resists the downward motion of the descending plate. The zone of deep earthquakes shows compression within the descending slab of lithosphere, indicating that the mantle material at that depth resists the movement of the descending plate.

Intraplate Seismicity

Although most of the world's seismicity occurs along plate boundaries, the continental platforms also experience infrequent and scattered shallow-focus earthquakes. The zones of seismicity of east Africa and the western United States are most striking. They are probably associated with spreading centers, which extend into those regions. The minor shallow earthquakes in the eastern United States and Australia are more difficult to explain. Apparently the lateral motion of a plate across the asthenosphere is accompanied by slight vertical movements. Built-up stress can exceed the strength of the rocks within the lithospheric plate, causing infrequent faulting and earthquakes. In contrast, the ocean floors beyond the spreading centers are seismically inactive, except for isolated earthquakes that result from sudden rupture of the rock caused by oceanic volcanoes.

Plate Motion as Determined from Seismicity

The data points in Figure 16.5 outline the seven major lithospheric plates. The present directions of plate movement can be deduced from the inclined zone of seismicity at deep-sea trenches and from the zone of shallow seismicity along the oceanic ridge. The Pacific plate, consisting almost entirely of oceanic crust, is moving northwestward, away from the spreading center along the eastern Pacific rise. The direction of movement along the convergent plate margins varies locally because the Pacific plate is bordered by several different plates (Cocos, Nazca, Antarctic, Australian, and so on). The width of the deep-focus earthquake zone indicates that the plate is descending into the mantle at an angle of roughly 45°. The slabs of lithosphere extending to depths of

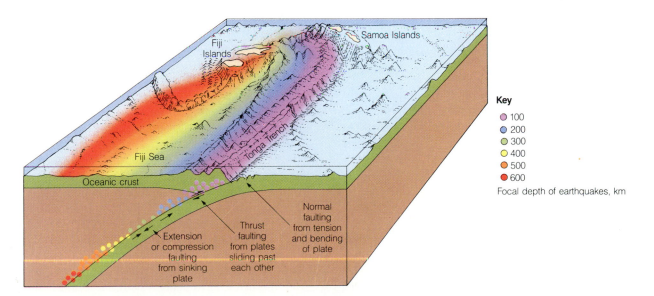

Key

- 🔵 100
- 🔵 200
- 🟢 300
- 🟡 400
- 🟠 500
- 🔴 600

Focal depth of earthquakes, km

Figure 16.7
Earthquake foci in the Tonga region of the South Pacific occur in a zone inclined from the Tonga Trench toward the Fiji Islands. The top of the diagram shows the aerial distribution of foci, with focal depths represented by different-colored dots. The cross section on the front of the diagram shows how the seismic zone is inclined away from the trench. This seismic zone accurately marks the boundary of the descending plate in the subduction zone.

700 km at convergent boundaries indicate that the present plate motion has continued long enough for at least 1,000 km of new crust to have been generated at the oceanic ridge.

The American plates move westward from the mid-Atlantic ridge, encountering the Pacific and adjacent plates along the trench on the West Coast. The African plate is moving north, toward the convergent boundary in the Mediterranean region, driven by spreading from the ridge, which essentially surrounds it in the Atlantic and Indian oceans. There are no subduction zones between the spreading centers of the Atlantic and Indian oceans; therefore, the plate boundaries surrounding Africa are apparently moving in relation to the African plate and to each other. The same is probably true of the Antarctic plate, which is completely surrounded by the spreading center of the oceanic ridge.

The Himalayas and the Tibetan Plateau define a wide belt of shallow earthquakes. In this area the convergence of plates has produced a collision of two continents. India moved in from the south and collided with Asia, which rode up and over the Indian plate to form a double thickness of continental rocks in the area. This convergence produced the wide zone of exceptionally high topography in the Himalayas and the Tibetan Plateau.

SEISMIC WAVES AS PROBES OF EARTH'S INTERIOR

How can seismic waves "X-ray" the internal structure of Earth?

Statement

Seismic waves passing through the Earth are refracted in ways that show distinct discontinuities in the material and structure of Earth's interior. They provide the basis for the belief that the Earth has a solid inner core, a liquid outer core, a soft asthenosphere, and a rigid lithosphere.

Discussion

Speculations about the interior of Earth have stimulated the imagination of humans for centuries. However, only after we learned how to use seismic waves to obtain an "X-ray" picture of Earth were we able to probe the deep interior of the Earth and formulate models of its structure and composition. Seismic waves—both P waves and S waves—travel faster through rigid material than through soft or plastic

material. The velocities of these waves traveling through a specific part of Earth thus give an indication of the type of rock there. Abrupt changes in seismic wave velocities indicate significant changes in the Earth's interior.

Seismic waves are similar in many respects to light waves, and their paths are governed by laws similar to those of optics. Both seismic waves and light waves move in a straight line through a homogeneous body. If they encounter a boundary between different substances, however, they are reflected or refracted. Familiar examples are light waves reflected from a mirror or refracted (bent) as they pass from air to water.

If Earth were a homogeneous solid, seismic waves would travel through it at a constant speed. A seismic ray (a line perpendicular to the wave front) would then be a straight line like those shown in Figure 16.8. Early investigations, however, found that seismic waves arrive progressively sooner than was expected at stations progressively farther from an earthquake's source. The rays arriving at a distant station travel deeper through Earth than those reaching stations closer to the epicenter. Obviously, then, if the travel times of long-distance waves are progressively shortened as they go deeper into Earth, they must travel more rapidly at depth than they do near the surface. The significant conclusion drawn from these studies is that Earth is not a homogeneous, uniform mass but has physical properties that change with depth. As a result, seismic rays are believed to follow curved paths through Earth (Figure 16.9).

In 1906 scientists recognized that whenever an earthquake occurs, there is a large region on the opposite side of the planet where the seismic waves are not detectable. To understand better the nature and significance of this shadow zone, refer to Figure 16.10. For an earthquake at a particular spot (labeled "0°"), the shadow zone for P waves invariably exists between 103 and 143° from the earthquake's focus. Evidently something deflects the waves from a linear path. The best explanation for this shadow zone is that Earth has a core through which P waves travel relatively slowly. Seismic rays traveling through the mantle follow a curved path from the earthquake's focus and emerge at the surface between 0 and 103° from the focus (slightly more than a quarter of the distance around the Earth). In Figure 16.10, ray 1 just misses the core and is received by a station located 103° from the focus. Ray 2, however, being steeper than ray 1, encounters the core's boundary where it is refracted. It travels through the core, is refracted again at the core's boundary, and is finally received at a station on the opposite side of Earth. Ray 3 is similarly refracted and emerges on the opposite side, 143° from the focus. Other rays that are steeper than ray 1 are also refracted through the core

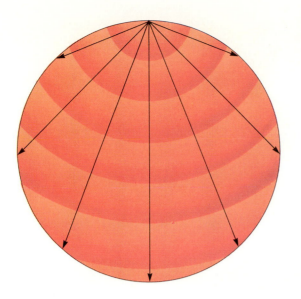

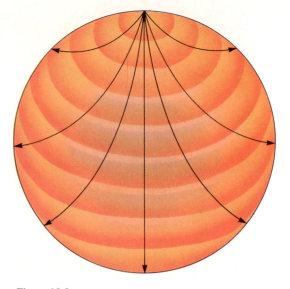

Figure 16.8
Seismic waves in a homogeneous planet would be neither reflected nor refracted. Lines drawn perpendicular to the wave fronts (rays) would follow a linear path.

Figure 16.9
Seismic waves in a differentiated planet would pass through material that gradually increases in density with depth. As a result, wave velocities would increase steadily with depth, and rays would follow a curved path.

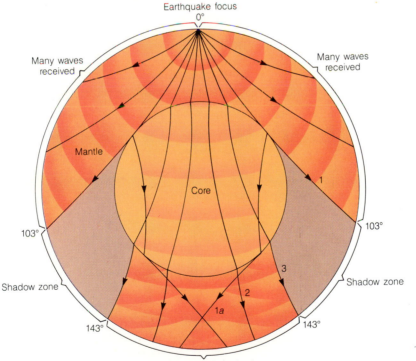

Figure 16.10
The shadow zone of P waves occurs in the area between 103 and 143° from an earthquake's focus. The best way to explain the P wave shadow zone is to postulate that Earth has a central core through which P waves travel relatively slowly. Ray 1 just misses the core and is received at a station located 103° from the earthquake's focus. A steeper ray, such as ray 2, encounters the boundary of the core and is refracted. It travels through the core, is refracted again at the core's boundary, and is received at a station less than 180° from the focus. Similarly, ray 3 is refracted and emerges at the surface 143° from the focus. Other rays that are steeper than ray 1 are severely bent by the core, so that no P waves are directly received in the shadow zone. From shadow zones, seismologists calculate that the boundary of the core is 2,900 km below the surface.

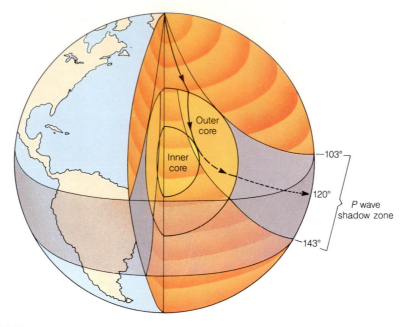

Figure 16.11
P waves are deflected by the inner core and are received in the shadow zone as weak indirect signals. This deflection suggests that the inner core is solid.

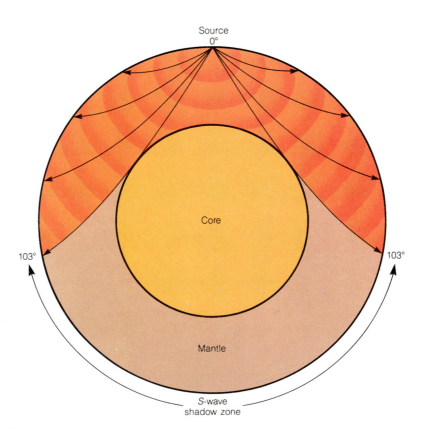

Figure 16.12
The shadow zone of S waves extends almost halfway around Earth, opposite the earthquake's focus. This can be explained if the outer core of Earth is liquid. Because S waves cannot travel through liquid, they do not pass through the core.

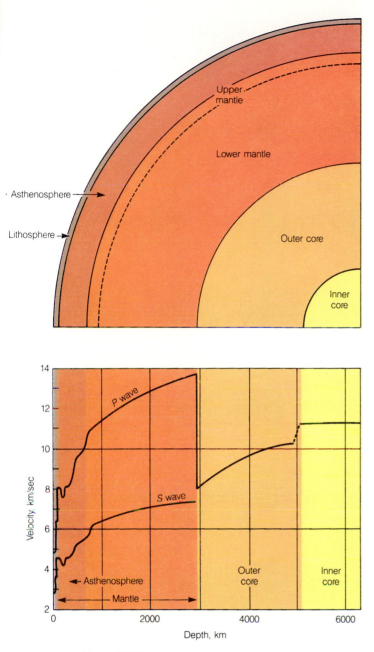

Figure 16.13
The internal structure of Earth is deduced from variations in the velocity of seismic waves at depth. The velocity of both P and S waves increases until they reach a depth of approximately 100 km. There the waves abruptly slow down, and their velocities continue to decrease until they have traveled to a depth of 200 km. This low-velocity layer is called the asthenosphere (see Figure 16.12). Below 200 km beneath the surface, the velocity of P and S waves increases, until the waves reach a depth of about 2,900 km, where both velocities change abruptly. S waves do not travel through the central part of Earth, and the velocity of the P waves decreases drastically. This variation is the most striking discontinuity and is considered to signify the boundary between the core and the mantle. Another discontinuity in P wave velocity, at a depth of 5,000 km, indicates the surface of the inner core.

and emerge between 143 and 180° from the focus. Refraction at the boundary between the core and the mantle thus causes a P wave shadow zone over part of the Earth's surface—that is, a zone where no P waves are received. The shadow zone extends as a band across the planet.

In 1914 Beno Gutenberg, a German seismologist, calculated that the depth to the surface of the core is 2,900 km. Later analysis of more numerous and more reliable seismic data showed that Gutenberg's original estimate was remarkably accurate, with a probable error of less than two-thirds of 1 percent. More recent studies of the P wave shadow zone show that some weak P waves of low amplitude are received in this zone. This suggests the presence of a solid inner core, which deflects the deep, penetrating P waves in the manner shown in Figure 16.11.

The surface of the core has an even more pronounced effect on S waves, but this effect cannot be explained by reflection or refraction. S waves simply do not pass through the core at all. They produce a huge shadow zone extending almost halfway around the Earth, opposite the earthquake's focus (Figure 16.12). One difference between P and S waves is particularly significant. P waves pass through any substance—solid, liquid, or gas. S waves, however, are transmitted only through solids that have enough elastic strength to return to their former shape after being distorted by the wave motion. They cannot be transmitted through a liquid. The fact that S waves will not travel through the core, therefore, is generally taken as evidence that the outer core is liquid.

With the present worldwide network of recording stations, even minor variations in seismic wave velocities, called seismic discontinuities, can be determined with considerable accuracy. These data, summarized graphically in velocity-depth curves like the one in Figure 16.13, provide additional information about the Earth's interior. The most striking variation occurs at the core's boundary, at a depth of 2,900 km. There, S waves stop, and the velocity of P waves is drastically reduced.

Other discontinuities are apparent but are less striking. The first occurs between 5 and 70 km below the surface. This is called the Mohorovicic discontinuity (or simply Moho), after Adrija Mohorovicic, the Yugoslavian seismologist who first recognized it. The discontinuity is considered to represent the base of the crust.

Perhaps the most significant discontinuity, however, is the low-velocity zone from 100 to 200 km below the surface (Figure 16.14). Gutenberg recognized this zone in the 1920s, but the discovery was viewed with skepticism by most other seismologists at that time. The normal trend is for seismic wave velocities to increase with depth. In this low-velocity zone,

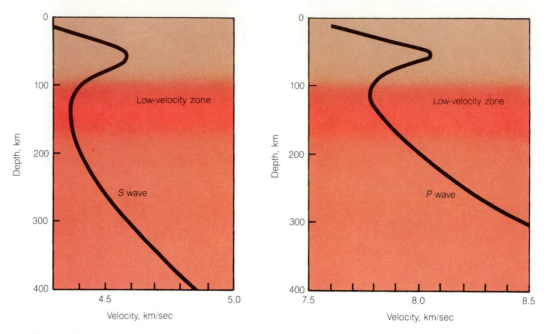

Figure 16.14
The rapid decrease in velocities of both S waves and P waves suggests a zone of low strength in the upper mantle, at depths between 100 and 200 km below the surface. This zone, the asthenosphere, is commonly referred to as the low-velocity layer of the upper mantle.

however, the trend is reversed, and seismic waves travel about 6 percent more slowly than they do in adjacent regions. More recent seismic data again confirmed Gutenberg's observations. Most seismologists are now convinced that this zone is more plastic than the areas above and below it and that it is capable of slow flow under long-term stress.

The generally accepted explanation for the low seismic wave velocities in this zone is that temperature and pressure cause part of the material, perhaps from 1 to 10 percent, to melt, so that a crystal-liquid mixture is produced. A small amount of liquid film around the mineral grains serves as a lubricant, increasing the plastic nature of the material. This low-velocity zone, as we will see in later chapters, plays a key role in the theories of motion of the material at the Earth's surface.

Other rapid changes in seismic wave velocities occur in the mantle at depths of about 350 km and 700 km. These are interpreted as the results of phase changes in the minerals, in which the atomic packing is rearranged in denser and more compact units.

Seismic Tomography

In the last few years, scientists have used a new analytic technique called seismic tomography, which promises to enhance greatly our knowledge of the Earth's internal structure, including the pattern of flow in the mantle.

Seismic tomography is like its medical analogue, the CAT scan (computer-assisted tomography; tomograph is based on a Greek word, *tomos,* meaning "section"). In a CAT scan, X-rays that penetrate the body from all directions are used to construct an image of a slice (cross section) through the body. Bones, organs, and tumors are identified because they have different densities and absorb X-rays differently. With the aid of a computer, these images are stacked side by side to produce a three-dimensional view. In seismic tomography seismic waves are used like X-rays. Seismologists analyze the velocity of hundreds of thousands of seismic waves as they pass through Earth in different directions. The results are images showing regions where the waves travel faster or slower than normal (Figure 16.15). Geophysicists know from laboratory studies and from observations near volcanoes that seismic waves are slowed down by unusually hot rock and are speeded up by cooler rock. The tomographs can thus be interpreted as temperature maps. The next step is to assume that hot parts of the mantle, being less dense than their surroundings, will rise, whereas the cool mantle rock will sink. Thus, the tomograph can be used to outline the patterns of flow in the mantle.

The three-dimensional view of the mantle obtained

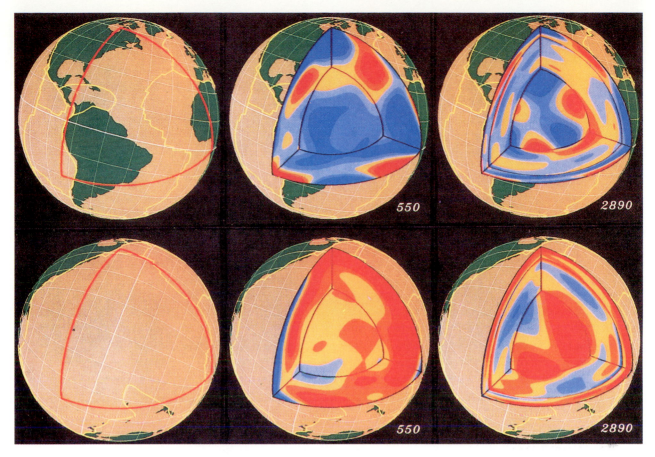

Figure 16.15
The internal structure of Earth is seen here by seismic tomography. This illustration consists of three-dimensional plots of velocity anomalies under two oceans: the Atlantic (top) and the Pacific (bottom). Depth to the base of the section is indicated by kilometers. The middle diagrams (550 km) show only the upper mantle. Scale is exaggerated by a factor of 5. The diagrams to the right show the structure of the Earth's core-mantle boundary, 2,890 km below the surface.

from seismic tomography confirms what one might expect from the plate tectonic theory. At a depth of 150 km, slow seismic zones occur under most of the volcanic regions, including the midocean ridges. In contrast, the shields of Canada, Brazil, Siberia, Africa, and Australia are all fast. At depths of 350 km, the midocean ridge system is no longer continuous but is broken up into isolated segments. At depths of 550 km, there is even less relation between mantle and surface features. This indicates that the midocean ridge system is not simply the surface expression of vertical upwelling currents. Instead it must be fed by the lateral transport of hot material from a few high-temperature zones in the upper mantle.

Seismic tomography is also providing a more sophisticated view of the core. Early maps of the core-mantle boundary show that the surface of the core is not smooth; it is marked by broad swells and depres-

sions with a difference in height of up to 20 km. The presence of any kind of topography on the core-mantle boundary bears on a number of fundamental geologic questions. A rough boundary would presumably disturb the flow of the liquid iron in the outer core, much like a mountain influences the flow pattern of winds. The topography of the core also has important implications on the flow of energy in the Earth. A rough core boundary would imply that a lot of energy escapes the core as heat, which might be the source of heat for convection in the mantle and the resulting tectonic plates.

Many other basic geologic questions must look to the Earth's interior for answers. Projects planned for the future include increasing the number of seismic stations so that we might obtain sharper global images of the mantle and the hidden flow that shapes the surface of Earth.

SUMMARY

1. Earthquakes are vibrations in the earth caused by the sudden rupture of rocks that have been strained beyond their elastic limits.
2. Three types of seismic waves are generated by an earthquake shock: P waves, S waves, and surface waves.
3. The magnitude of an earthquake is a measure of the total energy released.
4. The primary effect of earthquakes is ground motion. Secondary effects include (a) landslides, (b) tsunamis, and (c) regional uplift and subsidence.
5. The establishment of a worldwide network of sensitive seismic stations has enabled seismologists to monitor the thousands of earthquakes that occur each year. This new insight into Earth's seismicity confirms, in a most remarkable way, the theory of plate tectonics, because it shows the present-day movement of plates.
6. The distribution pattern of earthquakes dramatically outlines the plate margins.
7. Shallow earthquakes develop in a narrow zone along divergent plate margins, where the mantle rises and pulls the plate apart.
8. Where plates converge and one is thrust under the other, a zone of shallow, intermediate, and deep earthquakes is produced as the oceanic plate moves down into the mantle.
9. In the central parts of the plates there is little differential movement. Few earthquakes occur in these stable areas.
10. The velocities of P and S waves through the Earth indicate that it has a solid inner core, a liquid outer core, a thick mantle, a soft asthenosphere, and a rigid lithosphere.

KEY TERMS

deep-focus earthquake (p. 281)

earthquake (p. 280)

earthquake magnitude (p. 282)

elastic-rebound theory (p. 280)

epicenter (p. 281)

focus (p. 281)

intermediate-focus earthquake (p. 281)

P wave (primary wave) (p. 281)

S wave (secondary wave) (p. 281)

seismic discontinuity (p. 288)

seismic ray (p. 288)

seismic tomography (p. 293)

seismic wave (p. 280)

seismograph (p. 281)

shadow zone (p. 288)

shallow-focus earthquake (p. 281)

surface wave (p. 280)

REVIEW QUESTIONS

1. Explain the elastic-rebound theory of the origin of earthquakes.
2. Describe the motion and velocity of the three major types of seismic waves.
3. Explain how the location of the epicenter of an earthquake is determined.
4. What secondary effects commonly accompany earthquakes?
5. Describe the global pattern of earthquakes.
6. How does the depth of earthquakes indicate (a) convergent plate margins and (b) divergent plate margins?
7. Draw a diagram showing the paths that would be followed by seismic rays through the Earth if its core were only half the size shown in Figure 16.9.
8. What do seismic velocity-depth diagrams tell us about the internal structure of the Earth?

Multiple-Choice Questions

1. Which of the following is associated with most of Earth's major earthquakes?
   a. folding of rock sequences
   b. faulting of rock sequences
   c. igneous intrusions
   d. tides
   e. Earth's rotation

2. The epicenter of an earthquake is
   a. the shape of the wave front.
   b. a point directly above the focus.
   c. the point of initial movement.
   d. the direction in which the wave front moves.
   e. the direction of the earthquake's center.

3. The focus of an earthquake is
   a. the shape of the wave front.
   b. a point directly above the epicenter.
   c. the point of initial movement.
   d. the direction in which the wave front moves.
   e. the direction toward the earthquake's center.

4. The distribution of earthquakes on the planet Earth
   a. delineates plate boundaries.
   b. occurs parallel to all folded mountain belts.
   c. is at random throughout the interior of plates.
   d. is restricted to areas of fault zones in the shield.
   e. delineates continental margins.

5. A liquid core in the Earth is indicated by the fact that
   a. surface waves cannot pass through the core.
   b. P waves do not travel through a liquid and do not travel through the core.
   c. S waves will not travel through a liquid and do not travel through the core.
   d. all of the above.
   e. both a and b are correct.

6. In P (primary) waves, the particles in a rock body move
   a. back and forth at right angles to the direction in which the wave travels.
   b. back and forth parallel to the direction in which the wave travels.
   c. in a circle like sea waves.
   d. in an elliptical pattern.
   e. at a 45° angle to the direction the wave travels.

7. Which seismic wave cannot move through a liquid?
   a. P waves
   b. L waves
   c. S waves
   d. X waves
   e. T waves

8. If the interior of a planet is homogeneous, the paths followed by seismic waves will
   a. curve at depth.
   b. be linear.
   c. be reflected from the core.
   d. cause a shadow on the opposite side of the planet.
   e. be both reflected and refracted.

9. The surface of Earth's core is indicated by the shadow zone for
   a. surface waves.
   b. P waves.
   c. S waves.
   d. all of the above.
   e. both P and S waves.

10. A liquid core in Earth is indicated by the fact that
    a. surface waves cannot pass through the core.
    b. P waves do not travel through a liquid and do not travel through the core.
    c. S waves will not travel through a liquid and do not travel through the core.
    d. all of the above are true.
    e. both a and b are correct.

True/False Questions

1. *T F* Intermediate and deep-focus earthquakes are confined to convergent plate margins.
2. *T F* The areas where earthquakes are most likely in the near future are those areas where major tremors have occurred in the past but have not occurred within the last one hundred years.
3. *T F* The worldwide distribution of earthquakes delineates plate boundaries.
4. *T F* Abrupt changes in seismic wave velocity indicate significant changes in the structure and composition of the Earth's interior.
5. *T F* S waves simply do not pass through Earth's core at all.

Fill in the Missing Terms

1. The point on Earth's surface directly above the earthquake focus is called the _____.
2. The _____ of an earthquake is a measure of the amount of energy released.
3. Intermediate- and deep-focus earthquakes generally occur at _____ plate margins.
4. _____-focus earthquakes coincide with the crest of the oceanic ridge and with transform faults between ridge segments.
5. _____ waves do not pass through a liquid.

17

VOLCANISM

Volcanic eruptions are one of the most spectacular of all geologic phenomena, and for centuries they have caused dismay and terror to people who live nearby them. Although they have always attracted attention because of their sometimes violent and catastrophic destruction, the significance of volcanic eruptions was only partly understood. It is now clear that volcanism is a key process in Earth's dynamics and not a rare or abnormal event. It has occurred on Earth throughout most of geologic history and undoubtedly will continue far into the future.

To appreciate the importance of volcanic activity in the dynamics of Earth, consider the volume of rock it produces. More than two-thirds of the face of Earth—the ocean floors—consist entirely of rocks derived from lava during the last 200 million years. Volcanic activity is also the most significant process in building island arcs. In addition, not only is volcanism important in many mountain chains, but vast floods of lava have constructed large lava fields along the margins of many continents.

Volcanic activity is significant for more than just the quantity of lava it produces. It is closely associated with the movements of tectonic plates, movements that also produce earthquakes, ocean basins, continents, and mountain belts. Moreover, volcanism provides a window into the Earth's interior, giving us tangible evidence of processes operating far below the planet's surface.

MAJOR CONCEPTS

1. Most volcanism corresponds to active seismic zones and is clearly associated with plate boundaries.
2. The type of volcanic activity depends on the type of plate boundary.
3. Basaltic magma is generated at divergent plate boundaries by the partial melting of the asthenosphere. It is extruded mostly in quiet fissure eruptions.
4. Granitic magma is generated at convergent plate margins by the partial melting of the oceanic crust.
5. Volcanic activity within plates is probably caused by local hotspots in the upper mantle.

GLOBAL PATTERNS OF VOLCANISM

What determines the location of active volcanoes on Earth?

Statement

Volcanic eruptions occur in many areas of the world, including the islands and floors of the oceans and the young mountain ranges and plateaus of the continents. The distribution of volcanic activity is not random, however. If the locations of active or recently active volcanoes are plotted on a map that shows tectonic plates, two important facts stand out:

1. Most volcanic activity coincides with the active seismic regions of the world and is clearly associated with plate boundaries.
2. The type of volcanic activity depends on the type of plate boundary.

Discussion

The distribution of recently active volcanoes and their relationships with the major plate boundaries are shown in Figure 17.1. A worldwide belt of volcanic activity occurs along divergent plate margins, but it is largely concealed beneath the ocean. Locally enough lava is sometimes extruded along this zone to build a volcanic pile that rises above sea level. Iceland, the best example, is built entirely of volcanic rock and continues to be volcanically active. Surtsey, a new island off Iceland's southern coast, was built by eruptions that started in 1963. Ten years later, another volcano suddenly erupted on the tiny coastal island of Heimaey and buried much of Iceland's largest fishing port. More volcanism in this area can be expected in the future. South of Iceland, only a few submarine volcanoes rise above sea level, but submarine eruptions of gas and ash rising above the surface of the water have been seen by sailors in the area of the mid-Atlantic Ridge. Other volcanoes located along divergent plate margins include those associated with the East African rift valleys and perhaps those along the margins of the Basin and Range province in the western United States.

The most notable belt of volcanic activity occurs along subduction zones and coincides with the intense seismicity that also occurs there. A continuous belt of volcanism practically surrounds the Pacific Basin and has long been referred to as the Ring of Fire. The volcanoes in this belt are unquestionably among the Earth's great physical features. Their distribution is paralleled by the subduction zones of the

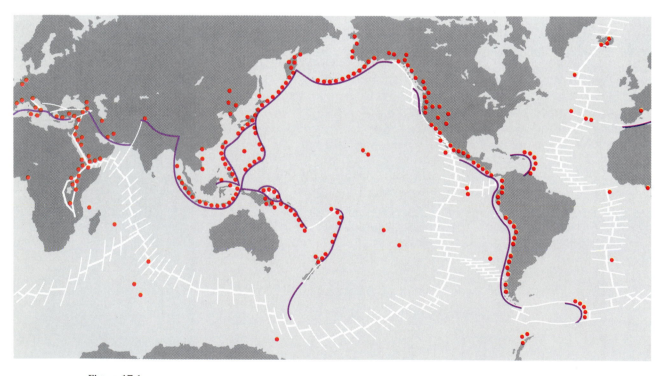

Figure 17.1
Active volcanoes are concentrated along plate boundaries. Basaltic volcanism and shallow intrusions occur all along the oceanic ridge, where plates are moving apart. Except for Iceland and a few small islands, however, the volcanic activity at spreading centers is concealed beneath the ocean. The most conspicuous volcanic activity occurs in the chains of andesitic volcanoes that form over subduction zones, at convergent plate margins. These include the "Ring of Fire" around the Pacific Ocean and the volcanoes of the Mediterranean and the Near East. Intraplate volcanism occurs mostly in the Pacific, where the movement of plates over hotspots in the mantle produces volcanic islands and seamounts.

three major plates that make up the Pacific Basin and the associated minor ones (such as the Caribbean and Philippine plates). Another similar belt of volcanic activity follows the convergent margin of the African Plate. It extends through southern Europe to the Middle East. Volcanic activity along the Australian Plate boundary is concentrated in the island arcs of Indonesia.

Although most of the world's famous volcanoes are located along plate margins and correspond to the belt of intense seismic activity, some volcanoes occur in the middle of tectonic plates. Most of these are in the Pacific.

VOLCANISM AT DIVERGENT PLATE MARGINS

Why is volcanic activity at divergent plate margins different from that at convergent margins?

Statement

Direct observations of the seafloor from submarines, together with geologic studies in areas where parts of the diverging plates are elevated above sea level, indicate that most volcanic activity along divergent plate margins takes the form of quiet fissure eruptions of basaltic magma, and that such eruptions have occurred there throughout all of geologic time. Magma extruded along divergent plate boundaries is believed to be generated by the partial melting of the mantle.

Discussion

The Generation of Magma at Divergent Plate Margins

Why and how is magma generated along spreading centers rather than in some other place? The answer lies in the special characteristics of temperature and pressure in the asthenosphere and in their relationship to the melting of minerals in the mantle. The generation of magma along divergent plate margins is not the result of high temperature alone; it is also related to the effects of pressure and the temperature at which melting occurs. The balance of temperature, pressure, and composition in the asthenosphere (between 100 and 200 km below the surface) allows some melting to occur there. The temperature in the overlying lithosphere is too low for it to melt. In the underlying mantle, the confining pressure is so great that the rocks are kept well under their melting points. The balance of temperature and pressure in the asthenosphere is just right for some minerals to melt.

The physical characteristics of the asthenosphere where part of the material is melted can be compared to those of slushy snow. It is a mixture of solid crystals and liquid. One factor particularly enhances the generation of magma along spreading centers. As the magma moves upward, the pressure is reduced; the decrease in pressure lowers the temperature at which melting occurs. Magma is thus generated along spreading centers, in contrast to other zones, largely because of reduced pressure. The diagram in Figure 17.2 illustrates the basic processes.

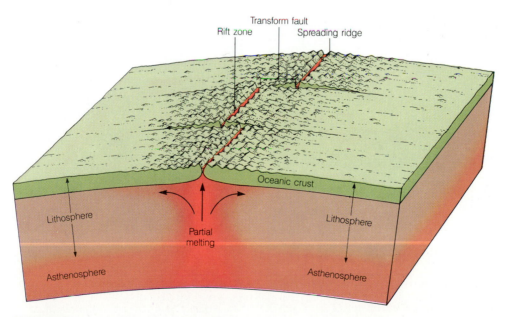

Figure 17.2
The origin of basaltic magma can be explained by the partial melting of mantle material that rises up at the spreading zone between two divergent plates. As mantle material moves upward along the spreading center, hot peridotite (the major rock in the mantle) partially melts because of the decrease in pressure. The resulting liquid forms a basaltic magma.

It should be reemphasized that each mineral in a rock has its own melting point. Magma produced by the partial melting of the asthenosphere does not have the composition of the original peridotite, which is a rock composed of the minerals olivine and pyroxene. Laboratory experiments on melting peridotite at high pressure indicate that if 10 to 30 percent of the rock melts, the resulting magma has a basaltic composition. Basaltic magma, because it is less dense than the solid mantle material, moves upward, into the fractures between the spreading plates, and is extruded on the ocean floor.

Examples

Direct Observations of the Seafloor

In recent years extensive studies of dredged samples and cores obtained from deep-sea drilling, together with observations from deep-diving vessels, have improved our understanding of the processes operating along the oceanic ridge. Thousands of photographs such as those shown in Figures 17.3 and 17.4 show that the oceanic ridge, as expected, is composed of innumerable structures of pillow lava. Instead of forming a single unit, as basalt flows do on land, submarine flows characteristically form multitudes of bubblelike structures that resemble a jumbled mass of pillows. Pillow basalt forms because underwater flows are chilled as soon as the liquid rock comes in contact with the water. This produces rounded, frozen skins on the lava, while the interior remains liquid. A succession of swelling, pillow-shaped bodies forms at breaks in the crust of the parent flow. These bodies can become detached and come to rest on the seafloor while they are still hot and plastic. Accumulations of pillow basalt thus show unmistakable evidence of underwater formation.

Another highly significant discovery was the presence of numerous open fissures or fractures in the crust along the oceanic ridge. More than 400 fractures, some as wide as 3 m (Figure 17.4) were observed in an area of 6 km². These are considered conclusive evidence that the oceanic crust is being pulled apart.

Geologic Studies of Iceland

Iceland is the best modern example of an area where the oceanic ridge rises above sea level. There geologists can examine in detail the surface expression of an active spreading center. The island is a plateau of basalt with a well-marked, troughlike rift extending through the center (see Figure 15.10). Although some cinder cones develop, the great floods of basalt have been extruded quietly along the fissures—that is, with little violent explosive phenomena. The youngest rocks are located along the rift, with progressively older basalts occurring toward the east and west coasts. Of special importance is the presence of innumerable vertical basalt dikes, called sheet dikes. The aggregate thickness of these vertical dikes is about 400 km, which represents the total thickness of new crust created in Iceland during the last 65 million years. This information—together with data obtained from seismic studies, deep-sea drilling, and seafloor observations made—confirms that volcanism along the divergent plate margins occurs largely as fissure eruptions in which great floods of basaltic magma are extruded to form new crust.

Figure 17.3
Pillow basalt along the mid-Atlantic Ridge was photographed at close range by scientists in the deep-diving submersible Alvin. Little or no sediment covers the basalt because this part of the seafloor is very young.

Figure 17.4
Open fissures along the mid-Atlantic Ridge were photographed by scientists in the deep-diving submersible Alvin. Hundreds of fissures such as this were mapped. They clearly indicate that the rift zone in the oceanic ridge is pulling apart.

Studies of Plateau Basalts

Where a spreading center passes beneath a continent, the continental crust is split, and great volumes of basalt commonly are extruded and spread over large areas near the rift system. These great floods of lava fill lowlands and depressions in the existing topography, and with subsequent crustal uplift, they erode into basalt plateaus (Figure 17.5).

Among the best examples of plateau basalts are those of the Columbia Plateau, in eastern Washington and Oregon and western Idaho. These basalts lie along the northward extension of the eastern Pacific Ridge, and they may well represent the initial stages of the breakup of western North America. The Columbia River Basalt covers an area of nearly 5,000,000 km² with a total thickness of between 1 and 2 km (Figure 17.5). This great accumulation of lava was not fed by central eruptions associated with a single volcano. Instead, the lava flowed to the surface through numerous fissures. Vast swarms of dikes now mark some of the fissures through which the volumes of lava were extruded.

An important implication of these and other observations is that the style of volcanic activity along divergent plate margins is predominantly the extrusion of basaltic lava along fissures. The fluid lava flows readily from cracks and fissures in the rift zone and tends to spread out laterally, instead of building high volcanoes. Extrusions beneath the sea form pillow lava, and eruptions of basalt on land produce floods of aa or pahoehoe flows. Regardless of where the lava

Figure 17.5
Fissure eruptions in the Snake River Plain, in Idaho, show the style of this type of volcanic activity. The basalt was extruded quietly along the zone of fissures and flowed rapidly across the surface as floods of lava.

is extruded, fissure eruptions dominate volcanic processes along divergent plate margins.

The great volume of lava extruded along divergent plate margins is difficult to comprehend because most of it is hidden beneath the ocean. The spreading centers, however, are the sites of the most extensive volcanism on Earth. To appreciate this fact, consider the amount of new oceanic crust created during the last ten million years (Figure 17.6). Approximately 20 km³ of basalt is extruded each year along the oceanic ridge.

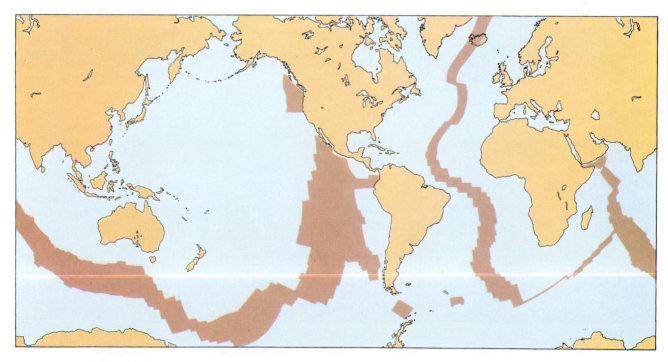

Figure 17.6
New crust created from volcanic activity along divergent plate margins during the last ten million years forms a broad swath along the crest and flanks of the oceanic ridge. This great volume of lava indicates that divergent plate margins are the site of the most extensive volcanic activity on Earth.

VOLCANISM AT CONVERGENT PLATE MARGINS

What is the origin of the "Ring of Fire"?

Statement

The type of volcanism along subduction zones is quite different from the basaltic fissure eruptions that characterize spreading centers. The magma generated at a subduction zone is largely andesitic. It is somewhat richer in silica than basalt is, and thus it is more viscous. Entrapped gas cannot escape easily. This composition results in violent, explosive eruptions from central vents, which commonly produce ash flows, stratovolcanoes, and collapse calderas. The geographic setting for volcanic activity along subducting zones depends on the type of plate interaction. Where two oceanic plates converge, an arcuate chain of volcanic islands forms on the edge of the overriding plate, parallel to the trench. Typical examples of this volcanic setting are Japan, the Aleutian Islands, and the Philippine Islands. Where a continent occurs on the active margin of the overriding plate, similar volcanic activity develops in the mountain belt. The Andes Mountains of South America are an example of this setting.

Discussion

The Generation of Magma at Convergent Plate Margins

Figure 17.7 is a quick visual summary of the major factors in the origin of magma in a subduction zone. There magma originates by the partial melting of the basalts and sediments of the oceanic crust as it plunges diagonally down into the hot asthenosphere. The key to understanding volcanic activity within this zone is the composition of the magma produced by the partial melting of the silica-rich sediment. The lithosphere, descending into a subduction zone, is subjected to progressively higher temperatures, and it begins to melt. The first material to melt is the layer of basaltic oceanic crust saturated with seawater. Na-plagioclase, amphibole, and, finally, pyroxene follow. The residue, containing minerals rich in magnesium and iron (olivine and some pyroxene), does not melt. It continues to sink and becomes assimilated in the mantle. The magma produced by the partial melting of the oceanic crust is thus enriched in silica, sodium,

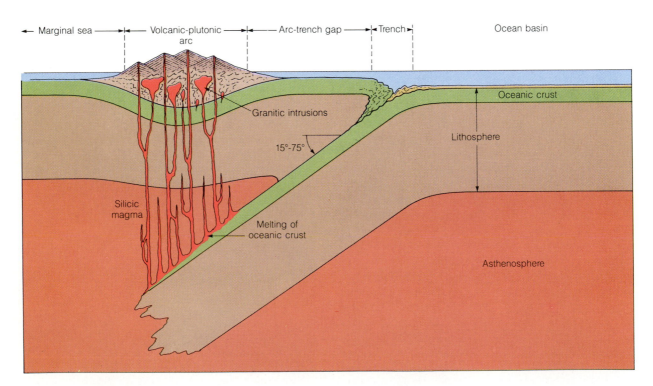

Figure 17.7
The generation of magma at convergent plate margins is primarily due to the slow heating of basalts and oceanic sediments as the descending plate slides down into the hot asthenosphere. At a critical depth, partial melting occurs in the descending slab of oceanic lithosphere. The melting generates andesitic magma, which rises buoyantly to form granitic batholiths or the volcanic rocks of an island arc.

and potassium (compared with the basaltic magma generated from the partial melting of the upper mantle) and typically produces andesites and associated rocks. Because of its lower density, it migrates upward through the crust and can accumulate as granitic batholiths or be extruded as ash flows or andesitic lava.

Note that at both spreading centers and subduction zones the generation of magma by partial melting differentiates and segregates the Earth's materials. At spreading centers, low-density material—enriched in silica, aluminum, sodium, potassium, and calcium—separates from the mantle and is concentrated in the oceanic crust. At subduction zones, magma is further enriched in silica by a second episode of partial melting and is concentrated in the island arcs or mountain belts of the continents. Because of its low density, this material cannot sink into the mantle but is concentrated in the continental crust. This magmatic differentiation is the principal method by which new material is added to the continents. We will say more about this process in Chapter 19.

Examples

There have been numerous historic volcanic eruptions, but most were not recorded, and many were not even observed. The record of volcanic activity is preserved in the fresh, unweathered, uneroded cones and

flows. A few eruptions, however, have had a major impact on human affairs and have therefore been described in minute detail. These accounts and studies of recent volcanic fields help us understand the nature of volcanic activity associated with converging plates.

Vesuvius

An extraordinarily vivid and accurate eyewitness account of the eruption of Mount Vesuvius in A.D. 79 was recorded by Pliny the Younger, then a seventeen-year-old boy, who related details of how his famous uncle, Pliny the Elder, died in the destruction of Pompeii and Herculaneum.

Beginning in A.D. 63 and continuing for 16 years, earthquakes shook the west coast of Italy around what is now Naples. Then, on the morning of August 24, A.D. 79, Mount Vesuvius exploded with a devastating eruption of white-hot ash and gas. Within a few hours, it asphyxiated and buried the population of Pompeii. Many people, suffocated by sulfurous fumes from ash clouds, died in their homes or on the streets. The entire town and most of its 20,000 inhabitants were buried by ash and forgotten for more than a thousand years, until Pompeii was excavated in 1748 (Figure 17.8).

The ash fall that buried Pompeii is a first-class example of the type of violent eruption that is common in volcanoes along converging plates. The once-

Figure 17.8
Pompeii, Italy, was completely covered with ash from the eruption of Vesuvius in A.D. 79. Left: An excavated part of the city of Pompeii reveals many details of Roman life. Right: The body of a Roman guard, asphyxiated by poison gas during the eruption, was cast in plaster by archaeologists.

smooth and symmetrical cone of Mount Vesuvius was shattered by the explosion, which created a large caldera where a peak once existed.

Mount Pelee (1902)

Ash flows are an important phenomenon associated with volcanism along converging plates. The great eruption of Mount Pelee, on the island of Martinique, in the West Indies, helped initiate an interest in and an understanding of this type of eruption. The eruption of Pelee was preceded by nearly a month of extrusions of steam and fine ash from the volcanic vent. Then, on May 8, 1902, a gigantic explosion blew ash and steam thousands of meters into the air. The denser hot ash moved as a body and swept down the slopes like an avalanche. In fewer than two minutes, the hot, incandescent ash flow moved 10 km from the side vent on Pelee and swept over the city of Saint Pierre. It annihilated the population of more than 30,000 people, except for one man, a prisoner, who was being held deep underground in the city jail. Every flammable object was instantly set aflame, and as the ash moved over the waterfront, all of the ships capsized.

The ash flow properly consisted of a mixture of hot glass shards, called pumice, and ash, which flowed at the base of a billowing cloud of gas. The fundamental force that caused the ash to flow so rapidly was simply the pull of gravity. A mixture of hot gas and fragments of ash and lava is highly mobile and practically frictionless. Each particle is separated from its neighbors and from the surface over which it moves by a cushion of expanding gas.

Intermittent ash-flow eruptions continued on Mt. Pelee for several months. By October a bulbous dome of lava too thick to flow had formed in the crater. A spire of solidified lava was then slowly pushed up from a vent in the dome, like toothpaste from a tube. This spire of Pelee was between 130 and 230 m in diameter, and it rose as much as 26 m per day, reaching a maximum height of 340 m above the crater floor. The spire repeatedly crumbled and grew again from the lava dome, which glowed red in the night.

The violent eruptions that characterized volcanoes at convergent plate margins result from the high silica content of the magma, which makes it thick and viscous. Dissolved gases cannot easily escape. As a result, tremendous pressure builds up in the magma, and when eruptions occur, they are highly explosive. The explosion produces huge quantities of ash, hot ash flows, and thick, viscous lava.

Mount Saint Helens, 1980

The best-documented example of the eruption of a stratovolcano is the eruption of Mount Saint Helens on May 18, 1980. Numerous indications of an impending eruption had drawn dozens of scientists to the site to monitor earthquakes, emissions of gas, and physical changes in the mountain. The explosion was estimated to have been roughly five hundred times the force of the atomic bomb that destroyed Hiroshima (Figure 17.9).

Mount Saint Helens, in the state of Washington, is part of the Cascade Range, which extends about 1,500 km in a north-south line from British Columbia to northern California and includes fifteen major composite volcanoes (Figure 17.10). This is the North American segment of the Ring of Fire that encircles the Pacific Ocean. All volcanoes in the Ring of Fire have the same origin. They result from the subduction of oceanic plates in the vicinity of the deep-sea trenches that surround the Pacific. Mount Saint Helens is the youngest volcano in the Cascade Range. Most of the existing mountain is only 2,500 years old, but it overlies an older volcanic center that has previously erupted on numerous occasions. The history of its eruptions extends back as far as 37,000 years.

Mount Saint Helens had been dormant for 123 years, but on March 20, 1980, it began to stir, with a series of small earthquakes. After a week of increasing number of earthquakes, it began to eject steam and ash. This was the first of a series of moderate eruptions that continued intermittently for the next six weeks. Within a few days after the first eruptions, warnings were issued by the U.S. Geological Survey. During the weeks to come, the U.S. Forest Service and state officials closed all the areas near the mountain, undoubtedly saving thousands of lives.

By the second week of activity, more than thirty geologists had gathered at the site to carry out a wide variety of studies. Much of their effort was directed toward monitoring the development of a large bulge on the north flank of the mountain. By the end of April, the bulge was 2 km long and 1 km wide and was expanding horizontally at a steady rate of 1.5 m per day. Clearly, the mountain was being inflated by magmatic intrusion. Sulfur dioxide (SO_2) was being released at a rate of 50 m tons per day from March to May 18. Volcanoes that are actively erupting give off as much as 1,000 metric tons of SO_2 per day, so an increase in the emission of SO_2 might suggest an increase in magmatic activity. Monitoring the occurrence and intensity of the earthquakes, the bulge, and the emission of gases, geologists believed they would detect some significant change to warn them of an impending large eruption, but no extraordinary activity occurred. In fact, the number of earthquakes decreased. Thirty-nine earthquakes were recorded on May 15, and only eighteen on May 17.

On Sunday morning, May 18, the mountain was silent. Only minor plumes of steam rose from two vents. At an observation post 8 km northwest of the

Figure 17.9
The eruption of Mount Saint Helens is well documented by aerial photographs and provides scientists with a rare opportunity to study details of volcanic eruptions. This photograph, taken about three hours after the initial blast on May 18, 1980, shows the ash cloud that reached an elevation of about 18 km. Steam and ash were blown to the northeast across the continental United States.

volcano's crater, David Johnston, a thirty-year-old geologist, was monitoring the emission of gases and reporting visual observations on his two-way radio. He cried, "Vancouver! Vancouver!" "This is it!" Moments later, Johnston vanished in an explosion of hot ash and gas, as more than 4 km^3 of material was blasted from the north side of the mountain.

The best way to understand the nature of the eruption is to study the sequence of diagrams in Figure 17.11. At 8.32 A.M. the mountain was shaken by an earthquake with a magnitude of approximately 5 on the Richter scale. The north slope began to undulate and then move downslope as a great debris flow. As the north side of the volcano gave way, it uncapped the bottled-up gas and magma, and an eruption cloud blasted laterally, above, and over the collapsing slope. This lateral blast of rock, ash, and gas caused most of

the destruction and loss of life. The blast wave leveled the forest in an area 35 km wide and 23 km outward on the north flank of the mountain. One man died in his truck still holding his camera. A young couple was buried in their tent by falling trees, their arms still around each other.

The eruption caused three separate, but somewhat interrelated, processes: (1) mudflows, (2) ash flows, and (3) ash falls. Mudflows originated largely from debris flows saturated with water on the upper slopes of the mountain. Most of the mudflows were hot. They moved rapidly downslope, and the most massive and devastating of them traveled many kilometers down the Toutle River. Their rush downslope killed several people. At the height of the flow, the lower Toutle River swelled to 1.5 km wide and was heated to 90°C. Mudflows swept up 123 homes, as

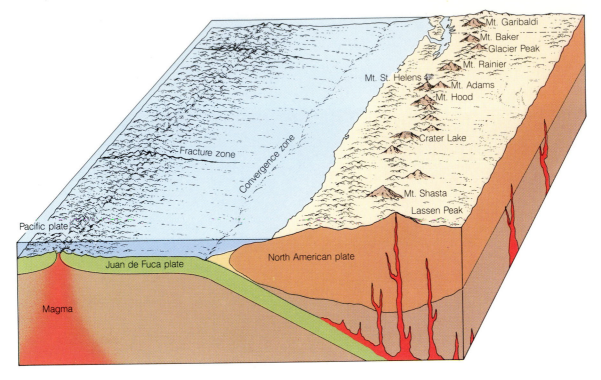

Figure 17.10
The Cascade Range contains fifteen large composite volcanoes extending in a line from British Columbia to northern California. The volcanoes are formed by the subduction of the Gorda Plate beneath the North American Plate. Note the location of Mt. St. Helens.

well as cars, logging trucks, and timber, and carried them downstream to batter and destroy bridges. So much sediment from the Toutle was carried into the Columbia River that its downstream depth was reduced from 12 m to 4 m within a day, and boats upstream were trapped.

An important part of the eruption was the extrusion of numerous ash flows. Traveling as fast as 130 km per hour, the incandescent ash and debris (at a temperature of about 500°C) extended north for a distance of 9 km. Some ash flows reached Spirit Lake, where, together with debris flows that were deposited earlier, they blocked the lake's outlet. Consequently, the level of water in the lake rose 60 m.

Immediately after the eruption, a vertical cloud of ash rose to heights of 18 km. The ash clouds then fanned out downwind (eastward), and ash began to settle like soft, gray snow. Within two hours, it began to fall on Yakima, Washington, and by the next day the ash cloud was over Cheyenne, Wyoming, approximately 1,400 km away. The cloud then moved in a broad arc across the United States. By June 5 it had completely circled the Earth.

The particles of ash were composed of fine-grained crystals, fragments of volcanic glass, and rock material that once formed the summit dome. The ash is slightly acidic and is often coated with sulfur from the eruption. In just a few hours, Mount Saint Helens had thrown up almost as much ash as Vesuvius did in A.D. 79, when it buried Pompeii and Herculaneum.

On May 25 another eruption sent a cloud of steam and ash to elevations of 13 km. This event caused no structural changes in the mountain. Ash flows were barely large enough to travel to the base of the mountain, and mudflows were minor. Similar activity occurred again on June 12, with the extrusion of ash flows that moved downslope as far as Spirit Lake. This eruption was preceded by a significant increase in harmonic tremors (constant, rhythmic vibrations). The mountain erupted again on July 22 and on August 7, ejecting ash and steam to heights of approximately 13 km and extruding new ash flows over the northern flank. Both events were preceded by harmonic tremors and a noticeable decrease in the ratio of CO_2 and SO_2 emissions the day before the eruption.

It is clear from the geologic record that Mount Saint Helens has had a history of spasmodic explosive activity separated by intervals of dormancy lasting two hundred or three hundred years. The most recent explosions demonstrate that the show is far from over. Future eruptions will undoubtedly produce lava flows, domes, volcanic ash, and ash-flows—the typical ejecta of stratovolcanoes. Whatever Mount Saint Helens does next, geologists stand only to gain.

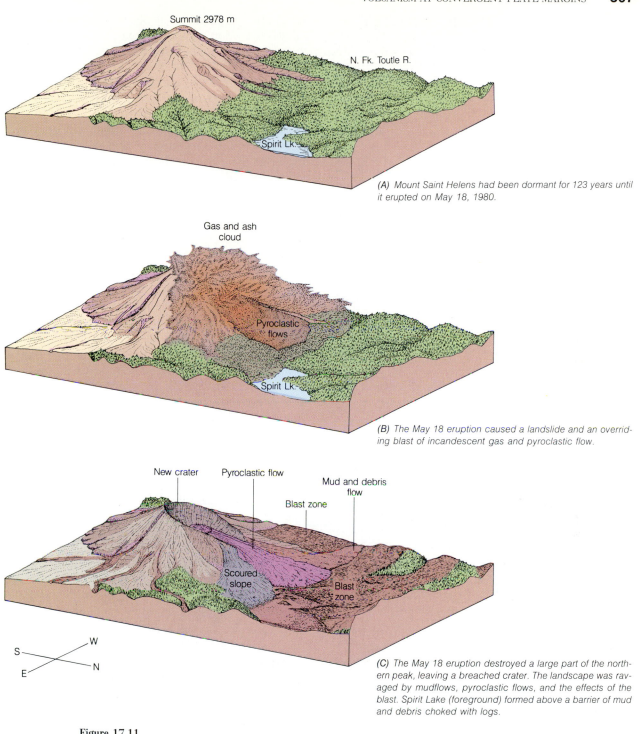

Summit 2978 m

N. Fk. Toutle R.

Spirit Lk.

(A) Mount Saint Helens had been dormant for 123 years until it erupted on May 18, 1980.

Gas and ash cloud

Pyroclastic flows

Spirit Lk.

(B) The May 18 eruption caused a landslide and an overriding blast of incandescent gas and pyroclastic flow.

New crater   Pyroclastic flow

Mud and debris flow

Blast zone

Scoured slope

Blast zone

(C) The May 18 eruption destroyed a large part of the northern peak, leaving a breached crater. The landscape was ravaged by mudflows, pyroclastic flows, and the effects of the blast. Spirit Lake (foreground) formed above a barrier of mud and debris choked with logs.

Figure 17.11
The sequence of events in the eruption of Mount Saint Helens, as seen from the northeast, is depicted in this series of diagrams.

By catching a volcanic eruption in the act, we are able to study many aspects of volcanism that are not preserved in the rocks from past eruptions. Geologists hope to learn how to assess potential hazards and even how to predict the time of major eruptions. A prime candidate as a predictive tool is the volcano's seismic activity, because harmonic tremors have preceded several eruptions. Other clues may be revealed by studies of the emission of gases, such as SO_2, CO_2 and H_2 (hydrogen).

INTRAPLATE VOLCANIC ACTIVITY

If plate margins are the site of most geologic activity, why do some volcanoes occur in the central parts of plates?

Statement

Volcanic eruptions in the central parts of plates, called intraplate volcanic activity, are trivial compared with those along spreading centers and subduction zones; however, they may be important as surface expressions of local thermal variations, or hotspots, in the mantle material. Most intraplate volcanic activity occurs on the floor of the South Pacific, producing numerous submarine volcanoes and volcanic islands, both as isolated features and in linear chains.

Discussion

Figure 17.12 shows the distribution of major centers of intraplate volcanic activity in the Pacific. At first glance the distribution of intraplate volcanoes may appear to be random, but obvious linear trends, or chains, are soon apparent. Excellent examples include the Hawaiian-Emperor, Tuamotu-Line, and Austral-Marshall-Gilbert chains. The best data available suggest that volcanic chains are formed as a lithospheric plate moves over a mantle plume, or hotspot (Figure 17.13). Volcanism occurring over a hotspot produces a submarine volcano, which can grow into an island. If the hotspot's position in the mantle remains fixed for a long time, the moving lithosphere carries the volcano beyond the magma source. This volcano then becomes dormant, and a new one forms over the fixed hotspot. A continuation of this process would build one volcano after another, producing a linear chain of volcanoes parallel to the direction of plate motion.

Isolated volcanoes can result from small hotspots that do not endure long enough to produce a volcanic chain, or they can develop from minor pockets of magma carried with the moving asthenosphere.

The type of volcanism in the interior of plates is similar, in most cases, to that along diverging plates. The products are largely basaltic lava extruded by quiet fissure eruptions. This basaltic lava is believed to be a derivative from the mantle, just like the lava that is found along the oceanic ridge.

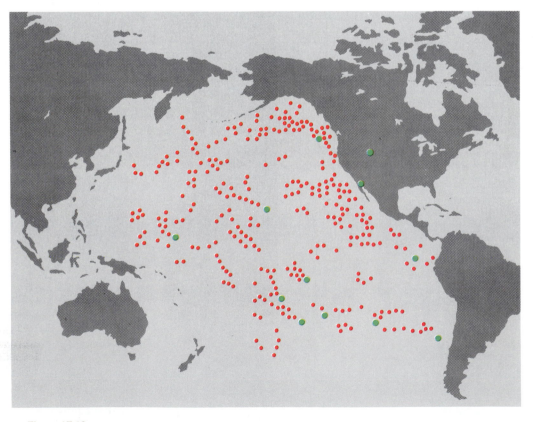

Figure 17.12
Intraplate volcanism occurs mostly in the Pacific Plate, where hundreds of islands and seamounts have been formed by basaltic volcanism. These are believed to be produced by masses of hot mantle material, called mantle plumes, rising beneath the plate. Linear chains of volcanoes can be produced as a plate moves over a hotspot, as indicated by red dots.

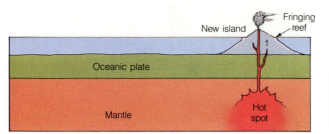

(A) A volcanic island or seamount is built up by extrusions from a fixed hotspot, or source of magma, in the mantle.

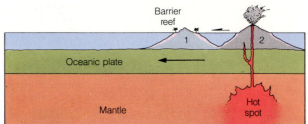

(B) As the plate moves, the volcano is carried away from the source of magma and becomes extinct. The surface of the island can then be eroded to sea level, and reefs can grow to form an atoll. A new island is then formed over the hotspot.

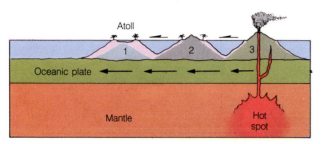

(C) Continued plate movement produces a chain of islands.

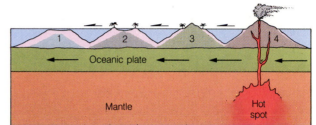

(D) The islands of the chain are progressively older away from the hotspot.

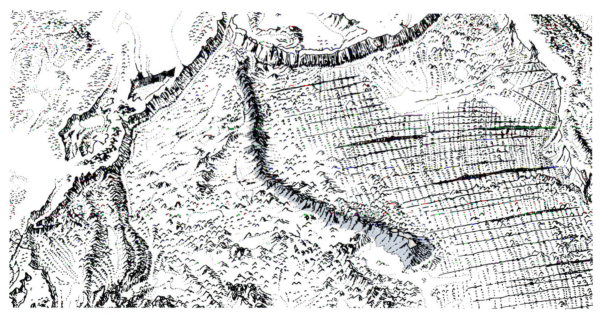

(E) The Hawaiian Islands are progressively older away from the area of recent volcanic activity. The numbers refer to the ages, in millions of years, of the volcanic basalts.

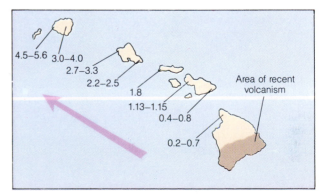

(F) The Hawaiian Islands are progressively older away from the area of recent volcanic activity. The numbers refer to the ages, in millions of years, of the volcanic basalts.

Figure 17.13
The origin of linear chains of volcanic islands and sea-mounts is explained by the plate tectonic theory.

SUMMARY

1. Most volcanic activity occurs in active seismic zones and is clearly associated with plate boundaries. The type of volcanism depends on the type of plate boundary.
2. At divergent plate margins, basaltic magma is generated by the partial melting of the asthenosphere and is extruded largely as fissure eruptions. Pillow lavas form on the seafloor, and plateau basalts are extruded where continents overlie rift zones.
3. At convergent plate margins, andesitic or granitic magma is generated in the subduction zone by the partial melting of the oceanic crust. It is intruded in the upper plate as batholiths or is extruded as composite volcanoes in island arcs or mountain belts. Andesitic magma is viscous and erupts violently, commonly in the form of an ash flow.
4. Minor intraplate volcanic activity probably indicates local hotspots in the mantle. The greatest intraplate volcanism occurs in the Pacific, where numerous shield volcanoes form islands and seamounts.

KEY TERMS

fissure (p. 300)

flood basalt (p. 301)

hotspots (p. 308)

island arc (p. 302)

magmatic differentiation (p. 302)

mantle plume (p. 308)

pillow lava (p. 300)

plateau basalt (p. 302)

sheet dikes (p. 300)

REVIEW QUESTIONS

1. What is the Ring of Fire?
2. Describe the type of volcanic activity in Iceland. What is the significance of sheet dikes?
3. Summarize the characteristics of volcanism along divergent plate margins.
4. Explain how magma is generated along divergent plate margins.
5. Describe the type of volcanic activity along convergent plate margins and cite several examples.
6. How does the composition of lava influence the style of volcanic eruption?
7. What type of volcanic activity occurs within the central parts of tectonic plates? Beyond the active margins?
8. Explain the origin of chains of volcanic islands and seamounts.

Multiple-Choice Questions

1. The type of volcanic activity on Earth
   a. depends upon the type of plate boundary.
   b. has changed in time from basaltic to andesitic.
   c. has evolved from andesitic to basaltic.
   d. depends upon the age of the volcanic activity.
   e. has changed from mostly explosive to mostly quiet.
2. The major volcanic region, called the Ring of Fire, is located
   a. along the mid-Atlantic Ridge.
   b. through the rift valleys of east Africa.
   c. within the Mediterranean Sea region.
   d. near the margins of the Pacific Ocean.
   e. along the east coast of North and South America.
3. Volcanic activity along the midoceanic ridge results from
   a. complete melting of parts of the upper mantle.
   b. convection in which material from the liquid core is brought to the surface.
   c. partial melting of the lithosphere.
   d. partial melting of the asthenosphere.
   e. partial melting of andesitic and granitic rock.
4. Photographs of the seafloor made from deep-diving vessels show that volcanism along the mid-Atlantic ridge
   a. is characterized by pillow basalt.
   b. is associated with open fissures.
   c. is relatively young because the lava flows are not covered by a significant amount of sediment.
   d. all of the above.
   e. only a and b are correct.
5. Which of the following features is least likely to be found on Iceland?
   a. composite cone
   b. flood basalt
   c. cinder cones
   d. sheet dikes
   e. rift valley
6. Volcanism at converging plate margins
   a. is violent because the magma is viscous and pressure builds up from trapped gas.
   b. results from partial melting of the oceanic crust.
   c. is largely andesitic.
   d. commonly produces ash flows.
   e. all of the above.

7. Intraplate volcanic activity
   a. probably results from local hot spots in the mantle.
   b. is restricted to the volcanic seamounts of the Pacific.
   c. is andesitic.
   d. represents old spreading centers.
   e. results from ancient subduction zones.
8. Chains of volcanic islands and seamounts result from
   a. extrusion along a fracture zone.
   b. extrusion along a rift system.
   c. a fixed hotspot in the mantle over which the moving lithosphere carries each volcano.
   d. a series of small hotspots in the mantle.
   e. extrusion along an incipient trench.
9. Which of the following is *not* true of chains of volcanic islands and seamounts?
   a. One end of the chain is older than the other.
   b. They appear to result from the movement of a plate over a "hotspot."
   c. They originate by the melting of the crust.
   d. They are composed of basalt.
   e. They indicate the direction of plate movement.
10. An example of intraplate volcanism is
    a. the Aleutian Islands.
    b. Hawaii.
    c. Iceland.
    d. Mt. St. Helens.
    e. Mt. Pelee.

True/False Questions

1. *T F* Most volcanic activity coincides with the active seismic regions of the world and is clearly associated with plate boundaries.
2. *T F* The type of volcanic activity is independent of the type of plate boundary.
3. *T F* The style of volcanic activity along the convergent plate margins is predominately the extrusion of basaltic lava along fissures.
4. *T F* The best data available suggest that chains or volcanic islands and seamounts such as the Hawaiian Chain are formed as a lithospheric plate moves over a mantle plume, or hotspot.
5. *T F* Most intraplate volcanic activity occurs on the stable platforms of continents.

Fill in the Missing Terms

1. Most volcanic activity along _____ plate margins takes the form of quiet eruptions of basaltic magma.
2. _____ is the best modern example of a geographic area where the oceanic ridge rises above sea level.
3. The magma generated at a subduction zone is largely _____ in composition.
4. Volcanic activity landward from the trench axis is mostly _____ in composition.
5. At convergent plate margins, granitic magma is generated by partial melting of the _____ _____.

18

EVOLUTION OF THE OCEAN BASINS

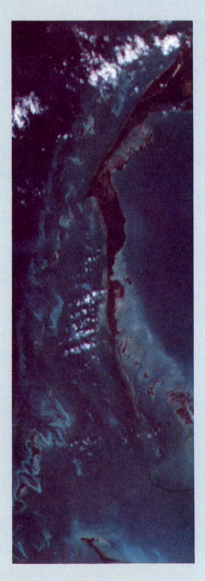

The ocean floor is an alien landscape, with features completely different from those we see on the continent. It was formed primarily by volcanic activity, faulting, and deposition of deep marine sediment, and much of its surface reflects these processes. In addition, sediment settling from the ocean blankets some areas to form large, flat, featureless surfaces. It is the ocean floor, not the continents, that is the most widespread surface of the planet.

Until recently we could not even dream of seeing and understanding the features on the ocean floor. They were almost as inaccessible as the stars. Now the surface of the ocean floor has been mapped from oceanographic vessels using special instruments that plot a profile of the seafloor. Most of our understanding of the ocean basin comes from these studies. The findings are as important as they are spectacular. A great "mountain range," called the oceanic ridge, extends as a continuous unit through all of the ocean basins, a distance of more than 64,000 km. The ridge is cut by long fracture zones, several thousand kilometers long. Flanking the oceanic ridge is the abyssal floor, above which rise numerous submarine volcanoes, called seamounts. The deepest parts of the ocean are the trenches, some of which descend to depths of 11,000 m. In addition, small ocean basins, commonly referred to as seas, have also been explored. These are mostly isolated from the major ocean basins.

This new knowledge of the landscape on the ocean basin helped revolutionize geological thinking. For the first time, we have been able to consider the geology of the entire planet. No amount of research on the continents alone could have revealed what we now know from studies of the ocean basin and the continents together.

MAJOR CONCEPTS

1. The major features of the ocean basins are (a) the oceanic ridge, (b) abyssal hills, (c) abyssal plains, (d) trenches, (e) islands and seamounts, and (f) continental margins.

2. The oceanic crust is composed of four major layers: (a) a surface layer of marine sediment, (b) a layer of pillow basalt, (c) a zone of sheeted dikes, and (d) a layer of gabbro.

3. Although the ocean basins originated by seafloor spreading, they are quite different in size, shape, and topography. These differences result largely from differences in age and stage of development.

4. The present oceans have evolved as a result of shifting tectonic plates during the last 200 million years.

THE LANDSCAPE OF THE SEAFLOOR

How does the theory of plate tectonics explain the location and origin of the major landforms on the ocean floor?

Statement

Within the past three decades, our knowledge of the ocean floor has increased enormously because of the new instruments with which we can make accurate images of its surface. The results indicate that the major landforms of the ocean basins are

1. the oceanic ridge.
2. the abyssal floor.
3. trenches.
4. seamounts.
5. continental margins.

Discussion

The surface features of the continents are familiar to everyone, but the landscape of the ocean floor remained largely a mystery until the development of modern oceanographic research. The regional landscape of the seafloor can now be studied with a variety of new instruments, including echo sounders, side scan sonar images, magnetometers, and seismic reflection profiles. All of these provide information that has been used to make maps and charts showing the topography of the ocean floor, in detail similar to that provided by early topographic maps of the land.

Seismic Reflection Profiles

The seismic reflection profile reproduced in Figure 18.1 illustrates the type of information obtained about the seafloor by seismic reflection. The method involves emitting pulses of sound from a ship that are subsequently reflected back from the seafloor. Some of the energy, however, penetrates the sediment and is reflected from the hard rock surface below. Automatic electronic equipment plots a continuous profile of the surface of the seafloor, the structure of the unconsolidated sediments, and in many cases the configuration of the surface of the consolidated bedrock of the oceanic crust. The profile in Figure 18.1 is a cross-sectional view of the geologic structure of the ocean basin flanking the mid-Atlantic Ridge. It shows that the oceanic sediments are relatively flat and undeformed, with bedrock protruding through them in irregular mounds. By combining a large number of profiles and manipulating them with a computer, we can obtain a perspective of the landforms on the seafloor, with different elevations shown in different colors (Figure 18.2).

From such records, we have been able to compile accurate maps of the seafloor (see the physiographic map inside the covers of this book) and can identify a variety of new and exciting geologic features not found on the continents. This is what we have found.

THE OCEANIC RIDGE

What geologic features characterize the oceanic ridge?

The oceanic ridge is the most pronounced tectonic feature on Earth. If the ridge were not covered with water, it would be visible from as far away as the Moon. It is essentially a broad, fractured swell, generally more than 1,500 km wide, with peaks rising as much as 3 km above the surrounding ocean floor. It covers nearly 23 percent of Earth's surface, almost as much as the surface of the continents. The remarkable characteristic of the ridge is that it extends as a continuous feature around the entire globe, like the seam of a baseball. It extends from the Arctic Basin, down through the center of the Atlantic, into the Indian Ocean, and across the South Pacific, terminating in the Gulf of California, a total length of more than 64,000 km. Without question, it is the greatest "mountain" system on Earth. The "mountains" of the oceanic ridge, however, are nothing like the mountains of the continents, which were built largely of folded and metamorphosed sedimentary rocks. By contrast, the ridge is composed entirely of basalt and is not deformed by folding.

Many of the characteristics of the ridge are apparent in the seismic-reflection profile reproduced in Figure 18.3. On a regional basis, the ridge is a broad segment of the ocean floor that is arched up and broken by numerous fault blocks that form linear hills and valleys. The highest and most rugged topography is located along the axis, and a prominent rift valley marks the crest of the ridge throughout most of its length. As is shown in Figure 18.3, oceanic sediments are thickest down the flanks of the ridge but thin rapidly toward the crest.

Throughout most of its length, the oceanic ridge is cut by a series of transform faults (Figure 18.4). These are sites of continued seismic activity. Beyond an active transform fault, the fracture zone is expressed by an abrupt, steep cliff that in places can be traced for several thousand kilometers.

The general character of the oceanic ridge seems to be a function of the rate of plate separation. Where the rate of spreading is relatively low (fewer than 5 cm per year), the ridge is higher and more rugged and mountainous than it is where rates of spreading are more rapid. Moreover, rift valleys on slow-spreading ridges are prominent, whereas rift valleys in areas with high rates of spreading are more subdued.

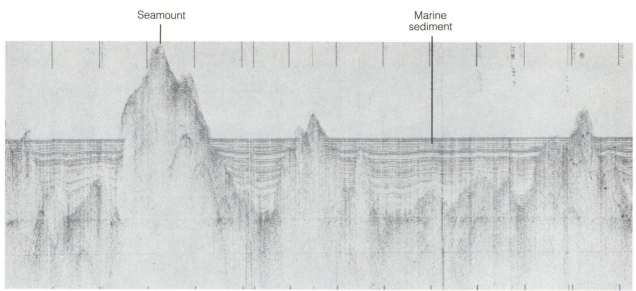

Seamount

Marine
sediment

Figure 18.1
A seismic-reflection profile provides a wealth of information about the seafloor. The profile shows not only
the morphology of the ocean floor, but also the configuration of unconsolidated marine sediments resting
on the solid bedrock.

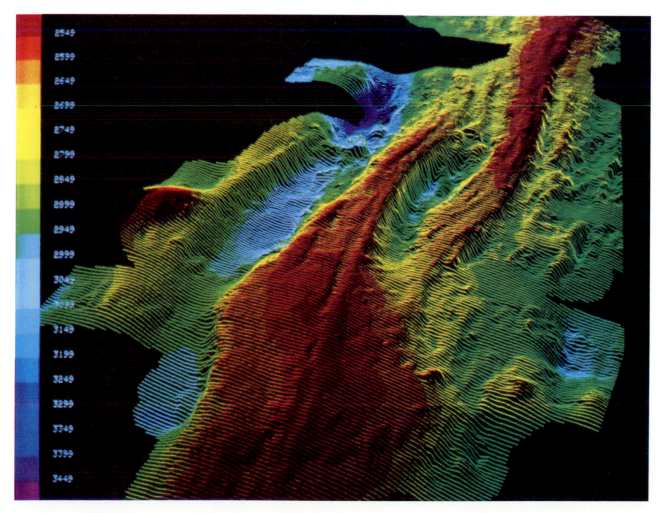

Figure 18.2
The topography of a part of the midoceanic ridge as revealed by computer-enhanced profiles.

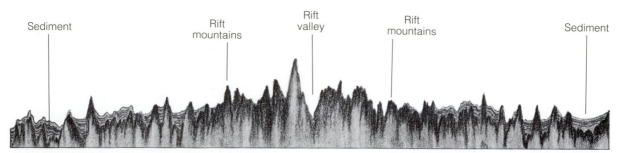

Sediment   Rift mountains   Rift valley   Rift mountains   Sediment

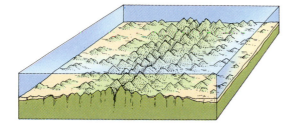

Figure 18.3
A seismic-reflection profile across the mid-Atlantic Ridge, at 44° north latitude, shows that the crest of the ridge is marked by a deep rift valley that can be traced along the entire length of the ridge. Sediment is thickest down the flanks of the ridge, but it thins rapidly near the crest. The idealized diagram of the ridge was based on a series of profiles.

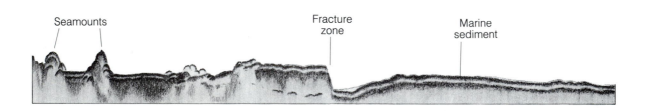

Seamounts   Fracture zone   Marine sediment

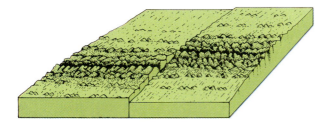

Figure 18.4
A seismic-reflection profile across the Murray Fracture Zone, in the eastern Pacific Ocean, shows that the fracture is expressed by a pronounced vertical cliff that separates areas of contrasting topography. On the left side of the fault, seamounts are abundant. To the right, the sea-floor is relatively smooth and featureless. Note how this contrast in topography on the block diagram is produced by strike-slip faults. Note the seismic profile is parallel to the front of the idealized block diagram.

THE ABYSSAL FLOOR

What controls the development of the abyssal floor?

Vast areas of the deep ocean consist of broad, relatively smooth surfaces known as the abyssal floor. In most areas the abyssal floor extends from the flanks of the oceanic ridge to the continental margins, generally lying at depths ranging from about 3 to 5.5 km.

In most ocean basins, the abyssal floor can be subdivided into two sections: the abyssal hills and the abyssal plains. The abyssal hills are relatively small hills, rising from 75 to 900 m above the ocean floor (Figure 18.5). They are circular or elliptical and range from 1 to 8 km in width at the base. The hills are found along the flanks of the oceanic ridge and occur in profusion in parts of the ocean floor separated from land by trenches. In the Pacific they cover between 80 and 85 percent of the ocean floor; thus, abyssal hills can be considered the most widespread landform on Earth. The abyssal plains are exceptionally flat areas of the ocean floor where the abyssal hills are completely buried by sediment (Figure 18.5). Commonly, they are located near the margins of a continent, where sediment from a continental mass is transported by turbidity currents and spreads over the adjacent ocean floor.

The origin of the abyssal hills and abyssal plains can be traced to the oceanic ridge and the development of new crust at the spreading center. New lithosphere, created at the ridge crest, slowly recedes from the rift zone and is gradually modified in several important ways. The new crust cools and contracts, deepening the ocean basin as it moves away from the ridge. Fine-grained pelagic sediments, consisting of dust and the shells of marine organisms, slowly, but continually, settle over all of the seafloor. Linear hills, formed by volcanic activity, intrusions of magma, and block faulting become the foundations of the abyssal hills. As plates move away from the spreading center, the superficial features of the landforms created at the ridge are gradually modified and concealed. Eventually, they may be completely buried in sediment, thus forming the flat abyssal plains. The outline of the buried rock surface can be traced in seismic-reflection profiles such as Figure 18.5.

The distribution of abyssal plains and abyssal hills substantiates this explanation of their origin. Abyssal plains occur only where the topography of the seafloor does not inhibit turbidity currents from spreading sediment from the continents over the seafloor. In the Atlantic Ocean, abyssal plains occur near the margins of the continents of North America, South America, Africa, and Europe. In the Pacific Ocean, by contrast, there are few abyssal plains because turbidity currents cannot flow past the deep trenches that lie along most of the continental margins. The inflowing sediment accumulates in the trenches, so that most of the Pacific floor lacks abyssal plains and is covered instead with abyssal hills.

Large cone-shaped or fan-shaped deposits of sediment derived from the continents lie on the abyssal plains offshore from most of the world's great rivers. These are called deep-sea fans (such as the areas of the seafloor east and west of India). They resemble alluvial fans and deltas in that they are fan-shaped accumulations of sediment located seaward from the mouth of major rivers and submarine canyons. They are different from the land deposits, however, because the sediment is transported and deposited primarily by turbidity currents. Most large fans are located at the base of the continental slope, with the head at the mouth of a submarine canyon cut into the edge of the shelf. The main source of sediment for the fan is mud, brought in by major rivers. Turbidity currents intermittently flush sediment through the canyon, building up depositional fans where the currents reach the lower gradient of the ocean floor.

The submarine fan of the Ganges River, in the northwestern Indian Ocean, is by far the largest deep-sea fan in the world. It is more than 2,800 km long and covers slightly more than 4,000,000 km_2. This accumulation represents about 70 percent of the debris derived from the erosion of the Himalayas. The remaining 30 percent is deposited on the floodplain and delta of the Ganges and by the Indus River, to the west.

Figure 18.5
A seismic-reflection profile across the abyssal floor of the Atlantic Ocean shows abyssal hills buried with sediment, which forms the smooth abyssal plains.

TRENCHES

What geologic features characterize the deep-sea trenches?

A subduction zone, where two plates converge and one slab of lithosphere plunges down into the mantle, is generally expressed topographically by a trench. We have seen in previous chapters that a subduction zone is characterized by intense volcanic activity and seismicity. It is also marked by a large gravity anomaly (abnormally high or low gravitational force within an area). Trenches, some of which reach nearly 11,000 m below sea level, are the deepest parts of the ocean. As is shown in the seismic-reflection profile in Figure 18.6, they typically are asymmetrical. A relatively steep slope lies on the landward side, along the continental landmass, and a gentler slope lies on the side of the ocean basin. Individual trenches fewer than 100 km wide can form continuous features extending for several thousand kilometers across the deep-ocean floor.

The most striking examples of trenches occur in the western Pacific. A trench system there extends from the vicinity of New Zealand to Indonesia to Japan, and then northeastward along the southern flank of the Aleutian Islands. Long trenches also occur along the western coast of Central America and South America, in the Indian Ocean west of Australia, in the Atlantic off the tip of South America, and in the Caribbean Sea.

ISLANDS AND SEAMOUNTS

What are the origin and significance of seamounts?

Literally thousands of submarine volcanoes occur on the ocean floor, with the greatest concentration in the western Pacific. Some risise above sea level and form islands, but most remain submerged and are called seamounts (Figure 18.7). They often occur in groups or chains, with individual volcanoes being as much as 100 km in diameter and 1,000 m high.

CONTINENTAL MARGINS

What are the major geologic features that characterize the margins of continents?

The continental margins are covered by the ocean but are not geologically part of the oceanic crust. They are composed of continental crust and sediment derived from erosion of the land. This part of the seafloor can be divided into three major sections: the continental shelf, the continental slope, and the continental rise. The continental shelf is simply a submerged part of the shield, or stable platform. The gently sloping shelf extends from the shoreline to the area where the continental margin begins its steep descent to the ocean floor. The shelf can be as much as 1,500 km wide; its depth ranges from 20 to 550 m at its outer edge. At present, the continental shelves comprise 18 percent of Earth's total continental area. At times in the geologic past, however, they were much larger because the oceans spread much farther over the continental platforms.

Characteristically, the continental shelf is smooth and flat, but its topography has been influenced greatly by changes in sea level. Because large areas were once exposed as dry land and thus were subjected to subaerial processes, the shelf topography can have features formed by both marine and nonmarine processes.

The continental slope descends from the outer edge of the continental shelf as a long, continuous slope to the deep-ocean basin (Figure 18.8). If the oceans were drained of water, the continental slope would appear as the most conspicuous boundary on Earth's surface. It would appear as a long, continuous slope rising from the abyssal floor to the high continental platform. It marks the edge of the continental granitic rock mass, the boundary between the continental crust and the oceanic crust.

The continental rise is the transition between the continent and the ocean basin. It has a gentle, inclined surface, rising from the abyssal plains to the continental slope. The continental rise is apparently formed by sediment deposited at the base of the continental slope.

Two types of continental margins can be recognized: active and passive. Each is subjected to different stresses, and thus each develops different characteristics. The passive margin of a continent is related to divergent plate motion and thus has only limited tectonic activity—for example, the Atlantic margin of North and South America. The shelves on the passive margin of a continent are typically wide and are not subjected to strong compressive stress.

In contrast, the active margin of a continent is associated with converging plates and is characterized by earthquakes, igneous activity, and mountain building. It is subjected to much more stress because it impinges on another moving plate. Crustal deformation results. Where a continent converges with an oceanic plate, the oceanic plate moves down and under the lighter continental crust, and a trench develops along the continental margin.

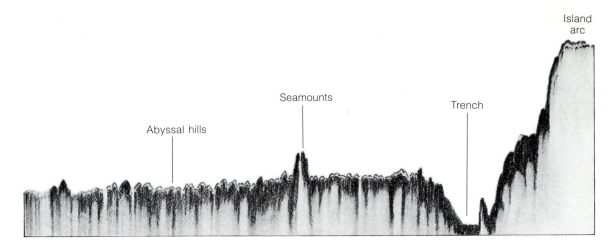

Abyssal hills

Seamounts

Trench

Island arc

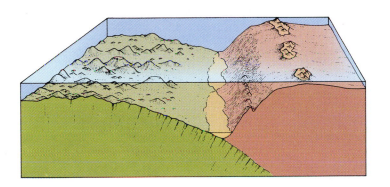

Figure 18.6
A seismic-reflection profile across the central part of the Aleutian Trench shows the steep flank of the trench alongside the Aleutian island arc (right) and the gentle slope toward the ocean basin. The trench is the surface expression of a subducting plate.

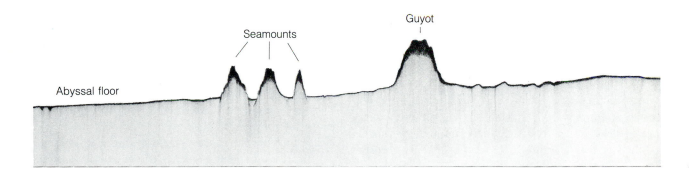

Seamounts

Guyot

Abyssal floor

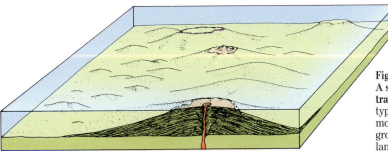

Figure 18.7
A seismic-reflection profile across seamounts in the central Pacific Ocean shows the general configuration of typical seamounts rising above the ocean floor. Seamounts are submarine volcanoes, which usually occur in groups or chains. Some rise above sea level to form islands.

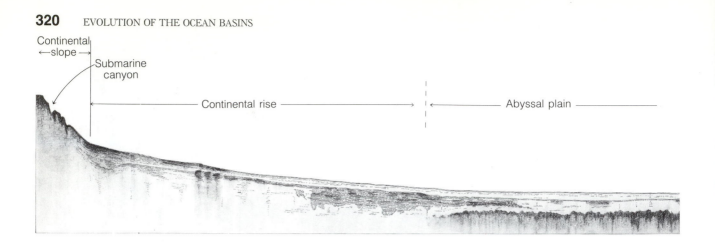

Continental
←slope→

Submarine
canyon

Continental rise

Abyssal plain

Figure 18.8
A seismic-reflection profile across the western continental slope and continental rise of Africa shows the profile of several submarine canyons near the upper part of the slope and the thick accumulation of undeformed sediments on the continental margin. The continental slope merges into the adjacent abyssal plains, which cover the abyssal hills.

SUBMARINE CANYONS

How do submarine canyons form?

Submarine canyons are common along the continental slope and have been studied for many years, long before the ocean floor was mapped. Some extend landward across the continental shelf for great distances. As is shown on the physiographic map on the inside covers of this book, they typically cut through the edge of the continental shelf and terminate on the deep abyssal floor, some 5,000 or 6,000 m below sea level. The profile in Figure 18.8 crosses three canyons near the upper part of the continental slope. Submarine canyons have a V-shaped profile and a system of tributaries; they thus closely resemble the great canyons cut by rivers on the continents (Figure 18.9). Many pioneer researchers therefore suggested that submarine canyons were also cut by rivers, but the problem of how this could happen remained unanswered for many years. Some canyons are 6,000 m below sea level, and it was difficult to understand how sea level could have changed enough for rivers to cut

to that depth. Some even suggested that the continents were uplifted thousands of meters, were dissected by streams to form canyons, and then subsided, so that the canyons became submerged. With our increased knowledge of the characteristics of the ocean floor and our better understanding of submarine processes, it now appears that submarine canyons are usually cut by turbidity currents flowing from the continental shelf to the abyssal floor. Turbidity currents can move at rates ranging up to 95 km per hour (Figure 6.8, page 95) and carry a large sediment load capable of vigorous erosion.

GRAVITY MAPPING OF THE OCEAN FLOOR

Why is the topography of the ocean floor reflected in the elevation of sea level?

A new map of the ocean floor has recently been created from satellite measurements and reveals many

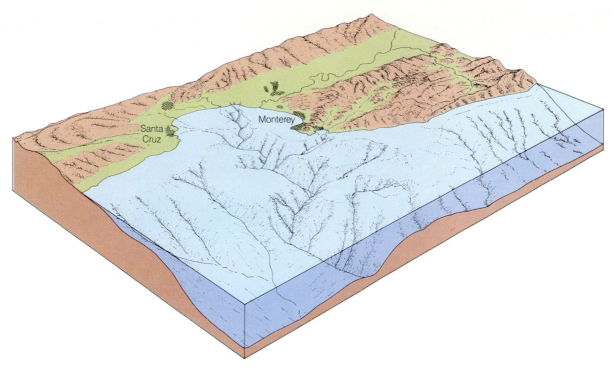

Figure 18.9
The Monterey submarine canyon, off the coast of southern California, shows many characteristics of canyons cut by rivers. A profile across the Monterey Canyon is similar to a profile across the Grand Canyon constructed with the same type of data.

details formerly either unknown or unconfirmed (Figure 18.10). Using a pencil-thin beam of microwaves, a radar altimeter on board NASA's Seasat satellite measured the distance between the satellite and the ocean surface with an accuracy of about 10 cm. This survey confirms that the ocean surface is as much as 200 m higher in some places than in others. The bulges and depressions occur because of gravitational differences that result from unequal distribution of mass in Earth's interior. Scientists also found much subtler variations on the ocean surface, and these corresponded with hills and valleys on the ocean floor. Where gravity is strong, such as around seamounts, water is attracted, creating a rise in sea level. This occurs because the added mass of rock in the seamount exerts a gravitational pull on the surrounding water. If, for example, the island of Hawaii were totally submerged, sea level would be about 30 m higher above the landmass than above the surrounding abyssal plains. In contrast, a trench has an ab-

sence of mass, so the ocean surface is depressed about 20 m above a deep trench. The ocean surface thus reflects the topography of the seafloor beneath it.

The map shown in Figure 18.10 was made by using a computer program to erase the regional bulges and depressions of the ocean surface caused by variations in Earth's internal mass. Only those variations resulting from the topography of the ocean floor remained. We are thus able for the first time to "see" a regional panorama of the ocean floor. We can recognize familiar features, originally discovered by sonar and plotted on physiographic maps, such as the mid-Atlantic Ridge, the Marianas Trench, long fracture zones, and other landforms. The new map has also revealed details of unexplored areas and shows previously unknown seamounts, continuous submarine ridges, and areas where the oceanic crust has been buckled by compression. Most intriguing are several swells on the Pacific floor, which may provide direct evidence of convection under the tectonic plates.

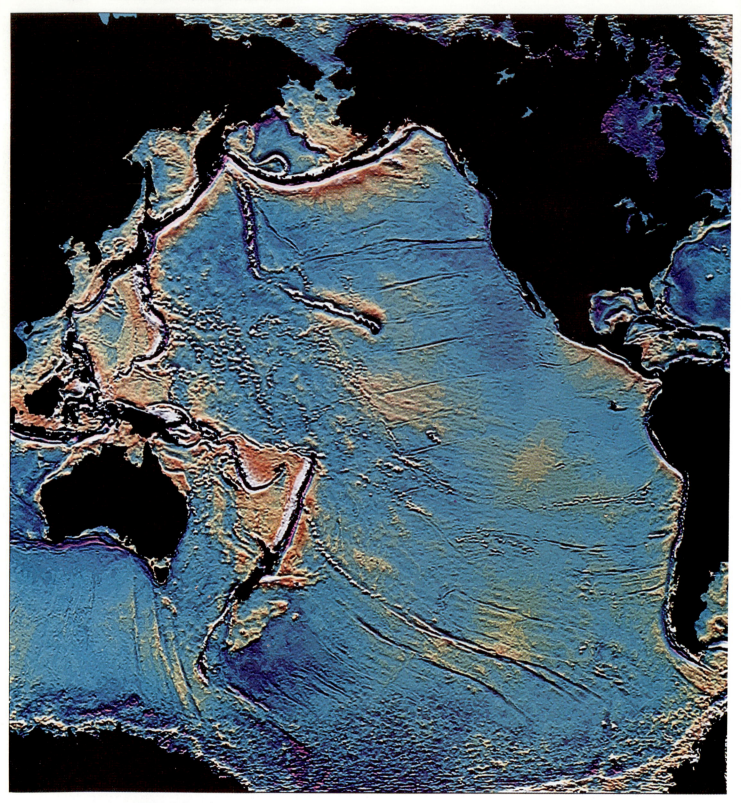

Figure 18.10
A map of the topography of the ocean floor was made from satellite measurements of the sea surface processed with sophisticated color computer graphics. This map not only confirms the data from sonar profiles, but reveals many details that were previously either unknown or unsubstantiated. For example, new seamounts and fractures were discovered in the Indian Ocean that will help geologists better understand how Africa, India, and Australia have drifted northward from Antarctica.

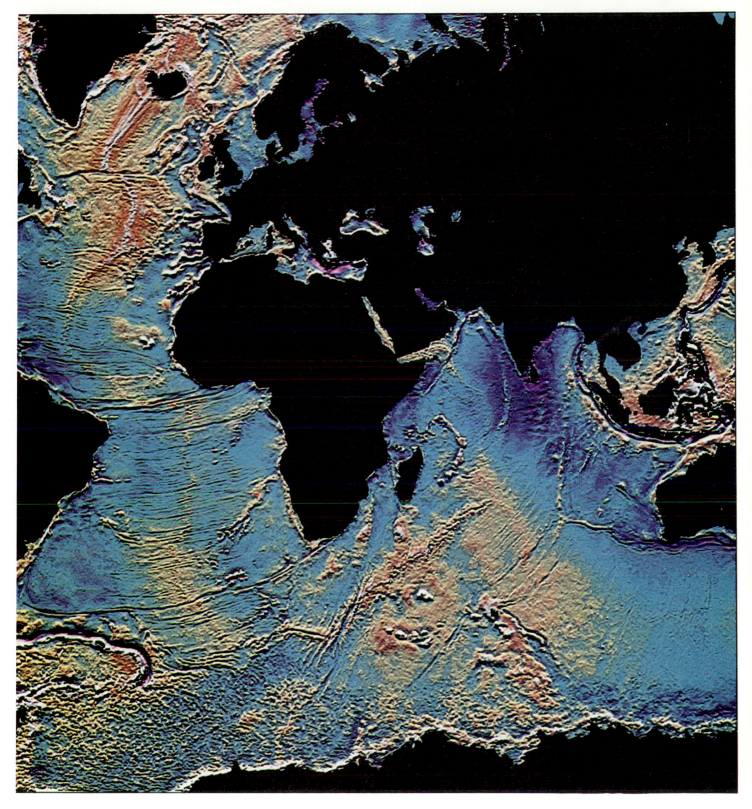

COMPOSITION AND STRUCTURE OF THE OCEANIC CRUST

How does the oceanic crust differ from the continental crust?

Statement

A variety of methods has been used to study the seismic composition and structure of the oceanic crust, such as seismic reflection, drilling, and studies of fragments of oceanic crust thrust up to the surface and plastered against the continent along convergent plate margins. Among the more significant facts we have learned about the oceanic crust are the following:

1. The oceanic crust is composed of four major layers: (a) a surface layer of marine sediment, (b) a layer of pillow basalt, (c) a zone of sheeted dikes, and (d) a layer of gabbro.
2. The oceanic crust and its topographic features are related in some way to igneous activity.
3. The rocks of the ocean basin have not been deformed by compression, so their simple structure stands out in marked contrast to the complex structure in the folded mountains and shields of the continents.
4. The rocks of the ocean basin are young in terms of geologic time. All appear to be fewer than 200 million years old, whereas the great bulk of continental rocks, the ancient rocks of the shields, are more than 700 million years old.

Discussion

The diagrams in Figure 18.11 illustrate the main elements of the composition and structure of the oceanic crust, as it is presently understood. At the top of the sequence (layer 1) is a relatively thin layer of sediment. This consists of shells of microscopic marine organisms together with red clay, which is derived from the continents. Layer 2 consists of pillow basalts, which were originally fed by numerous dikes and intruded into vertical fractures. The pillow lava represents extrusions of volcanic material on the seafloor, and the dikes are feeder vents, which were produced as the tectonic plates split and moved apart. Layer 3 consists almost entirely of the dikes, so that the individual basalt bodies are numerous vertical sheets. Layer 4 consists of rocks that are dominantly coarse-grained gabbro, which is believed to represent magma that was generated at a spreading center but cooled very slowly at some depth. Underlying the gabbro are peridotites, composed almost entirely of olivines and pyroxenes. This material is considered part of the mantle. The boundary between the gabbro and the peridotite is the Moho. Thus we see that the oceanic crust is strikingly different from the continental crust. Its general composition and structure are relatively simple, consisting of essentially flat layers of basalt and related rocks formed at the midoceanic ridge and partly covered with deep marine sediments.

Seismic reflection studies and drilling on the ocean floor provide important data concerning the upper layers of the oceanic crust. However, our most direct and detailed information is obtained from areas where large fragments of the oceanic crust have been incorporated in a folded mountain belt at a convergent plate margin and are available for direct observation. The most notable exposures are in Cyprus, Greece, New Guinea, Newfoundland, and California.

THE OCEAN BASINS

What is the difference in the size, shape, age, and history of the ocean basins?

Statement

Although all of the ocean basins originated in the same manner (that is, by seafloor spreading), they are quite different in size, shape, and topographic features. These differences are significant because they tell much about the ages and evolution of each individual ocean basin.

Discussion

The Atlantic Ocean

The structure and regional topography of the Atlantic floor are basically simple and reflected the initial opening of this young ocean basin. The Atlantic Basin also shows remarkable symmetry in the distribution of its major features (see the physiographic map inside the covers of this book). The dominant feature is the mid-Atlantic Ridge, which forms an S-shaped pattern down the center of the basin. It separates the ocean floor into two long, parallel subbasins, trending north and south, that are characterized by abyssal plains. Abyssal hills occur alongside the ridge, and the plains occur along the margins of the continental platforms. The symmetry of the Atlantic Basin extends to the continental margins: The outlines of Africa and Europe fit those of South America and North America.

The Arctic Ocean

Although the Arctic Ocean has been considered by some to be an extension of the Atlantic, recent explo-

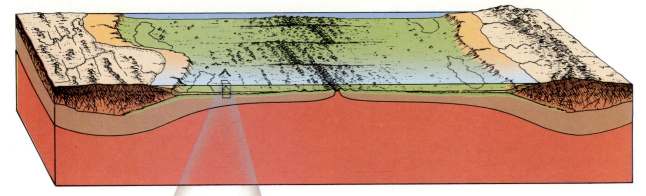

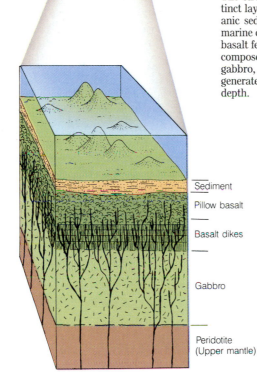

Figure 18.11
The structure of the oceanic crust consists of four distinct layers. Layer 1 is a relatively thin sequence of oceanic sediments, composed of the shells of microscopic marine organisms mixed with red clay. Layer 2 is pillow basalt fed by numerous dikes. Layer 3 is almost entirely composed of basalt dikes in vertical sheets. Layer 4 is gabbro, which is believed to represent magma that was generated at a spreading center and cooled slowly at depth.

Sediment

Pillow basalt

Basalt dikes

Gabbro

Peridotite
(Upper mantle)

ration shows that it is unique in several respects: (1) It is nearly landlocked. If the basin beyond the continental slope is considered, there is only one inlet, the Lena Trough (between Spitsbergen and Greenland). (2) The continental shelf north of Siberia is the widest shelf in the world (more than 1,100 km across). Most of the shelf is deeper than 200 m, but very little exceeds 600 m in depth. (3) Two prominent submarine ridges trend roughly 140° east and divide the deep Arctic Ocean into three major basins.

The Indian Ocean

The Indian Ocean is the smallest of the three great oceans. It connects with both the Atlantic and the Pacific through broad, open seas south of Africa and Australia. Like the Atlantic, its most conspicuous fea-

ture is the oceanic ridge, which continues from the Atlantic around southern Africa and splits near the center of the Indian Ocean to form a pattern similar to an inverted Y (see the physiographic map inside the covers). The northern segment of the ridge extends into the Gulf of Aden, apparently connecting with the African rift valleys and the Red Sea rift. The ridge thus divides the ocean basin into three major parts.

The topography of the Indian Ocean floor, unlike that of the other oceans, is dominated by scattered blocks and some remarkably linear plateaus, most of which are fragments of continental crust. Most are oriented in a north-south direction. Prominent parallel fracture zones are numerous. The striking northward trend of the parallel fracture zones, together with the trend of the linear microcontinents, imparts a linear structural fabric to the Indian Ocean floor.

The Pacific Ocean

The Pacific differs somewhat from the other oceans in that the oceanic ridge lies near its eastern margin. Its basin covers approximately half of the planet and is the largest region of oceanic crust. It is probably the oldest basin and lacks the symmetry of the Atlantic and Indian basins.

The oceanic ridge continues in a broad sweep from the Indian Ocean, trends eastward between New Zealand and Antarctica, and then turns northward along the American side of the ocean. The rifting associated with the ridge extends into the western United States at the head of the Gulf of California and is probably involved in the structure of the Basin and Range Province in Utah and Nevada and the San Andreas Fault. It reappears off the coast of Oregon.

The floor of the western Pacific is studded with more seamounts and atolls than all the other oceans combined. As is apparent from the physiographic map, many of the seamounts form linear chains that extend for considerable distances. The margins of the Pacific are also different from those of the other oceans. They generally are marked by a line of deep, arcuate trenches. In the eastern Pacific, the trenches lie along the margins of Central and South America, paralleling the great mountain systems of the Andes and the Rockies. Local relief from the top of the Andes to the bottom of the trench is 14,500 m (nearly nine times the depth of the Grand Canyon). In the western Pacific, a nearly continuous line of trenches extends from the margins of the Gulf of Alaska, along the margins of the Aleutians, Japan, and the Philippines, and down to New Zealand.

Small Ocean Basins

A number of small deep-ocean basins are nearly isolated from the major oceanic basins. Nevertheless, they are considered a direct consequence of plate tectonic processes. Most of them are nearly landlocked and connect to the main ocean by narrow gaps. The Mediterranean Sea, the Gulf of California, the Black Sea, the Red Sea, and the Sea of Japan are excellent examples. These small ocean basins originate in three ways: (1) by the growth of island arcs, (2) by the initial rifting of continental plates, and (3) by the convergence of continental plates and the destruction of larger ocean basins. The small ocean basins are temporary features. Like the major ocean basins, they grow but are ultimately destroyed.

The growth of island arcs created the small ocean basins of the Pacific. They formed along subduction zones that are not directly adjacent to a continent, so the associated volcanic arcs isolated segments of oceanic crust to form small independent basins. Several such basins extend from the Aleutian Islands to New Guinea. Most of them are shallower than the adjacent open ocean, partly because they trap the sediment washed from the continents.

The Red Sea and the Gulf of California are examples of another type of small ocean basin, which develops in the initial rift zone, where the continents split and begin to spread apart. Basalt from the upwelling mantle commonly fills the new opening during the early stages of rifting, and sediment eroded from the adjacent continents can help keep the basin filled to near sea level. If spreading is rapid, a long, narrow ocean basin develops.

As the sea floor continues to spread, the intervening sea develops the characteristics of a major ocean basin: an oceanic ridge, abyssal plains, and continental slopes. The Arctic Ocean is apparently in this transitional stage from sea to ocean.

The Mediterranean and Black seas, in contrast to those previously described, are basins formed by the closing of ocean basins as a result of converging continents, namely, Africa and Eurasia. These seas are remnants of the basin of the once vast Tethys Sea, which extended in an east–west direction and separated Pangaea into two large continental landmasses.

As is shown in Figure 18.12, the Mediterranean is almost landlocked, being connected with the Atlantic by a narrow gap through the Straits of Gibraltar. Evaporation removes more than 4000 km^3 of water each year from the Mediterranean Sea, but less than 500 km^3 are replaced by rain and surface runoff from Europe and Africa. Each year approximately 3500 km^3 of water flow in through the Straits of Gibraltar to maintain the Mediterranean at sea level. If the straits were closed, the Mediterranean Sea would evaporate in about 1000 years.

Total evaporation of the Mediterranean actually happened between 5 million and 8 million years ago. Core samples of windblown sand and salt deposits show that it was a stark desert region. Huge volcanoes (now islands) rose above the basin floor. The continental platforms of Africa and Europe surrounded the basin as huge plateaus. The area was undoubtedly devoid of life, since temperatures must have reached 65°C.

The salt deposits and windblown sand that formed in the dry basin are now covered throughout by the deep-sea organic oozes, which implies that the basin was flooded almost instantaneously about 5.5 million years ago. The inundation probably resulted from erosion of the barrier at the Straits of Gibraltar, so that water from the Atlantic flowed into the Mediterranean basin over an enormous waterfall. Estimates based on fossils preserved in the sediments indicate that the flow through the straits was 1000 times greater than the present-day flow over Niagara Falls and that the basin was filled in about 100 years.

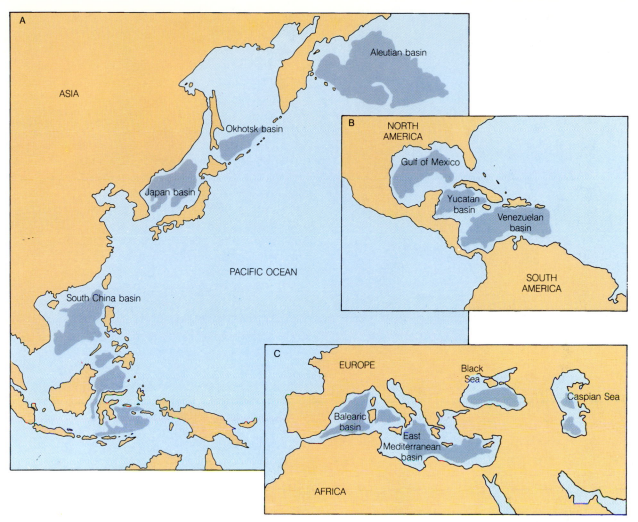

Figure 18.12
Small ocean basins originate in several ways. Those in the western Pacific and western Atlantic developed when island arcs isolated part of the sea from the main ocean basin. The Mediterranean basin and the Black and Caspian seas represent remnants of the ancient Tethys sea which was closed by the convergence of India and Africa with Europe and Asia.

HISTORY OF THE OCEANS

How have the oceans evolved? What major changes in the oceans and continents will likely occur in the next ten million years?

Statement

The considerable amount of data on plate motion accumulated during the last 200 million years enables us to reconstruct the positions of continents and to trace plate movement with considerable certainty. The variety of geologic data discussed in Chapter 15 suggest the same basic pattern of plate motion. They indicate that a large continental mass (Wegener's Pangaea) began to break up and drift apart about 200 million years ago. Dispersal and collision of the fragments have continued to the present time. The reconstruction of Pangaea in terms of absolute coordinates is now possible, and the directions and rates of plate movement have been determined.

Discussion

The history of relative plate movement during the last 200 million years in shown in Figure 18.13. These maps are adapted from those prepared by C. Denham and C. Scotese, which were constructed with cartographic precision. A computer was used to synthesize all available evidence, including the geologic fit of continents, paleomagnetic pole positions, and patterns of seafloor spreading.

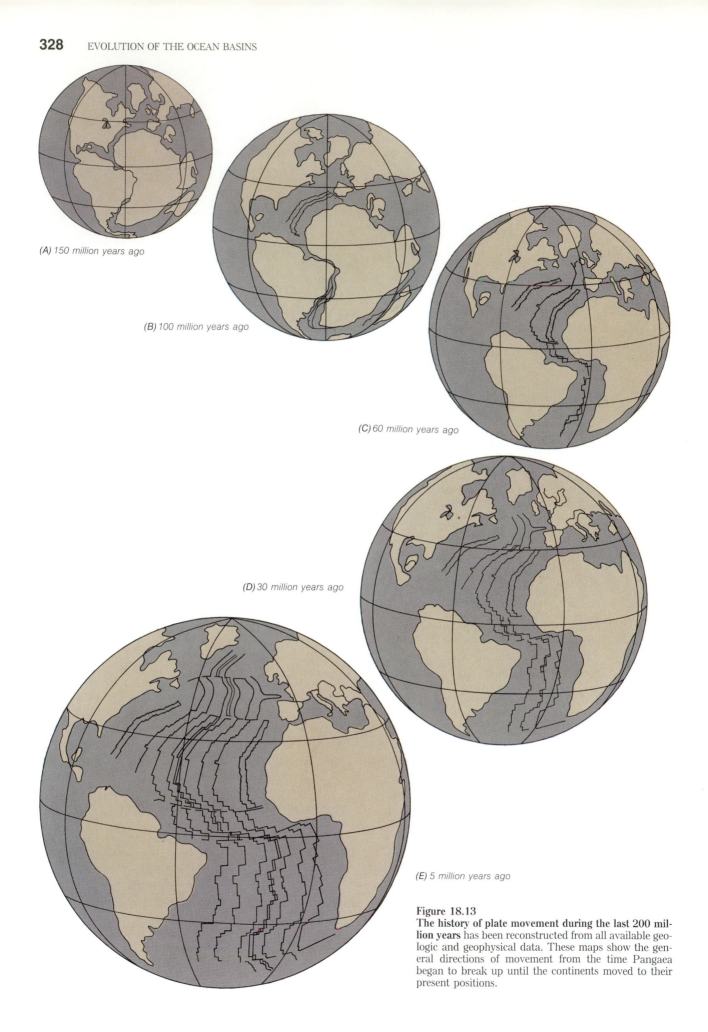

(A) 150 million years ago

(B) 100 million years ago

(C) 60 million years ago

(D) 30 million years ago

(E) 5 million years ago

Figure 18.13
The history of plate movement during the last 200 million years has been reconstructed from all available geologic and geophysical data. These maps show the general directions of movement from the time Pangaea began to break up until the continents moved to their present positions.

The initial event in the splitting of the continents was the extrusion of large volumes of basalt along the initial rift zone. Remnants of these basalts are found in the Triassic basins of the eastern United States and the plateau basalts of southwestern Africa, western India, and eastern Brazil. A northern rift split Pangaea along an east-west line, slightly north of the equator; it separated Laurasia from Gondwanaland, rotating Laurasia clockwise. A southern rift split South America and Africa away, and India was severed from Antarctica and moved rapidly northward. The plate containing Africa converged toward Eurasia, forming a subduction zone in the Tethyan Sea. By the end of the Cretaceous period, 65 million years ago, after 135 million years of movement, the separation of South America from Africa was complete, and the South Atlantic ocean had widened to at least 3,000 km. All the major continents were blocked out by this time, except for the connection between Greenland and Europe and between Australia and Antarctica. A new rift separated Madagascar from Africa, and India continued moving northward.

A north-south trench system must have existed in the Pacific (the South American-Central American trench) and consumed oceanic crust on the western edge of the rapidly westward-moving plates carrying North and South America. North America probably encountered this trench in late Jurassic time. The trench eventually was overridden with the continued westward drift of North America, causing the deformation of the Rocky Mountains. About the same time, the same trench was encountered by South America. This encounter developed the early Andean folded mountain belt.

Throughout the Cenozoic period (Figure 18.13), the mid-Atlantic Ridge extended into the Arctic and finally detached Greenland from Europe. During that time, the two Americas were joined by the Isthumus of Panama, which was created by volcanism along the subduction zone. The Indian landmass completed its northward movement and collided with Asia, creating the Himalayas. Australia drifted northward from Antarctica.

Finally, a branch of the Indian rift system split Arabia away from Africa, creating the Gulf of Aden and the Red Sea. Then a spur of the rift meandered west and south to create the East African rift valleys. Less pronounced changes induced the partial closing of the Caribbean region and the continued widening of the South Atlantic.

Figure 18.14 shows the anticipated position of plates fifty million years from now. Through the extrapolation of present directions and rates of plate movement, certain changes seem likely. The Atlantic and Indian oceans will continue to grow at the expense of the Pacific. Australia will move northward and encounter the Eurasian plate. The eastern part of Africa will split off along the rift valleys, but the northern part of the continent of Africa probably will move northward and close the Mediterranean Sea. New land will be created in the Caribbean by compressional uplift. Baja California and part of western California will be severed from North America and drift to the northwest.

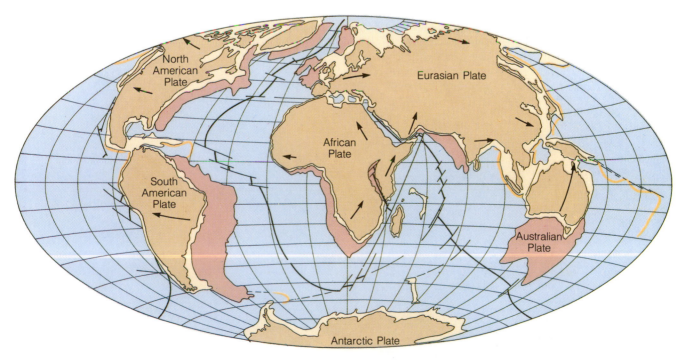

Figure 18.14
Anticipated plate movement during the next fifty million years.

SUMMARY

1. The oceanic ridge is a broad, fractured swell, extending as a continuous feature for more than 64,000 km. It has a narrow, axial rift valley and is cut by numerous transverse fractures.

2. Abyssal hills are formed at or near the ridge and move away from it with the drifting plates. As they move, they become covered with sediment. They can eventually be buried to form abyssal plains.

3. Trenches form at subduction zones, where two plates converge and one is thrust down into the mantle. Typically, they are bordered by island arcs or young mountain ranges.

4. Volcanic islands, or seamounts, can originate near a spreading center and move with the spreading plates. They may also form over local hotspots in the mantle, producing linear chains of islands as the plates move over the hotspots.

5. The continental shelf and slope are not parts of the oceanic crust but are submerged parts of the continental platform. The continental slope is dissected by many deep submarine canyons. These are similar in size and depth to the great canyons cut on the continents by rivers.

6. The oceanic crust is relatively thin (from 7 to 8 km) and is mostly basalt. It is composed of four major layers: (1) a surface layer of marine sediment, (2) a layer of pillow basalt, (3) a zone of sheeted dikes of basalt, and (4) a base layer of gabbro. The oceanic crust is not deformed by tight folding or thrust faulting. It is geologically young: essentially all of it was formed during the last 200 million years.

7. The three major ocean basins—the Atlantic, Indian, and Pacific—are believed to have originated from seafloor spreading. They differ from each other in size, shape, seafloor topography, and age. The small ocean basins in the western Pacific, the Red Sea, the Caribbean, and the Mediterranean resulted from one of three processes: the growth of island arcs, initial rifting, or the convergence of continental plates.

8. We can outline in some detail the history of the ocean basins during the last 200 million years. The Atlantic, Indian, and Arctic oceans were formed by the rifting of a supercontinent (Pangaea). The major fragments of the supercontinent moved with the spreading plates; as the ocean basins enlarged, the continents eventually arrived at their present positions.

KEY TERMS

abyssal floor (p. 316)
abyssal hills (p. 316)
abyssal plains (p. 316)
bathymetry (p. 314)
continental
  margin (p. 318)

continental rise (p. 318)
continental shelf (p. 318)
continental slope (p. 318)
deep-sea fan (p. 317)
gravity anomaly (p. 321)
island (p. 318)

oceanic ridge (p. 314)
pelagic sediment (p. 317)
rift valley (p. 314)
seismic-reflection
  profile (p. 314)
seamount (p. 318)

small deep-ocean
  basin (p. 326)
submarine
  canyon (p. 320)
trench (p. 318)

REVIEW QUESTIONS

1. Sketch a map of the major types of landforms found on the ocean floor.
2. Study the seismic-reflection profile in Figure 18.1 (page 315) and label the following: (a) oceanic sediment, (b) bedrock, and (c) probable faults.
3. Explain the origin of the following: (a) the oceanic ridge, (b) fracture zone, (c) abyssal hills, (d) abyssal plains, (e) trenches, (f) seamounts, and (g) submarine canyons.
4. Compare and contrast the Atlantic Ocean and the Pacific Ocean basins with respect to shape, age, major structural features, and history.
5. Briefly outline the history of plate movement during the last 200 million years.

Multiple-Choice Questions

1. Deep-sea trenches are most common in the
   a. Arctic Ocean.
   b. Atlantic Ocean.
   c. Indian Ocean.
   d. Pacific Ocean.
   e. Antarctic Ocean.
2. In the oceanic crust, the major layer beneath the pillow basalts is typically
   a. oceanic sediments.
   b. basalt dikes.
   c. gabbro.
   d. mantle material.
   e. andesite.
3. Abyssal plains in the Atlantic Ocean are located
   a. in the central part of the ocean.
   b. adjacent to the continents.
   c. at random in the deepest parts of the basin.
   d. adjacent to the crest of the mid-Atlantic Ridge.
   e. in the deep trenches.
4. Submarine canyons are most numerous and best developed on the
   a. abyssal hills.
   b. continental slope.
   c. abyssal plains.
   d. oceanic ridge.
   e. continental shelf.
5. The midoceanic ridge is
   a. a folded mountain belt.
   b. an uparched segment of the ocean floor split by rifting and transverse fractures.
   c. a zone where crust plates collide and are uparched.
   d. a zone where plates converge and descend back into the mantle.
   e. isolated segments of continental crust.
6. Seismic-reflection profiles of the ocean floor show
   a. the depth of the seafloor.
   b. a cross-sectional view of the geologic structure of the upper oceanic crust.
   c. a topographic profile of the ocean floor.
   d. all of the above.
   e. only (a) and (c).

7. The best explanation for the formation of submarine canyons is
   a. faulting.
   b. subduction.
   c. differential sedimentation.
   d. subaerial erosion modified by submarine erosion.
   e. submarine erosion only.
8. Seamounts are
   a. all flat-topped.
   b. atolls.
   c. andesitic.
   d. basaltic.
   e. of uniform height.
9. Most scientists believe that India
   a. has recently migrated north to its present location.
   b. was always part of the Asian landmass.
   c. rose from the Tethys Sea.
   d. is moving away from the Asian landmass.
   e. was built out as a delta from the rest of Asia.
10. An excellent example of the initial stage in the evolution of an ocean basin is the
    a. Indian Ocean.
    b. Atlantic Ocean.
    c. Mediterranean Sea.
    d. Red Sea.
    e. Black Sea.

True/False Questions

1. *T F* Oceanic sediments on the seafloor are relatively thin but are deformed into tight folds.
2. *T F* Oceanic sediments are thickest down the flanks of the oceanic ridge but thin rapidly toward the crest.
3. *T F* There are numerous large abyssal plains in the Pacific Ocean.
4. *T F* The topographic features of the continental shelf are formed by marine processes only.
5. *T F* A trench has an absence of mass, so the ocean surface is depressed about 20 m above a deep trench.

Fill in the Missing Terms

1. The most pronounced tectonic feature on Earth is the _____ _____.
2. A _____ is an isolated conical mount rising more than 1,000 m above the ocean floor. Most are probably submerged shield volcanoes.
3. The regional topography of the floor of the _____ Ocean shows remarkable symmetry in the distribution of the major features.
4. The floor of the western _____ Ocean is studded with more seamounts, volcanic islands, and atolls than all of the other ocean basins combined.
5. The main sources of sediment for deep-sea fans are the major _____.

19

EVOLUTION OF THE CONTINENTS

Although the theory of plate tectonics was developed largely from new studies of the seafloor, a record of plate motion during most of geologic time is preserved only in continental rocks. The oceanic crust is a temporary feature, continually created along the oceanic ridge and then destroyed at subduction zones. Plate movement can be studied and measured in the relatively young oceanic rocks, but essentially all of the oceanic crust is less than 190 million years old. By contrast, the rocks of the continents are as old as 3.8 billion years, and a record of plate movement during the early history of Earth is preserved only in the ancient metamorphic and igneous rocks of the continental shields.

On the continental shields, geologists find evidence of a long and complex history of Earth's dynamics. They also find evidence of how the continents originated and evolved. No fragments of original crust have ever been found. Why did continents evolve on our planet? Did Earth always have plate tectonics? Are continents still evolving?

In this chapter we will consider these questions and how deformation, metamorphism, igneous activity, erosion, sedimentation, and isostatic adjustment play a role in continental evolution.

MAJOR CONCEPTS

1. The continents are made up of three basic structural components: (a) shields, (b) stable platforms, and (c) young, folded mountain belts.
2. The major results of mountain building are (a) intensive crustal deformation, (b) metamorphism, and (c) igneous activity.
3. The characteristics of a mountain belt and the sequence of events that produced it depend on the types of interactions at convergent plate boundaries and the rock types that are involved.
4. Three major convergent plate interactions are recognized: (a) the convergence of two oceanic plates, (b) the convergence of a continental and an oceanic plate, and (c) the convergence of two continental plates.
5. Continents grow by accretion as new crustal material forms in a mountain belt.

THE CONTINENTAL CRUST

What makes the continental crust unique?
Why don't Mars, Mercury, and the Moon have continents?

Statement

The more important geologic characteristics of continents are:

1. Continental crust consists of huge slabs of "granitic" rock (30–50 km thick).
2. The continents contain the oldest rocks on Earth, which range in age up to 3.8 billion years.
3. Although each continent is unique, they all have three basic components:
   a. A large area of exposed basement complex, known as the shield, which consists of very old and highly deformed metamorphic and igneous rocks.
   b. Broad, flat stable platforms, where the basement complex is covered with a veneer of sedimentary rocks.
   c. Young folded mountain belts, which form along the continental margins.
4. Geologic differences among continents are mostly in the size, shape, and proportions of the three basic components.

Discussion

Why do we believe that the shields consist of a series of belts formed in the roots of ancient mountain systems?

The Basement Complex

The basement complex is the key to modern theories of the origin and evolution of continents. Take a moment to study Figure 19.1, which shows a part of the Canadian Shield, an area where the basement complex is exposed. The most striking characteristic in this photograph is the vast expanse of the low, relatively flat surface of the shield. Throughout an area of thousands of square kilometers, this surface lies within a few hundred meters of sea level. The only features that stand out in relief are resistant rock formations, which rise a few tens of meters above the surrounding, less resistant rocks. On a regional basis, shields are flat and almost featureless. A fundamental characteristic of the basement complex is that it is composed of a highly deformed sequence of metamorphic rocks and granitic intrusions. The structural complexities in Figure 19.1 are shown by patterns of

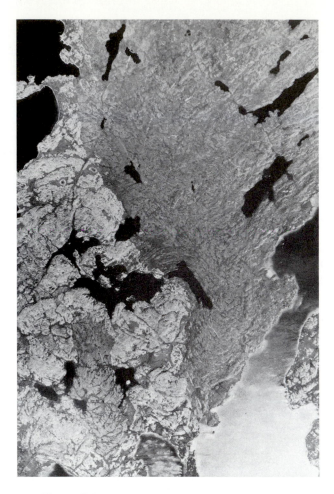

Figure 19.1
Complex metamorphic rocks (dark tones) are intruded by granitic rocks (light tones) to form the complex structural characteristics of a shield. These rocks were formed in the roots of an ancient folded mountain belt during Precambrian time, approximately 1.8 billion years ago. On a regional basis, the shield is a broad surface of low relief, eroded close to sea level.

erosion, the alignment of lakes, and differences in the tones of photographs. Faults and joints are common and are expressed at the surface by linear depressions, some of which can be traced for hundreds of kilometers. Most of the rocks are Precambrian, and most were formed under high temperature and highly compressive stresses several kilometers below the surface. Figure 19.1 shows evidence that the basement rocks have been intensely deformed, so that the deeper sedimentary and volcanic rocks were converted to complex metamorphic rocks. They were also intruded by granitic magmas. Subsequently, erosion removed the upper cover of the sedimentary and metamorphic terrain, exposing what we now see at the surface.

The belts of metamorphic rocks, together with igneous intrusions, indicate that the shields are composed of a series of zones that were once highly mobile and tectonically active. These facts have long been known. For many years, geologists have considered the shields to consist of a series of ancient mountain belts that have been eroded down to their roots and have since remained stable.

Stable Platforms

Large areas of the basement are covered with a series of horizontal sedimentary rocks, as is shown in and is known as the stable platform (Figure 19.2). The stable platforms form much of the broad, flat lowlands of the world and are known locally as plains, steppes, and low plateaus. The relationship between the sedimentary rocks of the stable platform and the underlying basement complex is known from thousands of wells that penetrate the sedimentary cover and from seismic studies, which reveal the rock structure beneath the surface. Although locally the rocks appear almost perfectly horizontal, on a regional basis they are warped into broad, shallow domes and basins.

The flat-lying sedimentary rocks that cover parts of the basement complex are predominantly sandstone, shale, and limestone, which were deposited in ancient shallow seas. These flat-lying marine sediments, preserved on all continents, show that large areas of the basement complex have periodically been flooded by the sea and have then reemerged as dry land.

There are several reasons for the periodic flooding of large areas of the shields during various periods of geologic time. The shields are broad, flat areas eroded down to within a few tens of meters of sea level. Uplifts or downwarps, which occur as the tectonic plates move across the globe, thus cause the shallow seas to expand and contract over the continents.

Mountain Belts

What is distinctive about mountain belts?

The great, linear folded mountain belts, which occur along the continental margins, are one of Earth's most distinctive tectonic features. The rocks within these relatively narrow zones are compressed and folded and in many places are broken by thrust faults. As a result, the crust is shortened by as much as 30 percent (Figure 19.3).

The young, active mountain belts of today occur along convergent plate margins, coinciding with zones of intense seismicity and andesitic volcanism. Two major mountain belts are still in the process of growing. They are (1) the Cordilleran belt, which includes the Rockies and the Andes, and (2) the Himalaya-Alpine belt, which extends across Asia and western Europe and into North Africa. Older mountain ranges can still be expressed by significant topographic relief, although deformation ceased long ago. Examples include the Appalachian Mountains of the eastern United States, the Great Dividing Range of eastern Australia, and the Ural Mountains of Russia.

The location of young mountains in long, narrow belts along continental margins is significant because it implies that mountains do not result from a uniform, worldwide force evenly distributed over Earth.

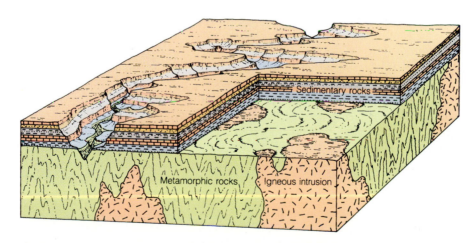

Figure 19.2

The general characteristics of the continental crust are depicted in this block diagram. The shield consists of a complex of metamorphic rocks and a variety of igneous intrusions. The upper surface is flat and is eroded down to near sea level. Throughout most of the interior of the United States, the basement complex is covered with a veneer of horizontal sedimentary rock. In some places, such as the Grand Canyon, erosion has cut through the sedimentary veneer to expose the basement complex below.

Figure 19.3
The structure of a folded mountain belt in western Pakistan is clearly visible in this satellite photograph. The ridges are plunging anticlines, many of which are in the initial stage of erosion. Some are partly breached, and exposures of older rocks can be seen in their cores. Young mountain belts such as this occur along continental margins as a result of the motion of converging plates.

They must be the result of forces concentrated along the margins of continents.

Another important aspect of their location is that many of the older mountain belts extend to the ocean and abruptly terminate at the continental margin. The northern Appalachians of the eastern United States, the Atlas Mountains of Africa, and the mountains of Great Britain are excellent examples. The abrupt termination of the folded structures suggests that the mountain systems were once much more continuous and have been separated by continental rifting.

From the point of view of geologic structure and mode of origin, all continents are similar. They are all composed of shields, stable platforms, and folded mountain belts (Figure 19.4). They differ mostly in the size, shape, and proportions of these structural features.

MOUNTAIN BUILDING

What is the origin of mountains?

Statement

The process of mountain building is called orogenesis. The factors that appear to be most important in this process are

1. structural deformation.
2. rock sequences.
3. igneous activity.
4. erosion and isostatic adjustment of the crust, which occur after mountain building and continue until the mountain belt is eroded to near sea level.

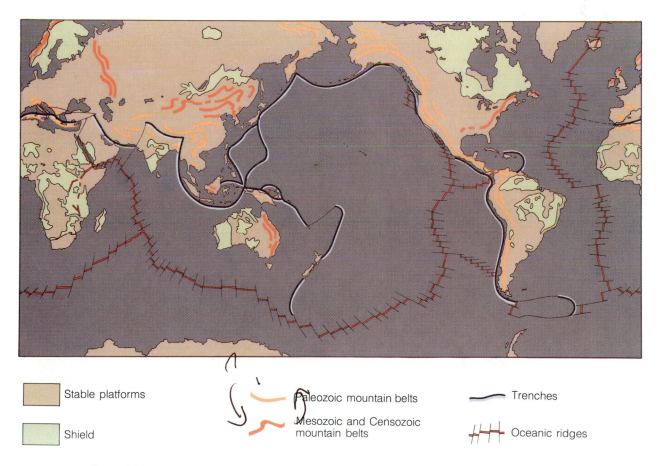

Stable platforms

Shield

Paleozoic mountain belts

Mesozoic and Censozoic mountain belts

Trenches

Oceanic ridges

Figure 19.4
Major tectonic features of the Earth provide a record of global dynamics during the last 3.8 billion years. The shields formed more than 600 million years ago and were eroded down to sea level. The stable platforms provide a history of sedimentation in the shallow seas that periodically covered the cratons during the past 600 million years. Paleozoic mountain belts represent ancient convergent plate margins, which existed more than 190 million years ago. Deep-sea trenches, young folded mountain belts, and the oceanic ridge outline the boundaries of the presently active plates.

Structural Deformation

Is there an organized structure in mountains?

Since the beginning of geologic studies more than 150 years ago, geologists have known that rocks in certain parts of the continents have been folded and fractured on a gigantic scale. Deformation of the crust is most intense in the great mountain belts of the world, where sedimentary rocks, which were originally horizontal, are now folded, contorted, fractured, and, in some places, completely overturned. In some mountains large bodies of rock have been thrust several tens of kilometers over younger strata. In the deeper parts of mountain belts, deformation is great enough to cause metamorphism.

In order for you to appreciate the magnitude and significance of the structural deformation of the rocks in the crust, you must first become familiar with some of the terms used to describe them. There are three general types of structures: folds, faults, and joints. Although we have defined these terms earlier, let us consider some of the details of their geometry and surface expression.

Folds

Folds are warps (wavelike contortions) in rock bodies. They are three-dimensional structures ranging in size from microscopic crinkles to large flexures that extended across several states. Small folds are common and can be seen in a mountainside, road cut, and even in hand specimens. Large folds cover thousands of square kilometers and are best recognized from air or space photographs. Three general types of folds are illustrated in Figure 19.5. Monoclines are folds in which horizontal or gently dipping beds are modified by simple steplike bends. Anticlines, in their simplest form, are uparched strata, with the limbs (sides) of the fold dipping away from the crest. Rocks exposed in an eroded anticline are progressively older toward the interior of the fold. Synclines, in their simplest form, are downfolds, or troughs, with the limbs dipping toward the center. Rocks exposed in an eroded syncline are progressively younger toward the center of the fold.

Fold geometry is not exceedingly complex. In many ways folds resemble the wrinkles in a rug, but many folds are large and we may see only isolated exposures of its various parts. The diagrams in Figure

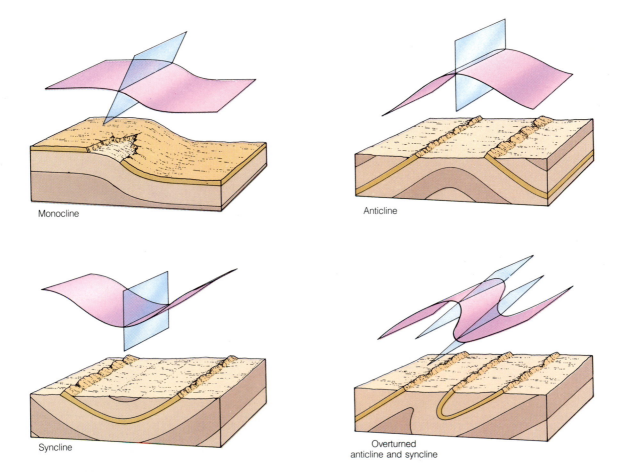

Monocline

Anticline

Syncline

Overturned
anticline and syncline

Figure 19.5
The nomenclature of folds is based on the three-dimensional geometry of the structure, although most exposures show only a cross-section or map view.

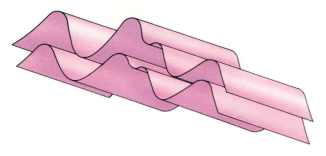

(A) The basic form of folded strata is similar to that of a wrinkled rug. In this diagram the strata are compressed and plunge toward the background.

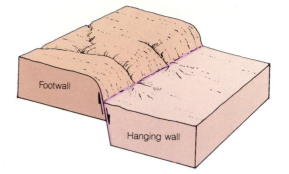

(A) In normal faults the hanging wall moves downward in relation to the footwall.

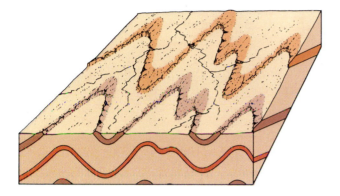

(B) If the tops of the folded strata are eroded away, a map of the individual layers shows a zigzag pattern at the surface. Rock units that are resistant to erosion form ridges, and nonresistant layers are eroded into linear valleys. In a plunging anticline, like the one shown here, the surface map pattern of the beds forms a "V" pointing in the direction of a plunge.

Figure 19.6
A series of plunging folds forms a zigzag outcrop pattern.

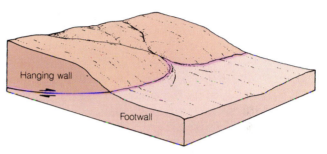

(B) In thrust faults the hanging wall moves upward in relation to the footwall.

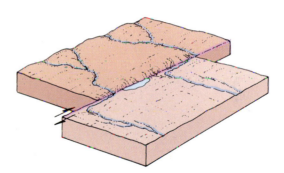

(C) In strike-slip faults the displacement is horizontal.

Figure 19.7
The three major types of faults are distinguished by the direction of their relative displacement.

19.6 illustrate the general configuration of folds and their surface expressions after they have been eroded. The basic form is shown in diagram A. Diagram B shows a fold system after the upper part has been removed by erosion. The outcrop of the eroded anticlines and synclines forms a characteristic zigzag pattern.

Faults

Faults are fractures along which slippage or displacement has occurred. The major types of faults are illustrated in the block diagram in Figure 19.7. In a road cut or in the walls of a canyon, a fault plane may be obvious, and the displaced, or offset, beds can easily be seen. Elsewhere, the surface expression of a fault can be very subtle, and detailed geologic mapping may be necessary before the precise location of such a fault can be established. In normal faults movement is mainly vertical, and the rocks above the fault plane move downward with respect to those beneath the fault plane.

Thrust faults are generally inclined at a low angle, and the rocks above the fault plane have moved up and over those below. They generally result from crustal shortening and are commonly associated with intense folding. They are prominent in all of the world's major mountain regions and commonly evolve from folds in a manner diagrammed in Figure 19.8.

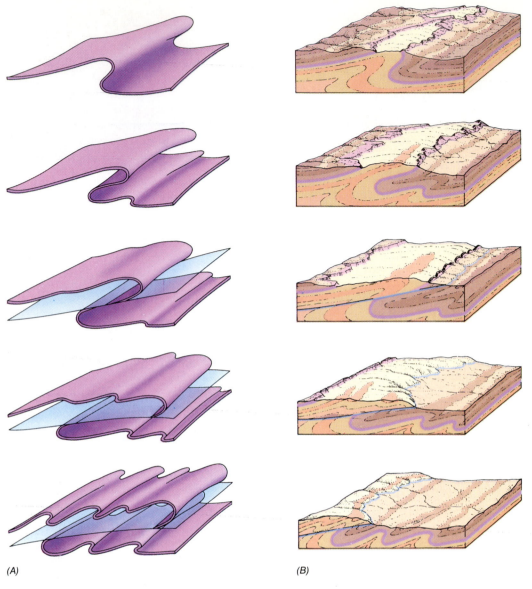

(A)   (B)

Figure 19.8
The evolution of thrust faults from folds is depicted in this sequence of diagrams. Diagram A shows the fault plane and the progressive development of folds into a thrust fault. Diagram B shows how the structure might be expressed at the surface, as a result of contemporaneous erosion.

Strike-Slip Faults

Strike-slip faults are high-angle fractures in which the displacement is horizontal, parallel to the strike of the fault plane. There is little or no vertical movement, so that high cliffs do not form along strike-slip faults. Instead, these faults are expressed topographically by a straight, low ridge extending across the surface, which commonly marks a discontinuity in the landscape.

Some of the topographic features produced by strike-slip faulting and subsequent erosion are shown in Figure 19.9. One of the more obvious is the offset of the drainage pattern. The relative movement is often shown by abrupt right-angle bends in streams at the fault line: A stream follows the fault for a short dis-

tance and then turns abruptly and continues down the regional slope. As the blocks move, some parts may be depressed to form sag ponds. Others buckle into low, linear ridges.

Joints

Joints, in contrast to faults, are fractures in rocks along which no appreciable displacement has occurred. They are very common and are found in almost every exposure. Usually they occur as two sets of fractures, which intersect at angles ranging from 45° to 90°. They thus divide the rock body into large, roughly rectangular blocks. Joints can be related to major faults or to broad upwarps of the crust (Figure 19.10).

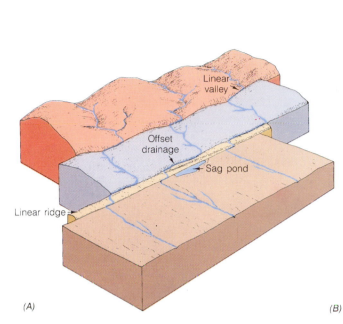

(A)

(B)

Figure 19.9
Strike-slip faults produce distinctive landforms such as streams are offset by recurrent movement, linear ridges and valleys form, and local sag ponds develop along the fault line. The San Andreas Fault, in California, is a major strike-slip fault. It is delineated by prominent, straight ridges and valleys. Recent movement has offset the drainage patterns on both sides of the fault. Relative movement between the fault blocks is evident from the direction in which the drainage is offset.

Figure 19.10
Joint systems in resistant sandstones in Arches National Park, Utah, have been enlarged by weathering to form deep, narrow crevasses. The set of intersecting joints reflects the orientation of the stress that deformed the rock body.

Structure of Mountains

From detailed field mapping, the structure of an entire mountain range can be determined and illustrated by geologic maps and cross sections. Figure 19.11 includes three well-known examples.

In the Canadian Rockies, a major type of deformation is thrust faulting, in which large blocks of rock have been thrust over others in a belt 60 km wide (Figure 19.11A). The orientation of the faults and the direction of displaced beds indicate that the rocks were thrust from the margin of the continent toward the interior.

A cross section of the Appalachian Mountains shows a different style and magnitude of deformation (Figure 19.11B). The major structural feature is a series of tight folds. Deformation is most intense near the continental margins, dying out toward the continental interior.

The structure of the Alps is even more complicated (Figure 19.11C). Great overturned folds called *nappes* (French for "tablecloths") show enormous amounts of crustal shortening. The rocks are so intensely deformed that spherical pebbles have been stretched into rods as much as thirty times longer than the original diameters of the pebbles. Most of these structures can be explained only in terms of compressive forces.

Note that, in each mountain belt, the internal structures result from strong horizontal compressive forces. Similar deformation exists in older rocks in places where erosion has reduced the topographic relief.

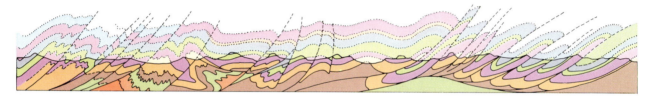

(A) The Canadian Rockies contain both folds and thrust faults.

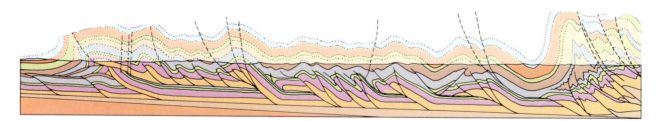

(B) The Appalachian Mountains consist of tight folds and thrust faults that have been eroded down to within 1,000–3,000 m of sea level. Resistant sandstones form the mountain ridges.

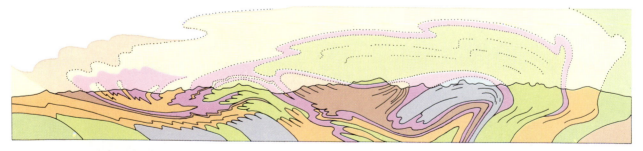

(C) The Alps are complex folds, many of which are overturned.

Figure 19.11
The structure of folded mountain belts reflects intense compression at convergent plate boundaries, but each range can have its own structural style.

Metamorphism

In the deeper parts of the orogenic belt, intensive plastic deformation and recrystallization at elevated temperatures and pressures transform the original sedimentary and volcanic rocks into schists and gneiss. The horizontal stress generated by converging plates causes the recrystallization of many minerals and develops foliation that is perpendicular to the direction of stress. Slaty cleavage, schistosity, and gneissic layering in the deeper parts of a mountain range are thus characteristically vertical or dip at a high angle.

In the deeper parts of a mountain belt, metamorphism can become intense enough to melt the rock and produce granitic magma.

Rock Sequences and Mountain Building

Studies of mountain belts in different parts of the world indicate that two roughly parallel belts of different types of thick sediments commonly exist along continental margins. One belt, called a miogeocline, is adjacent to the stable platform and is underlain by continental crust. It consists of clean, well-sorted, shallow-water sandstone, limestone, and shale, with no volcanic rocks. The other belt, called a eugeocline, consists of sediments deposited in deep-marine environments and typically includes volcanic rocks (Figure 19.12). A third rock assemblage, present in some mountain belts, is called an ophiolite sequence. It consists of peridotite, gabbro, pillow basalt, and the deep-marine sediments that form on the oceanic crust.

The theory of plate tectonics proposes that mountain building occurs along convergent plate margins and that mountain belts are formed by deformation of the geoclinal sediments. When they are deformed at convergent plate margins, different sequences produce different styles of mountain belts.

Igneous Activity

Why do andesitic and granitic magmas form in mountain belts?

Igneous activity associated with mountain building is part of the fundamental process of differentiation by which Earth's materials are separated and concentrated into layers according to density.

The concentration of Earth's low-density material in the continental crust occurs in two steps. The first phase begins at a spreading center, where partial melting of peridotite in the upper mantle generates a basaltic magma that rises to form oceanic crust. Basalt is richer than peridotite in the lighter elements, particularly silicon and oxygen. In the second phase, partial melting of the oceanic crust forms a silica-rich magma, which is then emplaced in the mountain belt as granitic intrusions and andesitic volcanic products. This process further separates the lighter elements, especially silicon and oxygen, and concentrates them in the continental crust. The granitic continental crust is less dense than the mantle and the oceanic crust, and its buoyancy prevents it from being consumed at subduction zones. Once it is formed, the continental crust remains on the outer surface of

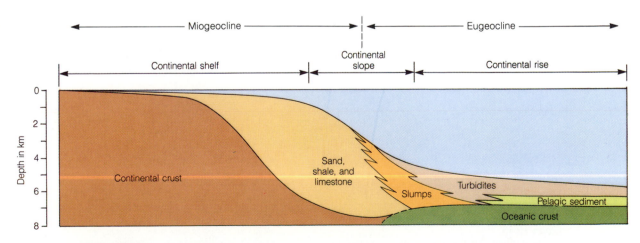

Figure 19.12
Two types of rock sequences accumulate along continental margins. Clean, well-sorted sandstone, shale, and limestone, deposited in shallow water, accumulate on the continental shelf. Poorly sorted, dirty sandstone and shale are deposited by turbidity currents in the deep water beyond the continental margins.

Earth. The generation of silica-rich magma at a subduction zone is shown in Figure 19.13. The magma conceivably can result from partial melting in three different regions of the subduction zone: (1) in the subducting oceanic crust, (2) in the overlying mantle, as hot fluids percolate upward, and (3) near the base of the continental crust, as upward-migrating magma raises the temperature. Water in the pore spaces of the rock and in chemical combination in many minerals of the oceanic crust plays an important role in the type of igneous activity at convergent plate margins. As the crust is subducted and heated, this water is driven out. It percolates upward and enhances melting at higher levels in the crust.

The Evolution of a Mountain Belt

The general concept of how a mountain belt evolved into a new region of the continental basement complex is shown in Figure 19.14. Folding and thrusting occur at relatively shallow depths, metamorphism occurs deeper, and partial melting occurs at still greater depths. Granitic magma initially forms deep within the crust, at points where the rock begins to melt. The magma is then injected into the foliation of the adjacent metamorphic rock. Much of the magma migrates upward because it is less dense than the solid rock. As it rises, it forms teardrop-shaped bodies, which collect into larger and larger masses. The boundaries of the body of magma are generally parallel to the broad zones of foliated metamorphic rock. The rising granitic magma cuts across the upper folded strata, which are not metamorphosed but are only deformed by folding and faulting. The magma can cool within a few kilometers of the surface, forming a batholith, or it can be extruded as andesitic volcanic materials.

Erosion and Isostatic Adjustment

The history of a mountain belt does not end with deformation, metamorphism, and igneous activity. After the orogenic activity is terminated (presumably because of shifts of the convection currents in the mantle), deformation ceases, but erosion and isostasy combine to modify the orogenic belt. As is illustrated in Figure 19.14, the entire crust is deformed by mountain building, so that a mountain root, which is composed of the most intensely deformed rocks, ex-

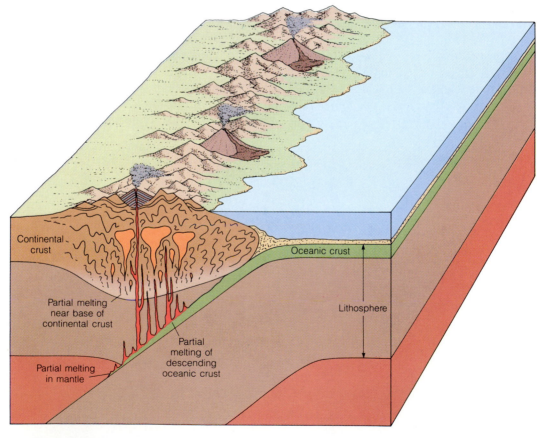

Figure 19.13
The generation of silica-rich magmas can occur by partial melting in three different regions of a subduction zone. The process can involve (1) partial melting of the descending oceanic crust, (2) partial melting of the upper mantle, and (3) partial melting of the lower part of the continental crust.

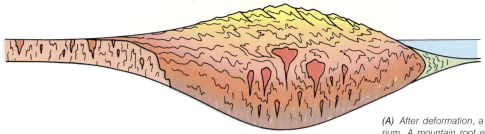

(A) After deformation, a mountain belt is in isostatic equilibrium. A mountain root extends down into the lithosphere to compensate for the high mountainous topography. Thrust faults and folds are the dominant structures exposed at the surface.

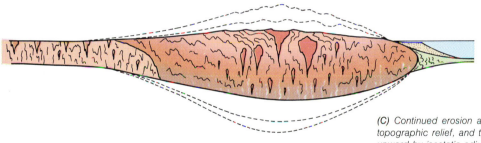

(B) As erosion removes rock from the mountain belt, isostatic equilibrium is upset, and the mountain root rebounds, so that some topographic relief is maintained. Tight folds, formed in the deeper part of the mountain belt, then become the dominant structures exposed at the surface.

(C) Continued erosion and isostatic adjustment reduce the topographic relief, and the root of the mountain belt moves upward by isostatic adjustment. Complex folds and igneous intrusions, originally formed deep in the mountains, are the dominant structures exposed at the surface.

(D) Ultimately, isostatic equilibrium is reached again. The mountain root, containing metamorphic and igneous rocks, is exposed at the surface, and the topography is eroded to sea level. At this stage the mountain belt constitutes a new segment of the basement complex.

Thrust faulting and folds

Intense folding

Large batholiths, discordant contacts

Small intrusions, concordant contacts with metamorphic rocks

Metamorphic rocks, small pods of intrusive rocks, concordant contacts

Figure 19.14
Erosion and isostatic adjustment combine to transform a newly deformed mountain belt into a segment of the basement complex of the continent.

tends down into the mantle. The high mountain range and its deep roots are in isostatic balance. As erosion wears away the high mountains, this balance is destroyed. Isostasy then causes the mountain belt to rise in a broad upwarp to reestablish the balance. In the early stages of erosion, the removal of 500 m of rock is accompanied by an isostatic uplift of approximately 400 m, with a net lowering of the surface of only 100 m. The rate of net lowering is believed to decrease as the mountain belt is eroded and approaches a new state of isostatic equilibrium. Uplift continues as long as erosion removes material from the mountain range. Eventually, however, a balance is reached, when the mountainous topography is eroded to near sea level and the deep mountain root is isostatically adjusted.

It is important to note that the rocks exposed at the surface when isostatic balance is established are the metamorphic and igneous rocks formed at great depths in the orogenic belt. These rocks are tectonically stable and become part of the basement complex.

TYPES OF MOUNTAIN BUILDING

Why are there three different types of mountain building?

Statement

There are three fundamentally different types of convergence, and therefore there are three different types of mountain building:

1. Convergence of two oceanic plates

2. Convergence of continental and oceanic plates
3. Convergence of two continental plates

Discussion

Convergence of Two Oceanic Plates

The main types of orogenic activity involving the convergence of two oceanic plates are shown in Figure 19.15. The major result is an arc of volcanic islands above the subduction zone. In the first stages of convergence, the process can be relatively simple and restricted to volcanic activity on the overriding plate. It does not involve widespread metamorphism or granitic intrusion. The Tongan Islands are an example of a simple island arc in the present oceans.

In other island arcs (such as Japan), crustal deformation, metamorphism, and igneous activity occur and produce more complex rock associations and deformational patterns.

Mountain building caused by the convergence of two oceanic plates is distinctive in (1) rock sequence (andesitic and volcanic sediments), (2) metamorphism (blueschist), and (3) zones of ophiolite and volcanic rock types. It is significant in that it represents the initial stage in the development of new continental crust in areas where only oceanic crust previously existed. This process provides important insight into the origin and evolution of the continents.

Convergence of Continental and Oceanic Plates

A schematic cross section showing the major features produced by the convergence of continental and oceanic plates is shown in Figure 19.16. Before the continental crust arrives at the subduction zone, a considerable thickness of sandstone, shale, and lime-

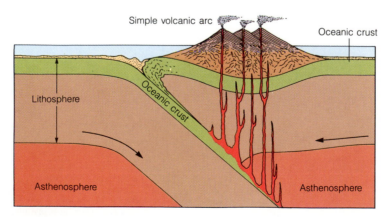

Figure 19.15
Ocean-to-ocean mountain building is restricted largely to volcanic activity and does not involve widespread metamorphism or granitic intrusions. An example is the Tongan island arc in the South Pacific.

stone can accumulate along the continental margins. During this same period, deep-marine sediments accumulate in the deep-ocean basin. When the plates converge, the buoyant granitic mass of the continental crust always overrides the adjacent oceanic plate. As the continental plate approaches the subduction zone, some of the deep-marine sediments on the oceanic plate can be crumpled and deformed. Slabs of oceanic crust shear off and are incorporated in the chaotic mass. This material, like that developed by the convergence of two oceanic plates, is subjected to high-pressure, low-temperature metamorphism.

The thick sequence of geoclinal sediments along the continental margin is then compressed and deformed. Thrust faults occur in the shallow zone of the mountain belt, where brittle rocks rupture. At intermediate depths, plastic flow develops tight folds and nappes. Intense metamorphism occurs within the deeper zones, where temperature and pressure are relatively high. The partial melting of the descending lithosphere generates silica-rich magma, which rises to form granitic intrusions in the deformed sediments in the upper plate or is extruded as volcanic material at the surface.

Convergence of Two Continental Plates

Although the generation of mountain belts by continental collision is exceedingly complex in detail, the major events are similar to those outlined in Figure 19.17.

1. Geoclinal sedimentation occurs along the margin of each continent (Figure 19.17A).
2. Before the actual continental collision, the wedge of sediments along the margin of the continent above the subduction zone is deformed (Figure 19.17B). The oceanic lithosphere is consumed at the subduction zone, and the ocean basin decreases in size.

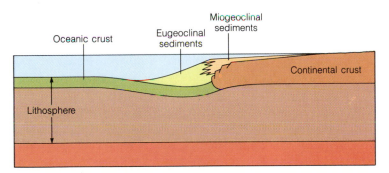

(A) Rock sequences before mountain building.

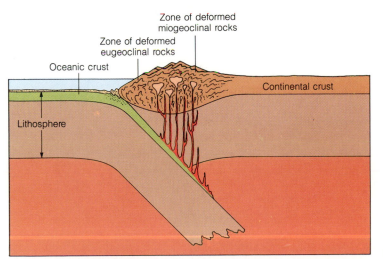

(B) Mountain building.

Figure 19.16
Continent-to-ocean mountain building involves the deformation of the thick sequence of sediment that accumulates along continental margins. This material is deformed into a folded mountain belt with miogeoclinal sediments (clean, well-sorted, shallow-marine sand, shale, and limestone) located near the continent interior and the eugeoclinal deep-marine turbidities located near the ocean margin. Granitic batholiths and metamorphosed sediments occur in the deeper zones of the orogenic belt. High topography is produced, and andesitic volcanism is common.

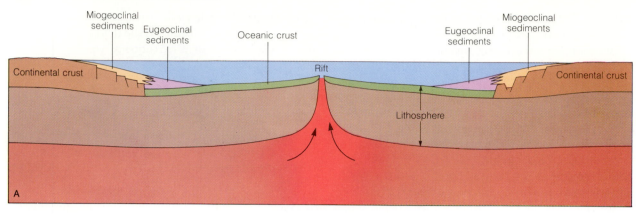

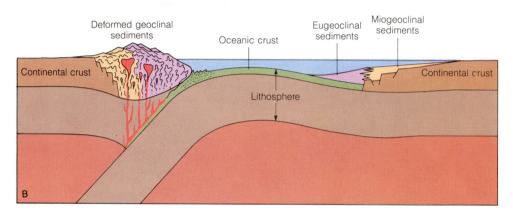

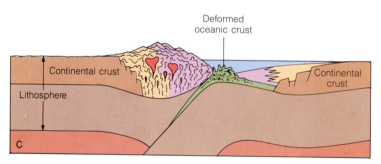

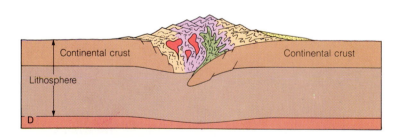

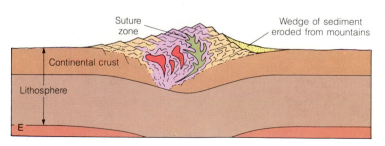

Figure 19.17
Continent-to-continent mountain building involves the deformation of oceanic and geoclinal sediments and commonly produces a complex, high mountain range. As two continents converge, the oceanic crust is caught between them and deformed. A double layer of continental crust can be produced, resulting in abnormally high topography. The continents are welded together. The descending oceanic plate becomes detached in the subduction zone and sinks independently. When this slab is consumed, volcanic activity and deep earthquakes end.

3. As the continents approach collision, segments of the remaining oceanic crust are deformed by overthrusting and are eventually squeezed between the converging plates (Figure 19.17C).

4. As the continental crust moves into the subduction zone, its buoyancy prevents it from descending into the mantle more than perhaps 40 km below its normal depth. It can be thrust under the overriding plate, however, so that a double layer of low-density continental crust is produced. This layer rises buoyantly to create a wide belt of deformed rock with an adjacent high plateau (Figure 19.17D).

5. Alternatively, the continental masses can become welded together, and fragments of ophiolite assemblage (oceanic crust) can be caught between them and squeezed upward.

6. The oceanic slab of lithosphere, descending into the mantle, ultimately becomes detached and sinks independently. When the slab has been consumed, the volcanic activity and earthquakes it generated cease.

7. Eventually convergence stops, as resisting forces build up, and the mountain belt is eroded and adjusts isostatically.

8. The welding together of two continents produces a single large continental mass with an internal mountain range (Figure 19.17E).

The Himalaya Mountains are an example of mountain building as a result of continental collision. They were formed during the last 100 million years as India moved northward and destroyed the oceanic lithosphere that formerly separated it from Asia. As the two continents collided, India was thrust under the Asian plate, and the Himalaya Mountains and the extensive highlands of the Tibetan Plateau were formed. Earthquakes are frequent in the region, but they are shallow and occur in a broad, diffuse zone because there is no descending oceanic plate.

The Ural Mountains are another example. They were formed much earlier, during late Paleozoic time, when the Siberian continental mass collided with Europe. They are not tectonically active, but erosion and isostatic adjustments maintain a mountainous topography.

In summary, mountain building is a fundamental process of the differentiation of Earth that occurs along convergent plate boundaries. Processes of compressional deformation, metamorphism, and igneous activity are involved in mountain building. The style of mountain building and the specific events can vary significantly, depending on the interactions that occur along the convergent plate boundaries. The three fundamental types of convergence include ocean-to-ocean, ocean-to-continent, and continent-to-continent collisions. Each produces its own distinctive style of mountain building.

Accretionary Terrains

Recent studies in western North America reveal a number of segments, or blocks, of diverse origins within the Rocky Mountain belt that appear to be juxtaposed in a disorganized manner. (Figure 19.18) The blocks or terrains (the term *terrain* is used with reference to a region or group of rocks that has a common age structure, stratigraphy, and origin) are variable in size, and their rocks, fossils, history, and magnetic properties contrast sharply with adjacent

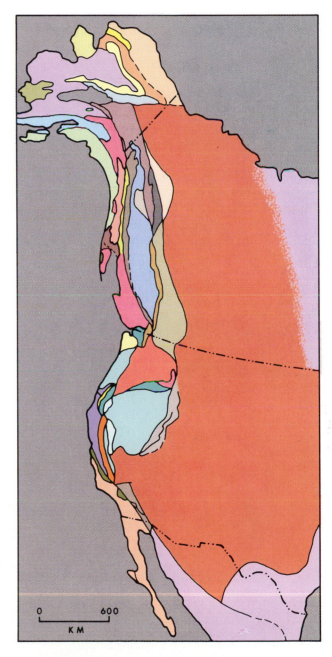

Figure 19.18
Accretionary terranes in western North America are composed of pre-mesozoic rocks. Prior to reaching their present position; these rocks were micro-continents or oceanic islands thousands of miles away in the south Pacific.

provinces. Each terrain is unique and completely unrelated to the others, except that they are now together in one place. These exotic segments of the mountain belt are referred to as accretionary terrains. They are believed to be small crustal fragments, island arcs, or seamounts that have been transported thousands of miles by the moving oceanic plate and added to a large continental mass at the subduction zone. The present oceans contain many such features that move with the plates and that will resist subduction and eventually be sutured onto a continent. These features stand out on a physiographic map of the oceans (see the inside covers). For example, in the Indian Ocean, Madagascar, the Seychelles Island platform, and Ninetyeast Ridge are small fragments of continental crust (microcontinents) that will eventually become a suspect terrain. New Zealand, the Philippines, Taiwan, and Japan, plus the numerous chains of seamounts, are similar examples in the Pacific. Others are being produced today as Baja, California, is split away from Mexico, and southern California is moving northward along the San Andreas fault.

Accretionary terranes were first recognized in the orogenic belt of western North America (Figure 22.16). In this region numerous independent segments of the crust, each with its own internal structure, rock types, and fossils, contrast sharply with adjacent provinces. One accretionary terrane contains remnants of ancient seamounts; another, segments of limestone platforms like the Bahamas; still others are

pieces of old volcanic arcs. Some are crustal fragments of metamorphosed basement complexes. Based on paleomagnetic properties and fossils, it is known that many of these crust segments originated in what is now the south Pacific and have traveled thousands of miles to eventually become part of the Rocky Mountains. The present distribution of oceanic plateaus (Figure 19.19) gives some idea of the source of accretionary terranes which will likely be welded onto continental blocks in the future.

ORIGIN AND EVOLUTION OF CONTINENTS

Statement

Continents are believed to originate and evolve through the plate tectonic processes operating along convergent plate boundaries. A continent begins as an island arc, composed of silica-rich volcanic material produced by the partial melting of a descending plate. This low-density material resists subduction because of its buoyancy. It remains on the surface of the lithosphere, continually drifting with the moving plates. The small, embryonic continent grows by accretion, accumulating new granitic material with each subsequent mountain-building event. Erosion and isostatic adjustment reduce the mountainous topography to a surface of low relief and develop a stable shield. Subsequent rifting can split and separate a

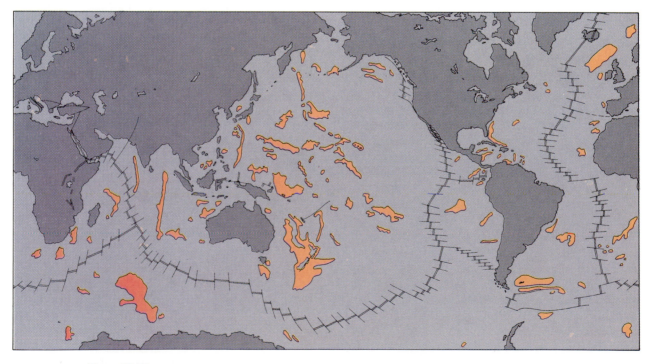

Figure 19.19
Oceanic Plateaus are concentrated in the western Pacific and Indian oceans. They are composed of fragments of continental crust, chains of seamounts, and segments of island arcs.

continent, and each segment then acts as a separate center for future continental growth.

Discussion

The origin and evolution of continents are shown graphically in the series of schematic diagrams in Figure 19.20. The initial stage of development is the formation of a subduction zone (Figure 19.20A). Magmatic differentiation occurs there by the partial melting of the descending plate. The silica-rich minerals melt first, producing a granitic magma that erupts as andesitic volcanoes or forms granitic intrusions. Sediment derived from erosion of the island arc further separates the low-density materials and concentrates them by sedimentary differentiation. The new igneous and sedimentary rocks produced in the island arc are less dense than the basaltic oceanic crust, and their buoyancy prevents any significant consumption in a subduction zone.

The volcanic material and sediment in the island arc are deformed into a mountain belt by subsequent collision with another plate. Metamorphism occurs in the mountain roots, and granitic intrusions result from partial melting in the subduction zone (Figure 19.20A). Erosion and isostatic adjustment produce a small, stable shield, in which igneous and metamorphic rocks formed in the mountain roots are exposed at the surface (Figure 19.20B). Sediment derived from the erosion of the mountains is deposited along the margins of the new continent. Repetition of this process deforms the geoclinal sediments along the continental margins into a new orogenic belt. New igneous material is added from the partial melting of the oceanic crust in the subduction zone (Figure 19.20C). The continent continues to grow by accretion of material along its margins during each subsequent orogenic event (Figure 19.20D).

The major processes of continental growth are (1) mountain building along convergent plate margins, involving deformation and metamorphism of geoclinal sediments, (2) igneous activity and emplacement of new material in the mountain belt, and (3) erosion and isostatic adjustment of the mountain belt to produce a new zone of the basement complex eroded to a flat surface near sea level. Broad upwarps or slight vertical movements of the craton permit the ocean to expand and contract over the continent, thus depositing a sediment cover, which forms the stable platform.

At some stage, rifts can form within the stable continental block as a new spreading center develops, presumably as a result of shifting convection currents in the mantle. The continent then splits, and its fragments drift apart with the spreading plates. Each fragment then acts as a separate center for future continental growth.

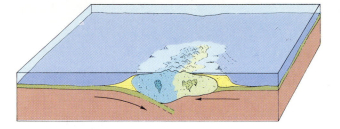

(A) *The sediment and andesitic volcanics are deformed by compression at converging plates. Mountain belts and metamorphic rocks form an embryonic continent.*

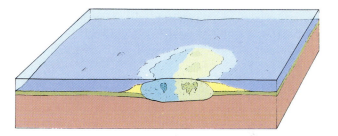

(B) *The erosion of mountains concentrates the light minerals (quartz, clay, and calcite) as geoclinal sediments along the margins of the small continent.*

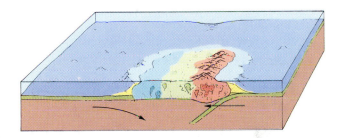

(C) *The geoclinal and orogenic cycle is repeated. New material is added to the continental mass in the form of granitic batholiths and andesitic flows.*

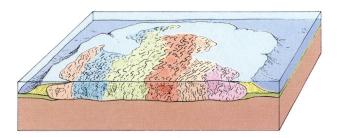

(D) *The continent continues to grow by accretion.*

Figure 19.20
Continents evolve by accretion of low-density rock material during a series of orogenic events. They begin as island arcs and grow by the addition of new material, which is metamorphosed into a segment of continental crust with each orogenic event.

Because Earth is a sphere, the moving plates carrying continental crust sometimes converge, and continents collide and become welded together.

SUMMARY

1. Geologic mapping shows that continents consist of three major components: (1) shields, (2) stable platforms, and (3) young mountain belts.

2. The characteristics of a mountain belt and the sequence of events that produced it depend largely on the type of crust carried by converging plates and the type of sediments involved in the deformation. Three distinctive types of convergence are recognized: (1) convergence of two oceanic plates, (2) convergence of a continental plate and an oceanic plate, and (3) convergence of two continental plates.

3. Convergence of two oceanic plates produces island arcs with andesitic volcanism, metamorphism of ophiolite rock sequences, and emplacement of granitic batholiths in the deeper parts of the mountain belt.

4. Convergence of a continental plate and an oceanic plate involves deformation of a thick geoclinal sequence of sediment, andesitic volcanism, and emplacement of granitic batholiths in the deformed mountain belt.

5. Convergence of two continental plates produces a continental suture or a double layer of continental crust with high, mountainous topography and limited igneous activity.

6. After mountain building, erosion and associated isostatic adjustment reduce the mountainous topography to a flat surface near sea level. These processes expose the igneous and metamorphic rock of the deep mountain roots as a new segment of the shield.

7. In mountain-building processes, silica-rich, low-density material is concentrated on the outer part of the crust through the partial melting of the oceanic plate in the subduction zone.

8. Continents thus grow by accretion, as new silica-rich material is added to a mountain belt through igneous activity at convergent plate margins. This low-density crust resists any further subduction and remains on the outer lithosphere. Subsequent rifting can split and separate continental blocks, but each block then acts as a separate center for future continental growth.

9. Continental growth is part of the process of planetary differentiation, in which the lighter elements are separated and concentrated in the outer layers of Earth. The process of differentiation has produced the ocean basins, the continental platforms, and the surface fluids (air and water). Differentiation thus is directly or indirectly responsible for the origin of all surface features and of the internal structure of the planet.

KEY TERMS

basement complex (p. 334)

continental accretion (p. 351)

eugeocline (p. 343)

folded mountain belt (p. 335)

geocline (p. 343)

isostasy (p. 344)

magmatic differentiation (p. 343)

miogeocline (p. 343)

ophiolite (p. 343)

shield (p. 334)

stable platform (p. 335)

REVIEW QUESTIONS

1. What are the composition and structure of the continental crust?
2. How does a stable platform differ from a shield?
3. How are young mountain belts related to island arcs?
4. What factors are most important in mountain building?
5. Describe and diagram the two types of rock sequences that accumulate along continental margins.
6. Explain the processes of metamorphism and igneous activity that accompany mountain building.
7. What role does isostatic adjustment play in the evolution of a mountain belt and the development of a new segment of the shield?
8. Compare and contrast mountain building involving two oceanic plates and mountain building involving a continental and an oceanic plate.
9. Describe mountain building involving two continental plates. How can this process be recognized in the rock record?

Multiple-Choice Questions

1. Which of the following is *not* characteristic of continental shields?
   a. presently being subjected to mountain-building forces
   b. high-grade metamorphic rocks
   c. granitic intrusions
   d. generally low relief
   e. belts of old intensely deformed rocks of distinctly different ages

2. Which of the following is *not* true about a stable platform?
   a. composed of sedimentary rocks underlain by the basement complex
   b. considered to be tectonically active
   c. may have numerous broad structural domes and basins
   d. the sedimentary rocks that compose the stable platform are mainly shallow marine
   e. at times it may be (or may have been) part of the continental shelf

3. Sediment on the continental shelf is predominantly
   a. the result of turbidity currents.
   b. shallow marine shale and limestone.
   c. the result of deep-sea fan sedimentation.
   d. nonmarine fluvial deposits.
   e. reefs and reef debris.

4. Most mountain belts were once
   a. geoclines.
   b. stable platforms.
   c. shields.
   d. seamount chains.
   e. rift systems.

5. Deep-water sediments deposited just beyond the continental margin are characterized by
   a. clean sand, shale, and limestone.
   b. poorly sorted sand and shale deposited by turbidity currents.
   c. coarse sand and gravel.
   d. pillow basalt and pelagic sediment.
   e. organic ooze.

6. An example of a mountain belt involving two continental plates would be the
   a. Ural Mountains.
   b. Andes Mountains.
   c. Rockie Mountains.
   d. Alps.
   e. Dividing Range of Australia.

7. Which of the following areas best describes the convergence of an oceanic plate and a continental plate?
   a. New Zealand
   b. western margin of South America
   c. eastern margin of Africa
   d. Himalayan region
   e. Ural Mountains

8. Which of the following statements about the continental crust is *not* true?
   a. Continental crust ranges in thickness from 30 to 50 km.
   b. The structure of continental crust is complex.
   c. The continents contain the oldest rocks on Earth.
   d. Continents are composed of huge slabs of granite rock.
   e. Each continent is unique and fundamentally different from the others in its basic features.

9. Erosion of a mountain belt is commonly accompanied by or followed by
   a. vertical uplift as a result of isostatic adjustment.
   b. subsidence of a geocline.
   c. rift systems and block faulting gravity sliding.
   d. gravity sliding.
   e. horizontal compression.

10. A mountain belt becomes part of the continental shield when
   a. the first mountain root is exposed.
   b. deformation at converging plate margins terminates.
   c. continents merge and are sutured together.
   d. erosion and isostatic adjustment form a stable surface eroded to near sea level.
   e. deformation is initiated.

True/False Questions

1. *T F* The volume of continental crust appears to have been constant throughout most of geologic time.
2. *T F* Many mountain belts extend to the ocean and terminate abruptly at the continental margin.
3. *T F* The major structural feature in the Appalachian Mountains is a series of tight folds.
4. *T F* Simple ocean-to-ocean mountain building involves deformation of a geocline.
5. *T F* Isostasy plays a critical role in the evolution of continents.

Fill in the Missing Terms

1. The thickest portions of the continental crust are beneath a(n) _____ _____.
2. A(n) _____ consists of sediments deposited in deep-marine environments and typically includes volcanic rocks.
3. In the Canadian Rockies, a major type of deformation is _____.
4. Mountain building is commonly restricted to plate _____.
5. Continents grow by _____ as new crustal material forms in an orogenic belt.

GLOSSARY

Aa flow A lava flow with a surface typified by angular, jagged blocks. Contrast with *pahoehoe flow*.

abrasion The mechanical wearing away of a rock by friction, rubbing, scraping, or grinding.

absolute time Geologic time measured in a specific duration of years (in contrast to relative time, which involves only the chronologic order of events).

abyssal floor The deep, relatively flat surface of the ocean floor located on both sides of the oceanic ridge. It includes the abyssal plains and the abyssal hills.

abyssal hills The part of the ocean floor consisting of hills rising as much as 1000 m above the surrounding floor. They are found seaward of most abyssal plains and occur in profusion in basins isolated from continents by trenches, ridges, or rises.

abyssal plains Flat areas of the ocean floor, having a slope of less than 1:1000. Most abyssal plains lie at the base of a continental rise and are simply areas where abyssal hills are completely covered with sediment.

active margin (plate tectonics) The leading edge of a lithospheric plate bordered by a trench.

aftershock An earthquake that follows a larger earthquake. Generally, many aftershocks occur over a period of days or even months after a major earthquake.

A **horizon** The topsoil layer in a soil profile.

alluvial fan A fan-shaped deposit of sediment built by a stream where it emerges from an upland or a mountain range into a broad valley or plain. Alluvial fans are common in arid and semiarid climates but are not restricted to them.

alpine glacier A glacier occupying a valley. Synonymous with *mountain glacier, valley glacier*.

amphibole An important rock-forming mineral group of ferromagnesian silicates. Amphibole crystals are constructed from double chains of silicon-oxygen tetrahedra. Example: hornblende.

andesite A fine-grained igneous rock composed mostly of plagioclase feldspar and from 25 to 40% amphibole and biotite, but no quartz or K-feldspar. It is abundant in mountains bordering the Pacific Ocean, such as the Andes Mountains of South America, from which the name was derived. Andesitic magma is believed to originate from fractionation of partially melted basalt.

angular unconformity An unconformity in which the older strata dip at a different angle (generally steeper) than the younger strata.

anticline A fold in which the limbs dip away from the axis. After erosion, the oldest rocks are exposed in the central core of the fold.

aphanitic texture A rock texture in which individual crystals are too small to be identified without the aid of a microscope. In hand specimens, aphanitic rocks appear to be dense and structureless.

aquifer A permeable stratum or zone below the Earth's surface through which groundwater moves.

arc-trench gap The geographic area in an island arc deep-sea trench system that separates the arc of volcanoes from the trench. In most cases, the gap is about 100 km wide.

arête A narrow, sharp ridge separating two adjacent glacial valleys.

arkose A sandstone containing at least 25% feldspar.

artesian basin A geologic structural feature in which groundwater is confined and is under artesian pressure.

artesian-pressure surface The level to which water in an artesian system would rise in a pipe high enough to stop the flow.

artesian water Groundwater confined in an aquifer and under pressure great enough to cause the water to rise above the top of the aquifer when it is tapped by a well.

ash Volcanic fragments the size of dust particles.

ash flow A turbulent blend of unsorted pyroclastic material (mostly fine-grained) mixed with high-temperature gases ejected explosively from a fissure or crater.

ash-flow tuff A rock composed of volcanic ash and dust, formed by deposition and consolidation of ash flows.

asthenosphere The zone in the Earth directly below the lithosphere, from 70 to 200 km below the surface. Seismic velocities are distinctly lower in the asthenosphere than in adjacent parts of the Earth's interior. The material in the asthenosphere is therefore believed to be soft and yielding to plastic flow.

asymmetric fold A fold (anticline or syncline) in which one limb dips more steeply than the other.

atmosphere The mixture of gases surrounding a planet. The Earth's atmosphere consists chiefly of oxygen and nitrogen, with minor amounts of other gases. Synonymous with *air*.

atoll A ring of low coral islands surrounding a lagoon.

atom The smallest unit of an element. Atoms are composed of protons, neutrons, and electrons.

axial plane With reference to folds, an imaginary plane that intersects the crest or trough of a fold so as to divide the fold as symmetrically as possible.

axis 1 (crystallography) An imaginary line passing through a crystal around which the parts of the crystal are symmetrically arranged. 2 (fold) The line where folded beds show maximum curvature. The line formed by the intersection of the axial plane with the bedding surface.

Backswamp The marshy area of a floodplain at some distance beyond and lower than the natural levees that confine the river.

backwash The return sheet flow down a beach after a wave is spent.

bajada The surface of a system of coalesced alluvial fans.

bar An offshore, submerged, elongate ridge of sand or gravel built on the sea floor by waves and currents.

barchan dune A crescent-shaped dune, the tips or horns of which point downwind. Barchan dunes form in desert areas where sand is scarce.

barrier island An elongate island of sand or gravel formed parallel to a coast.

barrier reef An elongate coral reef that trends parallel to the shore of an island or a continent, separated from it by a lagoon.

355

basalt A dark-colored, aphanitic (fine-grained) igneous rock composed of plagioclase (over 50%) and pyroxene. Olivine may or may not be present. Basalt and andesite represent 98% of all volcanic rocks.

basement complex A series of igneous and metamorphic rocks lying beneath the oldest stratified rocks of a region. In shields, the basement complex is exposed over large areas.

basin 1 (structural geology) A circular or elliptical downwarp. After erosion, the youngest beds are exposed in the central part of the structure. 2 (topography) A depression into which the surrounding area drains.

batholith A large body of intrusive igneous rock exposed over an area of at least 100 km^2.

bay (coast) A wide, curving recess or inlet between two capes or headlands.

baymouth bar A narrow, usually submerged ridge of sand or gravel deposited across the mouth of a bay by longshore drift. Baymouth bars commonly are formed by extension of spits along embayed coasts.

beach A deposit of wave-washed sediment along a coast between the landward limit of wave action and the outermost breakers.

beach drift The migration of sediment along a beach caused by the impact of waves striking the shore at an oblique angle.

bed A layer of sediment 1 cm or more in thickness.

bedding plane A surface separating layers of sedimentary rock.

bed load Material transported by currents along the bottom of a stream or river by rolling or sliding, in contrast to material carried in suspension or in solution.

bedrock The continuous solid rock that underlies the regolith everywhere and is exposed locally at the surface. An exposure of bedrock is called an outcrop.

B **horizon** The solid zone of accumulation underlying the *A* horizon of a soil profile. Some of the material dissolved by leaching in the *A* horizon is deposited in the *B* horizon.

biosphere The totality of life on or near the Earth's surface.

biotite "Black mica." An important rock-forming ferromagnesian silicate with silicon-oxygen tetrahedra arranged in sheets.

block faulting A type of normal faulting in which segments of the crust are broken and displaced to different elevations and orientations.

blowout A basin excavated by wind erosion.

blueschist A fine-grained schistose rock characterized by high-pressure, low-temperature mineral assemblages, and typically blue in color.

boulder A rock fragment with a diameter of more than 256 mm (about the size of a volleyball). A boulder is one size larger than a cobble.

braided stream A stream with a complex of converging and diverging channels separated by bars or islands. Braided streams form where more sediment is available than can be removed by the discharge of the stream.

breaker A collapsing water wave.

breccia A general term for sediment consisting of angular fragments in a matrix of finer particles. Examples: sedimentary breccias, volcanic breccias, fault breccias, impact breccias.

butte A somewhat isolated hill, usually capped with a resistant layer of rock and bordered by talus. A butte is an erosional remnant of a formerly more extensive slope.

Calcite A mineral composed of calcium carbonate ($CaCO_3$).

caldera A large, more or less circular depression or basin associated with a volcanic vent. Its diameter is many times greater than that of the included vents. Calderas are believed to result from subsidence or collapse and may or may not be related to explosive eruptions.

carbonaceous Containing carbon.

carbonate mineral A mineral formed by the bonding of carbonate ions (CO_3^{2-}) with positive ions. Examples: calcite ($CaCO_3$), dolomite [$CaMg(CO_3)_2$].

carbonate rock A rock composed mostly of carbonate minerals. Examples: limestone, dolomite.

carbon 14 A radioisotope of carbon. Its half-life is 5730 years.

catastrophism The belief that geologic history consists of major catastrophic events involving processes that were far more intense than any we observe now. Contrast with *uniformitarianism*.

cave A naturally formed subterranean open area, chamber, or series of chambers, commonly produced in limestone by solution activity.

cement Minerals precipitated from groundwater in the pore spaces of a sedimentary rock and binding the rock's particles together.

Cenozoic The era of geologic time from the end of the Mesozoic era (65 million years ago) to the present.

chalk A variety of limestone composed of shells of microscopic oceanic organisms.

chemical decomposition Synonymous with chemical weathering.

chemical weathering Chemical reactions that act on rocks exposed to water and the atmosphere so as to change their unstable mineral components to more stable forms. Oxidation, hydrolysis, carbonation, and direct solution are the most common reactions. Synonymous with *decomposition*.

chert A sedimentary rock composed of granular cryptocrystalline silica.

C **horizon** The zone of soil consisting of partly decomposed bedrock underlying the *B* horizon. It grades downward into fresh, unweathered bedrock.

cinder A fragment of volcanic ejecta from 0.5 to 2.5 cm in diameter.

cinder cone A cone-shaped hill composed of loose volcanic fragments.

cirque An amphitheater-shaped depression at the head of a glacial valley, excavated mainly by ice plucking and frost wedging.

clastic 1 Pertaining to fragments (such as mud, sand, and gravel) produced by the mechanical breakdown of rocks. 2 A sedimentary rock composed chiefly of consolidated clastic material.

clastic texture The texture of sedimentary rocks consisting of fragmentary particles of minerals, rocks, and organic skeletal remains.

clay Sedimentary material composed of fragments with a diameter of less than 1/256 mm. Clay particles are smaller than silt particles.

clay minerals A group of fine-grained crystalline hydrous silicates formed by weathering of minerals such as feldspar, pyroxene, or amphibole.

cleavage The tendency of a mineral to break in a preferred direction along smooth planes.

columnar jointing A system of fractures that splits a rock body into long prisms, or columns. It is characteristic of lava flows and shallow intrusive igneous flows.

composite volcano A large volcanic cone built by extrusion of alternating layers of ash and lava. Synonymous with *stratovolcano*.

compression A system of stresses that tends to reduce the volume of or shorten a substance.

conchoidal fracture A type of fracture that produces a smooth, curved surface. It is characteristic of quartz and obsidian.

cone of depression A conical depression of the water table surrounding a well after heavy pumping.

conglomerate A coarse-grained sedimentary rock composed of rounded fragments of pebbles, cobbles, or boulders.

contact metamorphism Metamorphism of a rock near its contact with a magma.

continent A large landmass, from 20 to 60 km thick, composed mostly of granitic rock. Continents rise abruptly above the deep-ocean floor and include the marginal areas submerged beneath sea level. Examples: the African continent, the South American continent.

continental accretion The theory that the continents have grown by incorporation of deformed sediments along their margins.

continental crust The type of crust underlying the continents, including the continental shelves. The continental crust is commonly about 35 km thick. Its maximum thickness is 60 km, beneath mountain ranges. Its density is 2.7 g/cm^3, and the velocities of primary seismic waves traveling through the crust are less than 6.2 km/sec. Synonymous with *sial*. Contrast with *oceanic crust*.

continental drift The theory that the continents have moved in relation to one another.

continental glacier A thick ice sheet covering large parts of a continent. Present-day examples are found in Greenland and Antarctica.

continental margin The zone of transition from a continental mass to the adjacent ocean basin. It generally includes a continental shelf, continental slope, and continental rise.

continental rise The gently sloping surface located at the base of a continental slope (see diagram for *abyssal plains*).

continental shelf The submerged margin of a continental mass extending from the shore to the first prominent break in slope, which usually occurs at a depth of about 120 m.

continental slope The slope that extends from a continental shelf down to the ocean deep. In some areas, such as off eastern North America, the continental slope grades into the more gently sloping continental rise.

convection Movement of portions of a fluid as a result of density differences produced by heating.

convection cell The space occupied by a single convection current.

convection current A closed system in which material is transported as a result of thermal convection. Convection currents are characteristic of the atmosphere and of bodies of water. They are believed also to be generated in the interior of the Earth. In the plate tectonic theory, convection within the mantle is thought to be responsible for the movement of tectonic plates.

convergent plate boundary The zone where the leading edges of converging plates meet. Convergent plate boundaries are sites of considerable geologic activity and are characterized by volcanism, earthquakes, and crustal deformation. See also *subduction zone*.

coquina A limestone composed of an aggregate of shells and shell fragments.

coral A bottom-dwelling marine invertebrate organism of the class Anthozoa.

core The central part of the Earth below a depth of 2900 km.

country rock A general term for rock surrounding an igneous intrusion.

covalent bond A chemical bond in which electrons are shared between different atoms so that none of the atoms has a net charge.

crater An abrupt circular depression formed by extrusion of volcanic material, by collapse, or by the impact of a meteorite.

craton The stable continental crust, including the shield and stable platform areas, most of which have not been affected by significant tectonic activity since the close of the Precambrian era.

creep The imperceptibly slow downslope movement of material.

crevasse 1 (glacial geology) A deep crack in the upper surface of a glacier. 2 (natural levee) A break in a natural levee.

cross-bedding Stratification inclined to the original horizontal surface upon which the sediment accumulated. It is produced by deposition on the slope of a dune or sand wave.

crosscutting relations, principle of The principle that a rock is younger than any rock across which it cuts.

crust (planetary structure) The outermost layer, or shell, of the Earth (or any other differentiated planet). The Earth's crust is generally defined as the part of the Earth above the Mohorovičić discontinuity. It represents less than 1% of the Earth's total volume. See also *continental crust, oceanic crust*.

crystal A solid, polyhedral form bounded by naturally formed plane surfaces resulting from growth of a crystal lattice.

crystal face A naturally formed smooth plane surface of a crystal.

crystal form The geometric shape of a crystal.

crystal lattice A systematic, symmetrical network of atoms within a crystal.

crystalline texture The rock texture resulting from simultaneous growth of crystals.

crystallization The process of crystal growth. It occurs as a result of condensation from a gaseous state, precipitation from a solution, or cooling of a melt.

crystal structure The orderly arrangement of atoms in a crystal.

cuesta An elongate ridge formed on the tilted and eroded edges of gently dipping strata.

Debris flow The rapid downslope movement of debris (rock, soil, and mud).

debris slide A type of landslide in which comparatively dry rock fragments and soil move downslope at speeds ranging from slow to fast. The mass of debris does not show backward rotation (which occurs in a slump) but slides and rolls forward.

decomposition Weathering by chemical processes. Synonymous with *chemical weathering*.

deep-focus earthquake An earthquake that originates at a depth greater than 300 km.

deep-marine environment The sedimentary environment of the abyssal plains.

deep-sea fan A cone-shaped or fan-shaped deposit of land-derived sediment located seaward of large rivers or submarine canyons. Synonymous with *abyssal cone, abyssal fan, submarine cone*.

deep-sea trench See *trench*.

deflation Erosion of loose rock particles by wind.

deflation basin A shallow depression formed by wind erosion where groundwater solution activity has left unconsolidated sediment exposed at the surface.

delta A large, roughly triangular body of sediment deposited at the mouth of a river.

dendritic drainage pattern A branching stream pattern, resembling the branching of certain trees, such as oaks and maples.

density The measure of concentration of matter in a substance; mass per unit volume, expressed in grams per cubic centimeter (g/cm^3).

density current A current that flows as a result of differences in density. In oceans, density currents are produced by differences in temperature, salinity, and turbidity (the concentration of material held in suspension).

denudation The combined action of all of the various processes that cause the wearing away and lowering of the land, including weathering, mass wasting, stream action, and groundwater activity.

deranged drainage A distinctively disordered drainage pattern formed in a recently glaciated area. It is characterized by irregular direction of stream flow, few short tributaries, swampy areas, and many lakes.

desert pavement A veneer of pebbles left in place where wind has removed the finer material.

detrital 1 Pertaining to detritus. 2 A rock formed from detritus.

detritus A general term for loose rock fragments produced by mechanical weathering.

differential erosion Variation in the rate of erosion on different rock masses. As a result of differential erosion, resistant rocks form steep cliffs, whereas nonresistant rocks form gentle slopes.

dike A tabular intrusive rock that occurs across strata or other structural features of the surrounding rock.

diorite A phaneritic intrusive igneous rock consisting mostly of intermediate plagioclase, feldspar, and pyroxene, with some amphibole and biotite.

dip The angle between the horizontal plane and a structural surface (such as a bedding plane, a joint, a fault, foliation, or other planar features).

disappearing stream A stream that disappears into an underground channel and does not reappear in the same, or even in an adjacent, drainage basin. In karst regions, streams commonly disappear into sinkholes and follow channels through caves.

discharge Rate of flow; the volume of water moving through a given cross section of a stream in a given unit of time.

disconformity An unconformity in which beds above and below are parallel.

discontinuity A sudden or rapid change in physical properties of rocks within the Earth. Discontinuities are recognized by seismic data. See also *Mohorovičić discontinuity*.

disintegration Weathering by mechanical processes. Synonymous with *mechanical weathering*.

dissolution The process by which materials are dissolved.

dissolved load The part of a stream's load that is carried in solution.

distributary Any of the numerous stream branches into which a river divides where it reaches its delta.

divergent plate boundary A plate margin formed where the lithosphere splits into plates that drift apart from one another. Divergent plate boundaries are areas subject to tension, where new crust is generated by igneous activity. Synonymous with

spreading center. See also *oceanic ridge.*

divide A ridge separating two adjacent drainage basins.

dolomite 1 A mineral composed of $CaMg(CO_3)_2$. 2 A sedimentary rock composed primarily of the mineral dolomite.

dolostone A sedimentary rock composed mostly of the mineral dolomite. Sometimes referred to simply as *dolomite.*

dome 1 (structural geology) An uplift that is circular or elliptical in map view, with beds dipping away in all directions from a central area. 2 (topography) A general term for any dome-shaped landform.

downwarp A downward bend or subsidence of a part of the Earth's crust.

drainage basin The total area that contributes water to a single drainage system.

drainage system An integrated system of tributaries and a trunk stream, which collect and funnel surface water to the sea, a lake, or some other body of water.

drift (glacial geology) A general term for sediment deposited directly on land by glacial ice or deposited in lakes, oceans, or streams as a result of glaciation.

drip curtain A thin sheet of dripstone hanging from the ceiling or wall of a cave.

dripstone A cave deposit formed by precipitation of calcium carbonate from groundwater entering an underground cavern.

drumlin A smooth, glacially streamlined hill that is elongate in the direction of ice movement. Drumlins are generally composed of till.

dune A low mound of fine-grained material that accumulates as a result of sediment transport in a current system. Dunes have characteristic geometric forms that are maintained as they migrate. Sand dunes are commonly classified according to shape. See also *barchan dune, longitudinal dune, parabolic dune, seif dune, star dune,* and *transverse dune.*

Earthquake A series of elastic waves propagated in the Earth, initiated where stress along a fault exceeds the elastic limit of the rock so that sudden movement occurs along the fault.

ecology The study of relationships between organisms and their environments.

elastic deformation Temporary deformation of a substance, after which the material returns to its original

size and shape. Example: the bending of mica flakes.

elastic limit The maximum stress that a given substance can withstand without undergoing permanent deformation either by solid flow or by rupture.

elastic-rebound theory The theory that earthquakes result from energy released by faulting; the sudden release of stored strain creates earthquake waves.

end moraine A ridge of till that accumulates at the margin of a glacier.

entrenched meander A meander cut into the underlying rock as a result of regional uplift or lowering of the regional base level.

environment of sedimentation See *sedimentary environment.*

eolian Pertaining to wind.

eolian environment The sedimentary environment of deserts, where sediment is transported and deposited primarily by wind.

epicenter The area on the Earth's surface that lies directly above the focus of an earthquake.

epoch A division of geologic time; a subdivision of a period. Example: Pleistocene epoch.

erosion The processes that loosen sediment and move it from one place to another on the Earth's surface. Agents of erosion include water, ice, wind, and gravity.

erratic A large boulder carried by glacial ice to an area far removed from its point of origin.

escarpment A cliff or very steep slope.

esker A long, narrow, sinuous ridge of stratified glacial drift deposited by a stream flowing beneath a glacier in a tunnel or in a subglacial stream bed.

estuary A bay at the mouth of a river formed by subsidence of the sand or by a rise in sea level. Fresh water from the river mixes with and dilutes seawater in an estuary.

eugeocline (plate tectonics) A geocline in which volcanism is associated with elastic sedimentation. Eugeoclines are usually associated with an island arc.

eugeosyncline A geosyncline situated seaward from a continent and characterized by sediments deposited by turbidity currents and derived in part from a volcanic arc.

eustatic change of sea level A worldwide rise or fall in sea level resulting from a change in the volume of water or the capacity of ocean basins.

evaporite A rock composed of minerals derived from evaporation of mineralized water. Examples: rock salt, gypsum.

exfoliation A weathering process by which concentric shells, slabs, sheets, or flakes are successively broken loose and stripped away from a rock mass.

exposure Bedrock not covered with soil or regolith; outcrop.

extrusive rock A rock formed from a mass of magma that flowed out on the surface of the Earth. Example: basalt.

Faceted spur A spur or ridge that has been beveled or truncated by faulting, erosion, or glaciation.

facies A distinctive group of characteristics within part of a rock body (such as composition, grain size, or fossil assemblages) that differ as a group from those found elsewhere in the same rock unit. Examples: conglomerate facies, shale facies, brachiopod facies.

fan A fan-shaped deposit of sediment. See also *alluvial fan, deep-sea fan.*

fault A surface along which a rock body has broken and been displaced.

fault block A rock mass bounded by faults on at least two sides.

fault scarp A cliff produced by faulting.

faunal succession, principle of The principle that fossils in a stratigraphic sequence succeed one another in a definite, recognizable order.

feldspar A mineral group consisting of silicates of aluminum and one or more of the metals potassium, sodium, or calcium. Examples: K-feldspar, Ca-plagioclase, Na-plagioclase.

felsite A general term for light-colored aphanitic (fine-grained) igneous rocks. Example: rhyolite.

ferromagnesian minerals A variety of silicate minerals containing abundant iron and magnesium. Examples: olivine, pyroxene, amphibole.

fiord A glaciated valley flooded by the sea to form a long, narrow, steep-walled inlet.

fissure An open fracture in a rock.

fissure eruption Extrusion of lava along a fissure.

flood basalt An extensive flow of basalt erupted chiefly along fissures. Synonymous with *plateau basalt.*

floodplain The flat, occasionally flooded area bordering a stream.

fluvial Pertaining to a river or rivers.

fluvial environment The sedimentary environment of river systems.

focus The area within the Earth where an earthquake originates.

fold A bend, or flexure, in a rock.

folded mountain belt A long, linear zone of the Earth's crust where rocks have been intensely deformed by horizontal stresses and generally intruded by igneous rocks. The great folded mountains of the world (such as the Appalachians, the Himalayas, the Rockies, and the Alps) are believed to have been formed at convergent plate margins.

foliation A planar feature in metamorphic rocks, produced by the secondary growth of minerals. Three major types are recognized: slaty cleavage, schistosity, and gneissic layering.

footwall The block beneath a dipping fault surface.

formation A distinctive body of rock that serves as a convenient unit for study and mapping.

fossil Naturally preserved remains or evidence of past life, such as bones, shells, casts, impressions, and trails.

fossil fuel A fuel containing solar energy that was absorbed by plants and animals in the geologic past and thus is preserved in organic compounds in their remains. Fossil fuels include petroleum, natural gas, and coal.

fracture zone 1 (field geology) A zone where the bedrock is cracked and fractured. 2 (oceanography) A zone of long, linear fractures on the ocean floor, expressed topographically by ridges and troughs. Fracture zones are the topographic expression of transform faults.

fringing reef A reef that lies alongside the shore of a landmass.

frost heaving The lifting of unconsolidated material by the freezing of subsurface water.

frost wedging The forcing apart of rocks by the expansion of water as it freezes in fractures and pore spaces.

Gabbro A dark-colored, coarse-grained rock composed of Ca-plagioclase, pyroxene, and possibly olivine, but no quartz.

gas The state of matter in which a substance has neither independent shape nor independent volume. Gases can readily be compressed and tend to expand indefinitely.

geocline An elongate prism of sedimentary rock deposited in a subsided part of the continental margins and adjacent oceanic crust. A modern example is the continental margin of the eastern United States. See also *eugeocline, miogeocline.*

geode A hollow nodule of rock lined with crystals; when separated from the rock body by weathering, it appears as a hollow, rounded shell partly filled with crystals.

geologic column A diagram representing divisions of geologic time and the rock units formed during each major period.

geologic cross section A diagram showing the structure and arrangement of rocks as they would appear in a vertical plane below the Earth's surface.

geologic map A map showing the distribution of rocks at the Earth's surface.

geologic time scale The time scale determined by the geologic column and by radiometric dating of rocks.

geosyncline A subsiding part of the lithosphere in which thousands of meters of sediment accumulate. See also *eugeosyncline, miogeosyncline.*

geothermal Pertaining to the heat of the interior of the Earth.

geothermal energy Energy useful to human beings that can be extracted from steam and hot water found within the Earth's crust.

geothermal gradient The rate at which temperature increases with depth.

geyser A thermal spring that intermittently erupts steam and boiling water.

glacial environment The sedimentary environment of glaciers and their meltwaters.

glacier A mass of ice formed from compacted, recrystallized snow that is thick enough to flow plastically.

glass 1 A state of matter in which a substance displays many properties of a solid but lacks crystal structure. 2 An amorphous igneous rock formed from a rapidly cooling magma.

glassy texture The texture of igneous rocks in which the material is in the form of natural glass rather than crystal.

global tectonics The study of the characteristics and origin of structural features of the Earth that have regional or global significance.

gneiss A coarse-grained metamorphic rock with a characteristic type of foliation (gneissic layering), resulting from alternating layers of light-colored and dark-colored minerals. Its composition is generally similar to that of granite.

gneissic layering The type of foliation characterizing gneiss, resulting from alternating layers of the constituent silicic and mafic minerals.

Gondwanaland The ancient continental landmass that is thought to have split apart during Mesozoic time to form the present-day continents of South America, Africa, India, Australia, and Antarctica.

graben An elongate fault block that has been lowered in relation to the blocks on either side.

gradation Leveling of the land due

to erosion by such agents as river systems, groundwater, glaciers, wind, and waves.

graded bedding   A type of bedding in which each layer is characterized by a progressive decrease in grain size from the bottom of the bed to the top.

gradient   (stream) The slope of a stream channel measured along the course of the stream.

grain   A particle of a mineral or rock, generally lacking well-developed crystal faces.

granite   A coarse-grained igneous rock composed of K-feldspar, plagioclase, and quartz, with small amounts of ferromagnesian minerals.

gravity anomaly   An area where gravitational attraction is greater or less than its normal value.

groundmass   The matrix of relatively fine-grained material between the phenocrysts in a porphyritic rock.

ground moraine   Glacial deposits that cover an area formerly occupied by a glacier; they typically produce a landscape of low, gently rolling hills.

groundwater   Water below the Earth's surface. It generally occurs in pore spaces of rocks and soil.

gypsum   An evaporite mineral composed of calcium sulfate with water ($CaSO_4 \cdot 2H_2O$).

Half-Life   The time required for half of a given sample of a radioactive isotope to decay to its daughter isotope.

halite   An evaporite mineral composed of sodium chloride (NaCl).

hanging valley   A tributary valley with the floor lying ("hanging") above the valley floor of the main stream or shore to which it flows. Hanging valleys commonly are created by deepening of the main valley by glaciation, but they can also be produced by faulting or rapid retreat of a sea cliff.

hanging wall   The surface or block of rock that lies above an inclined fault plane.

hardness   1 (mineralogy) The measure of the resistance of a mineral to scratching or abrasion.   2 (water) A property of water resulting from the presence of calcium carbonate and magnesium carbonate in solution.

headland   An extension of land seaward from the general trend of the coast; a promontory, cape, or peninsula.

headward erosion   Extension of a stream headward, up the regional slope of erosion.

heat flow   The flow of heat from the interior of the Earth.

high-grade metamorphism   Metamorphism that occurs under high temperature and high pressure.

hogback   A narrow, sharp ridge formed on steeply inclined, resistant rock.

horizon   1 (geologic) A plane of stratification assumed to have been originally horizontal.   2 (soil) A layer of soil distinguished by characteristic physical properties. Soil horizons generally are designated by letters (for example, *A* horizon, *B* horizon, *C* horizon).

horn   A sharp peak formed at the intersection of the headwalls of three or more cirques.

hornblende   A variety of the amphibole mineral group.

hornfels   A nonfoliated metamorphic rock of uniform grain size, formed by high-temperature metamorphism. Hornfelses typically are formed by contact metamorphism around igneous intrusions.

horst   An elongate fault block that has been uplifted in relation to the adjacent rocks.

hot spot   The expression at the Earth's surface of a mantle plume, or column of hot, buoyant rock rising in the mantle beneath a lithospheric plate.

hydraulic head   The pressure exerted by a fluid at a given depth beneath its surface. It is proportional to the height of the fluid's surface above the area where the pressure is measured.

hydrologic system   The system of moving water at the Earth's surface.

hydrolysis   Chemical combination of water with other substances.

hydrosphere   The waters of the Earth, as distinguished from the rocks (lithosphere), the air (atmosphere), and living things (biosphere).

hydrostatic pressure   The pressure within a fluid (such as water) at rest, exerted on a given point within the body of the fluid.

Ice sheet   A thick, extensive body of glacial ice that is not confined to valleys. Localized ice sheets are sometimes called *ice caps*.

ice wedging   A type of mechanical weathering in which rocks are broken by the expansion of water as it freezes in joints, pores, or bedding planes. Synonymous with *frost wedging*.

igneous rock   Rock formed by cooling and solidification of molten silicate minerals (magma). Igneous rocks include volcanic and plutonic rocks.

inclination, magnetic   The angle between the horizontal plane and a magnetic line of force.

inclusion   A rock fragment incorporated into a younger igneous rock.

intermediate-focus earthquake   An earthquake with a focus located at a depth between 70 and 300 km.

intermittent stream   A stream through which water flows only part of the time.

internal drainage   A drainage system that does not extend to the ocean.

intrusion   1 Injection of a magma into a preexisting rock.   2 A body of rock resulting from the process of intrusion.

intrusive rock   Igneous rock which, while it was fluid, penetrated into or between other rocks and solidifed. It can later be exposed at the Earth's surface after erosion of the overlying rock.

inverted valley   A valley that has been filled with lava or other resistant material and has subsequently been eroded into an elongate ridge.

ion   An atom or combination of atoms that has gained or lost one or more electrons and thus has a net electrical charge.

ionic bond   A chemical bond formed by electrostatic attraction between oppositely charged ions.

ionic substitution   The replacement of one kind of ion in a crystalline lattice by another kind that is of similar size and electrical charge.

island   A landform smaller than a continent and completely surrounded by water.

island arc   A chain of volcanic islands. Island arcs are generally convex toward the open ocean. Example: the Aleutian Islands.

isostasy   A state of equilibrium, resembling flotation, in which segments of the Earth's crust stand at levels determined by their thickness and density. Isostatic equilibrium is attained by flow of material in the mantle.

isotope   One of the several forms of a chemical element that have the same number of protons in the nucleus but differ in the number of neutrons and thus differ in atomic weight.

Joint   A fracture in a rock along which no appreciable displacement has occurred.

Kame   A body of stratified glacial sediment. A mound or an irregular ridge deposited by a subglacial stream as an alluvial fan or a delta.

karst topography   A landscape characterized by sinks, solution valleys, and other features produced by groundwater activity.

kettle   A closed depression in a deposit of glacial drift formed where a block of ice was buried or partly buried and then melted.

Laccolith A concordant igneous intrusion that has arched up the strata into which it was injected, so that it forms a pod-shaped or lens-shaped body with a generally horizontal floor.

lag deposit A residual accumulation of coarse fragments that remains on the surface after finer material has been removed by wind.

lagoon A shallow body of seawater separated from the open ocean by a barrier island or reef.

landform Any feature of the Earth's surface having a distinct shape and origin. Landforms include major features (such as continents, ocean basins, plains, plateaus, and mountain ranges) and minor features (such as hills, valleys, slopes, drumlins, and dunes). Collectively, the landforms of the Earth constitute the entire surface configuration of the planet.

landslide A general term for relatively rapid types of mass movement, such as debris flows, debris slides, rockslides, and slumps.

lateral moraine An accumulation of till deposited along the side margins of a valley glacier. It accumulates as a result of mass movement of debris on the sides of the glacier.

lava Magma that reaches the Earth's surface.

lee slope The part of a hill, dune, or rock that is sheltered or turned away from the wind. Synonymous with *slip face.*

levee, natural A broad, low embankment built up along the banks of a river channel during floods.

limb The flank, or side, of a fold.

limestone A sedimentary rock composed mostly of calcium carbonate ($CaCO_3$).

liquid The state of matter in which a substance flows freely and lacks crystal structure. Unlike a gas, a liquid retains the same volume independent of the shape of its container.

lithification The processes by which sediment is converted into sedimentary rock. These processes include cementation and compaction.

lithosphere The relatively rigid outer zone of the Earth, which includes the continental crust, the oceanic crust, and the part of the mantle lying above the softer asthenosphere.

load The total amount of sediment carried at a given time by a stream, glacier, or wind.

loess Unconsolidated, wind-deposited silt and dust.

longitudinal dune An elongate sand dune oriented in the direction of the prevailing wind.

longitudinal profile The profile of a stream or valley drawn along its length, from source to mouth.

longitudinal wave A seismic body wave in which particles oscillate along lines in the direction in which the wave travels. Synonymous with *P wave.*

longshore current A current in the surf zone moving parallel to the shore. Longshore currents occur where waves strike the shore at an angle. The waves push water and sediment obliquely up the beach, and the backwash returns straight down the beach face, so the water and sediment follow a zigzag pattern, with net movement parallel to the shore.

longshore drift The process in which sediment is moved in a zigzag pattern along a beach by the swash and backwash of waves that approach the shore obliquely.

low-grade (low-rank) metamorphism Metamorphism that is accomplished under low or moderate temperature and low or moderate pressure.

Magma A mobile silicate melt, which can contain suspended crystals and dissolved gases as well as liquid.

magmatic differentiation A general term for the various processes by which early-formed crystals or early-formed liquids are separated and removed from a magma to produce a rock with composition different from that of the original magma. Early-crystallized ferromagnesian minerals commonly are separated by gravitational settling, so that the parent magma is left enriched in silica, sodium, and potassium.

magmatic segregation Separation of crystals of certain minerals from a magma as it cools. For example, some minerals (including certain valuable metals) crystallize while other components of the magma are still liquid. These early-formed crystals can settle to the bottom of a magma chamber and thus become concentrated there, forming an ore deposit.

magnetic anomaly A deviation of observed magnetic inclination or intensity (as measured by a magnetometer) from a constant normal value.

magnetic reversal A complete 180-degree reversal of the polarity of the Earth's magnetic field.

mantle The zone of the Earth's interior between the base of the crust (the Moho discontinuity) and the core.

mantle plume A buoyant mass of hot mantle material that rises to the base of the lithosphere. Mantle plumes commonly produce volcanic activity and structural deformation in the central part of lithospheric plates.

marble A metamorphic rock consisting mostly of metamorphosed limestone or dolomite.

mass movement The transfer of rock and soil downslope by direct action of gravity without a flowing medium (such as a river or glacial ice). Synonymous with *mass wasting.*

matrix The relatively fine-grained rock material occupying the space between larger particles in a rock. See also *groundmass.*

meander A broad, looping bend in a river.

mechanical weathering The breakdown of rock into smaller fragments by physical processes such as frost wedging. Synonymous with *disintegration.*

medial moraine A ridge of till formed in the middle of a valley glacier by the junction of two lateral moraines where two valley glaciers converge.

melt A substance altered from the solid state to the liquid state.

mesa A flat-topped, steep-sided highland capped with a resistant rock formation. A mesa is smaller than a plateau but larger than a butte.

Mesozoic The era of geologic time from the end of the Paleozoic era (225 million years ago) to the beginning of the Cenozoic era (65 million years ago).

metaconglomerate A metamorphosed conglomerate.

metallic bond A chemical bond in which shared electrons move freely among the atoms.

metamorphic Pertaining to the processes or products of metamorphism.

metamorphic rock Any rock formed from preexisting rocks within the Earth's crust by changes in temperature and pressure and by chemical action of fluids.

metamorphism Alteration of the minerals and textures of a rock by changes in temperature and pressure and by a gain or loss of chemical components.

meteorite Any particle of solid matter that has fallen to the Earth, the Moon, or another planet from space.

mica A group of silicate minerals exhibiting perfect cleavage in one direction.

microcontinent A relatively small, isolated fragment of continental crust. Example: Madagascar.

mid-Atlantic ridge The mountain range extending from north to south down the central part of the Atlantic Ocean floor.

mineral A naturally occurring inorganic solid having a definite internal structure and a definite chemical composition that varies only within strict limits. Chemical composition and internal structure determine its physical properties, including the tendency to assume a particular geometric form (crystal form).

miogeocline (plate tectonics) A geocline situated near a continental margin containing a thick sequence of well-sorted clastic and chemical sediments derived from the continent.

miogeosyncline A geosyncline situated near a continental margin and receiving well-sorted clastic and chemical sediments from the continent, not associated with volcanism.

mobile belts Long, narrow belts in the continents that have been subjected to mountain-building processes.

Mohorovičić discontinuity The first global seismic discontinuity below the surface of the Earth. It lies at a depth varying from about 5 to 10 km beneath the ocean floor to about 35 km beneath the continents. Commonly referred to as the Moho.

monocline A bend or fold in gently dipping horizontal strata.

moraine A general term for a landform composed of till.

mountain A general term for any landmass that stands above its surroundings. In the stricter geological sense, a mountain belt is a highly deformed part of the Earth's crust that has been injected with igneous intrusions and the deeper parts of which have been metamorphosed. The topography of young mountains is high, but erosion can reduce old mountains to flat lowlands.

mud crack A crack in a deposit of mud or silt resulting from the contraction that accompanies drying.

mudflow A flowing mixture of mud and water.

Nappe Faulted and overturned folds.

natural arch An arch-shaped landform produced by weathering and differential erosion.

névé Granular ice formed by recrystallization of snow. Synonymous with *firm*.

nonconformity An unconformity in which stratified rocks rest on eroded granitic or metamorphic rocks.

normal fault A steeply inclined fault in which the hanging wall has moved downward in relation to the footwall. Synonymous with *gravity fault*.

Obsidian A glassy igneous rock with a composition equivalent to that of granite.

ocean basin A low part of the lithosphere lying between continental masses. The rocks of an ocean basin are mostly basalt with a veneer of oceanic sediment.

oceanic crust The type of crust that underlies the ocean basins. It is about 5 km thick, composed predominantly of basalt. Its density is 3.0 g/cm^3. The velocities of compressional seismic waves traveling through it exceed 6.2 km/sec. Compare with *continental crust*.

oceanic ridge The continuous ridge, or broad, fractured topographic swell, that extends through the central part of the Arctic, Atlantic, Indian, and South Pacific oceans. It is several hundred kilometers wide, and its elevation above the ocean floor is 600 m or more. It thus constitutes a major structural and topographic feature of the Earth.

offshore The area from low tide seaward.

oil shale Shale that is rich in hydrocarbon derivatives. In the United States, the chief oil shale is the Green River Formation, in the Rocky Mountain region.

olivine A silicate mineral with magnesium and iron but no aluminum $[(Mg,Fe)_2SiO_4]$.

oolite A limestone consisting largely of spherical grains of calcium carbonate in concentric spherical layers.

ooze (marine geology) Marine sediment consisting of more than 30% shell fragments of microscopic organisms.

ophiolite A sequence of rocks characterized by ultramafic rocks at the base and (in ascending order) gabbro, sheeted dikes, pillow lavas, and deep-sea sediments. The typical sequence of rocks constituting the oceanic crust.

outcrop An exposure of bedrock.

outlet glacier A tonguelike stream of ice, resembling a valley glacier, that forms where a continental glacier encounters a mountain system and is forced to move through a mountain pass in large streams.

outwash Stratified sediment washed out from a glacier by meltwater streams and deposited in front of the end moraine.

outwash plain The area beyond the margins of a glacier where meltwater deposits sand, gravel, and mud washed out from the glacier.

overturned fold A fold in which at least one limb has been rotated through an angle greater than 90 degrees.

oxbow lake A lake formed in the channel of an abandoned meander.

oxidation Chemical combination of oxygen with another substance.

Pahoehoe flow A lava flow with a billowy or ropy surface. Contrast with *aa flow*.

paleogeography The study of geography in the geologic past, including the patterns of the Earth's surface, the distribution of land and ocean, and ancient mountains and other landforms.

paleomagnetism The study of ancient magnetic fields, as preserved in the magnetic properties of rocks. It includes studies of changes in the position of the magnetic poles and reversals of the magnetic poles in the geologic past.

paleontology The study of ancient life.

paleowind An ancient wind, existing in the geologic past, the direction of which can be inferred from patterns of ancient ash falls, orientation of cross-bedding, and growth rates of colonial corals.

Paleozoic The era of geologic time from the end of the Precambrian (600 million years ago) to the beginning of the Mesozoic era (225 million years ago).

Pangaea A hypothetical continent from which the present continents originated by plate movement from the Mesozoic era to the present.

parabolic dune A dune shaped like a parabola with the concave side toward the wind.

partial melting The process by which minerals with low melting points liquefy within a rock body as a result of an increase in temperature or a decrease in pressure (or both) while other minerals in the rock are still solid. If the liquid (magma) is removed before other components of the parent rock have melted, the composition of the magma can be quite different from that of the parent rock. Partial melting is believed to be important in the generation of basaltic magma from peridotite at spreading centers and in the generation of granitic magma from basaltic crust at subduction zones.

passive margin (plate tectonics) A lithospheric plate margin at which crust is neither created nor destroyed. Passive plate margins generally are marked by transform faults. Contrast with *active margin*.

peat An accumulation of partly carbonized plant material containing approximately 60% carbon and 30% oxygen. It is considered an early stage, or rank, in the development of coal.

pebble A rock fragment with a diameter between 2 mm (about the size of a matchhead) and 64 mm (about the size of a tennis ball).

pediment A gently sloping erosion

surface formed at the base of a receding mountain front or cliff. It cuts across bedrock and can be covered with a veneer of sediment. Pediments characteristically form in arid and semiarid climates.

peninsula An elongate body of land extending into a body of water.

perched water table The upper surface of a local zone of saturation that lies above the regional water table.

peridotite A dark-colored igneous rock of coarse-grained texture, composed of olivine, pyroxene, and some other ferromagnesian minerals, but with essentially no feldspar and no quartz.

permafrost Permanently frozen ground.

permanent stream A stream or reach of a stream that flows continuously throughout the year. Synonymous with *perennial stream.*

permeability The ability of a material to transmit fluids.

phaneritic texture The texture of igneous rocks in which the interlocking crystals are large enough to be seen without magnification.

phenocryst A crystal that is significantly larger than the crystals surrounding it. Phenocrysts form during an early phase in the cooling of a magma when the magma cools relatively slowly.

physiographic map A map showing surface features of the Earth.

physiography The study of the surface features and landforms of the Earth.

pillar A landform shaped like a pillar.

pillow lava An ellipsoidal mass of igneous rock formed by extrusion of lava underwater.

pinnacle A tall, tower-shaped or spire-shaped pillar of rock.

placer A mineral deposit formed by the sorting or washing action of water. Placers are usually deposits of heavy minerals, such as gold.

plagioclase A group of feldspar minerals with a composition range from $NaAlSi_3O_8$ to $CaAl_2Si_2O_8$.

planetary differentiation The processes by which the materials in a planetary body are separated according to density, so that the originally homogeneous body is converted into a zoned or layered (shelled) body with a dense core, a mantle, and a crust.

plastic deformation A permanent change in a substance's shape or volume that does not involve failure by rupture.

plate (tectonics) A broad segment of the lithosphere (including the rigid upper mantle, plus oceanic and continental crust) that floats on the underlying asthenosphere and moves independently of other plates.

plateau An extensive upland region.

plateau basalt Basalt extruded in extensive, nearly horizontal layers, which, after uplift, tend to erode into great plateaus. Synonymous with *flood basalt.*

plate tectonics The theory of global dynamics in which the lithosphere is believed to be broken into individual plates that move in response to convection in the upper mantle. The margins of the plates are sites of considerable geologic activity.

platform reef An organic reef with a flat upper surface developed on submerged segments of a continental platform.

playa A depression in the center of a desert basin, the site of occasional temporary lakes.

playa lake A shallow temporary lake formed in a desert basin after rain.

Pleistocene The epoch of geologic time from the end of the Pliocene epoch of the Tertiary period (about 2 million years ago) to the beginning of the Holocene epoch of the Quaternary period (about 10,000 years ago). The major event during the Pleistocene was the expansion of continental glaciers in the Northern Hemisphere. Synonymous with *glacial epoch, ice age.*

plucking (glacial geology) The process of glacial erosion by which large rock fragments are loosened by ice wedging, become frozen to the bottom surface of the glacier, and are torn out of the bedrock and transported by the glacier as it moves. The process involves the freezing of subglacial meltwater that seeps into fractures and bedding planes in the rock.

plunge The inclination, with respect to the horizontal plane, of any linear structural element of a rock. The plunge of a fold is the inclination of the axis of the fold.

plunging fold A fold with its axis inclined from the horizontal.

plutonic rock Igneous rock formed deep beneath the Earth's surface.

pluvial lake A lake that was created under former climatic conditions, at a time when rainfall in the region was more abundant than it is now. Pluvial lakes were common in arid regions during the Pleistocene.

point bar A crescent-shaped accumulation of sand and gravel deposited on the inside of a meander bend.

polarity epoch A relatively long period of time during which the Earth's magnetic field is oriented in either the normal direction or the reverse direction.

polarity event A relatively brief interval of time within a polarity epoch; during a polarity event, the polarity of the Earth's magnetic field is reversed with respect to the prevailing polarity of the epoch.

polar wandering The apparent movement of the magnetic poles with respect to the continents.

pore fluid A fluid, such as groundwater or liquid rock material resulting from partial melting, that occupies pore spaces of a rock.

pore space The spaces within a rock body that are unoccupied by solid material. Pore spaces include spaces between grains, fractures, vesicles, and voids formed by dissolution.

porosity The percentage of the total volume of a rock or sediment that consists of pore space.

porphyritic texture The texture of igneous rocks in which some crystals are distinctly larger than others.

pothole A hole formed in a stream bed by sand and gravel swirled around in one spot by eddies.

Precambrian The division of geologic time from the formation of the Earth (about 4.5 billion years ago) to the beginning of the Cambrian period of the Paleozoic era (about 600 million years ago). Also, the rocks formed during that time. Precambrian time constitutes about 90% of the Earth's history.

pressure ridge An elongate uplift of the congealing crust of a lava flow, resulting from the pressure of underlying and still fluid lava.

primary coast A coast shaped by subaerial erosion, deposition, volcanism, or tectonic activity.

primary sedimentary structure A structure of sedimentary rocks (such as cross-bedding, ripple marks, or mud cracks) that originates contemporaneously with the deposition of the sediment (in contrast to a secondary structure, such as a joint or fault, which originates after the rock has been formed).

primary wave See *P wave.*

pumice A rock consisting of frothy natural glass.

P wave (primary seismic wave) A type of seismic wave, propagated like a sound wave, in which the material involved in the wave motion is alternately compressed and expanded.

pyroclastic Pertaining to fragmental rock material formed by volcanic explosions.

pyroclastic texture The rock texture of igneous rocks consisting of fragments of ash, rock, and glass produced by volcanic explosions.

pyroxene A group of rock-forming silicate minerals composed of single chains of silicon-oxygen tetrahedra. Compare with *amphibole,* which is composed of double chains.

Quartz An important rock-forming silicate mineral composed of silicon-oxygen tetrahedra joined in a three-dimensional network. It is distinguished by its hardness, glassy luster, and conchoidal fracture.

quartzite A sandstone recrystallized by metamorphism.

Radioactivity The spontaneous disintegration of an atomic nucleus with the emission of energy.

radiocarbon A radioactive isotope of carbon, ^{14}C, which is formed in the atmosphere and is absorbed by living organisms.

radiogenic heat Heat generated by radioactivity.

radiometric dating Determination of the age in years of a rock or mineral by measuring the proportions of an original radioactive material and its decay product. Synonymous with *radioactive dating.*

recessional moraine A ridge of till deposited at the margin of a glacier during a period of temporary stability in its general recession.

recharge Replenishment of the groundwater reservoir by the addition of water.

recrystallization Reorganization of elements of the original minerals in a rock resulting from changes in temperature and pressure and from the activity of pore fluids.

reef A solid structure built of shells and other secretions of marine organisms, particularly coral.

regolith The blanket of soil and loose rock fragments overlying the bedrock.

rejuvenated stream A stream that has had its erosive power renewed by uplift or lowering of the base level or by climatic changes.

relative age The age of a rock or an event as compared with some other rock or event.

relative dating Determination of the chronologic order of a sequence of events in relation to one another without reference to their ages measured in years. Relative geologic dating is based primarily on superposition, faunal succession, and crosscutting relations.

relative time Geologic time as determined by relative dating, that is, by placing events in chronologic order without reference to their ages measured in years.

relief The difference in altitude between the high and the low parts of an area.

reverse fault A fault in which the hanging wall has moved upward in relation to the footwall; a high-angle thrust fault.

rhyolite A fine-grained volcanic rock composed of quartz, K-feldspar, and plagioclase. It is the extrusive equivalent of a granite.

rift system A system of faults resulting from extension.

rift valley 1 A valley of regional extent formed by block faulting in which tensional stresses tend to pull the crust apart. Synonymous with *graben.* 2 The downdropped block along divergent plate margins.

rill A very small stream.

rip current A current formed on the surface of a body of water by the convergence of currents flowing in opposite directions. Rip currents are common along coasts where longshore currents move in opposite directions.

ripple marks Small waves produced on a surface of sand or mud by the drag of wind or water moving over it.

river system A river with all of its tributaries.

roche moutonnée An abraded knob of bedrock formed by an overriding glacier. It typically is striated and has a gentle slope facing the upstream direction of ice movement.

rock An aggregate of minerals that forms an appreciable part of the lithosphere.

rockfall The most rapid type of mass movement, in which rocks ranging from large masses to small fragments are loosened from the face of a cliff.

rock flour Fine-grained rock particles pulverized by glacial erosion.

rock glacier A mass of poorly sorted, angular boulders cemented with interstitial ice. It moves slowly by the action of gravity.

rockslide A landslide in which a newly detached segment of bedrock suddenly slides over an inclined surface of weakness (such as a joint or bedding plane).

runoff Water that flows over the land surface.

Sag pond A small lake that forms in a depression, or sag, where active or recent movement along a fault has impounded a stream.

saltation The transportation of particles in a current of wind or water by a series of bouncing movements.

salt dome A dome produced in sedimentary rock by the upward movement of a body of salt.

saltwater encroachment Displacement of fresh groundwater by salt water in coastal areas, due to the greater density of salt water.

sand Sedimentary material composed of fragments ranging in diameter from 0.0625 to 2 mm. Sand particles are larger than silt particles but smaller than pebbles. Much sand is composed of quartz grains, because quartz is abundant and resists chemical and mechanical disintegration, but other materials, such as shell fragments and rock fragments, can also form sand.

sandstone A sedimentary rock composed mostly of sand-size particles, usually cemented by calcite, silica, or iron oxide.

sand wave A wave produced on a surface of sand by the drag of air or water moving over it. Sand waves include dunes and ripple marks.

scarp A cliff produced by faulting or erosion.

schist A medium-grained or coarse-grained metamorphic rock with strong foliation (schistosity) resulting from parallel orientation of platy minerals, such as mica, chlorite, and talc.

schistosity The type of foliation that characterizes schist, resulting from the parallel arrangement of coarse-grained platy minerals, such as mica, chlorite, and talc.

scoria An igneous rock containing abundant vesicles.

sea arch An arch cut by wave erosion through a headland.

sea cave A cave formed by wave erosion.

sea cliff A cliff produced by wave erosion.

sea-floor spreading The theory that the sea floor spreads laterally away from the oceanic ridge as new lithosphere is created along the crest of the ridge by igneous activity.

seamount An isolated, conical mound rising more than 1000 m above the ocean floor. Seamounts are probably submerged shield volcanoes.

sea stack A small, pillar-shaped, rocky island formed by wave erosion through a headland near a sea cliff.

secondary coast A coast formed by marine processes or the growth of marine organisms.

secondary wave See *S wave.*

sediment Material (such as gravel, sand, mud, and lime) that is transported and deposited by wind, water, ice, or gravity; material that is precipitated from solution; deposits of organic origin (such as coal and coral reefs).

sedimentary differentiation The process in which distinctive sedimentary products (such as sand, shale, and lime) are generated and progressively separated from a rock mass by means of weathering, erosion, transportation, and deposition.

sedimentary environment A place where sediment is deposited and the physical, chemical, and biological conditions that exist there. Examples: rivers, deltas, lakes, shallow-marine shelves.

sedimentary rock Rock formed by the accumulation and consolidation of sediment.

seep A spot where groundwater or other fluids (such as oil) is discharged at the Earth's surface.

seif dune A longitudinal dune of great height and length.

seismic Pertaining to earthquakes or to waves produced by natural or artificial earthquakes.

seismic discontinuity A surface within the Earth at which seismic wave velocities abruptly change.

seismic ray The path along which a seismic wave travels. Seismic rays are perpendicular to the wave crest.

seismic reflection profile A profile of the configuration of the ocean floor and shallow sediments on the floor obtained by reflection of artificially produced seismic waves.

seismic wave A wave or vibration produced within the Earth by an earthquake or artificial explosion.

seismograph An instrument that records seismic waves.

settling velocity The rate at which suspended solid material subsides and is deposited.

shadow zone (seismology) An area where there is very little or no direct reception of seismic waves from a given earthquake because of refraction of the waves in the Earth's core. The shadow zone for *P waves* is between about 103 and 143 degrees from the epicenter.

shale A fine-grained clastic sedimentary rock formed by consolidation of clay and mud.

shallow-focus earthquake An earthquake with a focus less than 70 km below the Earth's surface.

shallow-marine environment The sedimentary environment of the continental shelves, where the water is usually less than 200 m deep.

sheeting A set of joints formed essentially parallel to the surface. It allows layers of rock to spall off as the weight of overlying rock is removed by erosion. It is especially well developed in granitic rock.

shield An extensive area of a continent where igneous and metamorphic rocks are exposed and have approached equilibrium with respect to erosion and isostasy. Rocks of the shield are usually very old (that is, more than 600 million years old).

shield volcano A large volcano shaped like a flattened dome and built up almost entirely of numerous flows of fluid basaltic lava. The slopes of shield volcanoes seldom exceed 10 degrees, so that in profile they resemble a shield or broad dome.

shore The zone between the waterline at high tide and the waterline at low tide. A narrow strip of land immediately bordering a body of water, especially a lake or an ocean.

silicate A mineral containing silicon-oxygen tetrahedra, in which four oxygen atoms surround each silicon atom.

silicon-oxygen tetrahedron The structure of the ion SiO_4^{-2}, in which four oxygen atoms surround a silicon atom to form a four-sided pyramid, or tetrahedron.

sill A tabular body of intrusive rock injected between layers of the enclosing rock.

silt Sedimentary material composed of fragments ranging in diameter from 1/265 to 1/16 mm. Silt particles are larger than clay particles but smaller than sand particles.

siltstone A fine-grained clastic sedimentary rock composed mostly of silt-size particles.

sinkhole A depression formed by the collapse of a cavern roof.

slate A fine-grained metamorphic rock with a characteristic type of foliation (slaty cleavage), resulting from the parallel arrangement of microscopic platy minerals, such as mica and chlorite.

slaty cleavage The type of foliation that characterizes slate, resulting from the parallel arrangement of microscopic platy minerals, such as mica and chlorite. Slaty cleavage forms distinct zones of weakness within a rock, along which it splits into slabs.

slope retreat Progressive recession of a scarp or the side of a hill or mountain by mass movement and stream erosion.

slump A type of mass movement in which material moves along a curved surface of rupture.

snowline The line on a glacier separating the area where snow remains from year to year from the area where snow from the previous season melts.

soil The surface material of the continents, produced by disintegration of rock. Regolith that has undergone chemical weathering in place.

soil profile A vertical section of soil showing the soil horizons and parent material.

solid The state of matter in which a substance has a definite shape and volume and some fundamental strength.

solution valley A valley produced by solution activity, either by dissolution of surface materials or by removal of subsurface materials such as limestone, gypsum, or salt.

sorting The separation of particles according to size, shape, or weight. It occurs during transportation by running water or wind.

spatter cone A low-steep-sided volcanic cone built by accumulation of splashes and spatters of lava (usually basaltic) around a fissure or vent.

specific gravity The ratio of the weight of a substance to the weight of an equal volume of water.

spheroidal weathering The process by which corners and edges of a rock body become rounded as a result of exposure to weathering on all sides, so that the rock acquires a spheroidal or ellipsoidal shape.

spit A sandy bar projecting from the mainland into open water. Spits are formed by deposition of sediment moved by longshore drift.

splay A small deltaic deposit formed on a floodplain where water and sediment are diverted from the main stream through a crevasse in a levee.

spreading axis The imaginary axis through the Earth about which a set of tectonic plates moves. The motion of a diverging plate can be described as rotation around a spreading axis.

spreading center A plate boundary formed by tensional stress along the oceanic ridge. Synonymous with *divergent plate boundary, spreading edge.*

spreading pole A pole of the imaginary axis about which a set of tectonic plates moves. The spreading poles are the two points at which a spreading axis intersects the Earth's surface.

spring A place where groundwater flows or seeps naturally to the surface.

stable platform The part of a continent that is covered with flatlying or gently tilted sedimentary strata and underlain by a basement complex of igneous and metamorphic rocks. The stable platform has not been extensively affected by crustal deformation.

stalactite An icicle-shaped deposit of dripstone hanging from the roof of a cave.

stalagmite A conical deposit of dripstone built up from a cave floor.

star dune A mound of sand with a high central point and arms radiating in various directions.

stock A small, roughly circular intrusive body, usually less than 100 km^2 in surface exposure.

strata Plural of *stratum.*

stratification The layered structure of sedimentary rock.

stratovolcano A volcano built up of alternating layers of ash and lava flows. Synonymous with *composite volcano.*

stratum (pl. *strata*) A layer of sedimentary rock.

stream load The total amount of sediment carried by a stream at a given time.

stream order The hierarchical number of a stream segment. The smallest tributary has the order number of 1, and successively larger tributaries have progressively higher numbers.

stream piracy Diversion of the headwaters of one stream into another stream. The process occurs by headward erosion of a stream having greater erosive power than the stream it captures.

stream terrace One of a series of level surfaces in a stream valley representing the dissected remnants of an abandoned floodplain, stream bed, or valley floor produced in a previous stage of erosion or deposition.

stress Force applied to a material that tends to change its dimensions or volume; force per unit area.

striation A scratch or groove produced on the surface of a rock by a geologic agent, such as a glacier or stream.

strike The bearing (compass direction) of a horizontal line on a bedding plane, a fault plane, or some other planar structural feature.

strike-slip fault A fault in which movement has occurred parallel to the strike of the fault.

strike valley A valley that is eroded parallel to the strike of the underlying nonresistant strata.

subaerial Occurring beneath the atmosphere or in the open air, with reference to conditions or processes (such as erosion) that occur on the land. Contrast with *submarine* and *subterranean.*

subaqueous sand flow A type of mass movement in which saturated sand or silt flows beneath the surface of a lake or an ocean.

subduction Subsidence of the leading edge of a lithospheric plate into the mantle.

subduction zone An elongate zone in which one lithospheric plate descends beneath another. A subduction zone is typically marked by an oceanic trench, lines of volcanoes, and crustal deformation associated with mountain building. See also *convergent plate boundary.*

submarine canyon A V-shaped trench or valley with steep sides cut into a continental shelf or continental slope.

subsidence A sinking or settling of a part of the Earth's crust with respect to the surrounding parts.

superposed stream A stream with a course originally established on a cover of rock now removed by erosion, so that the stream or drainage system is independent of the newly exposed rocks and structures. The stream pattern is thus superposed on, or placed upon, ridges or other structural features that were previously buried.

superposition, principle of The principle that, in a series of sedimentary strata that has not been overturned, the oldest rocks are at the base and the youngest are at the top.

surface creep Slow downwind movement of large sand grains by rolling or sliding along the surface due to the impact of smaller, saltating grains.

surface wave (seismology) A seismic wave that travels along the Earth's surface. Contrast with *P waves* and *S waves,* which travel through the Earth.

suspended load The part of a stream's load that is carried in suspension for a considerable period of time without contact with the stream bed. It consists mainly of mud, silt, and sand. Contrast with *bed load* and *dissolved load.*

swash The rush of water up onto a beach after a wave breaks.

S wave **(secondary seismic wave)** A seismic wave in which particles vibrate at right angles to the direction in which the wave travels. Contrast with *P wave.*

symmetrical fold A fold in which the two limbs are essentially mirror images of each other.

syncline A fold in which the limbs dip toward the axis. After erosion, the youngest beds are exposed in the central core of the fold.

Talus Rock fragments that accumulate in a pile at the base of a ridge or cliff.

tectonics The branch of geology that deals with regional or global structures and deformational features of the Earth.

tension Stress that tends to pull materials apart.

tephra A general term for pyroclastic material ejected from a volcano. It includes ash, dust, bombs, and other types of fragments.

terminal moraine A ridge of material deposited by a glacier at the line of maximum advance of the glacier.

terrace A nearly level surface bordering a steeper slope, such as a stream terrace or wave-cut terrace.

texture The size, shape, and arrangement of the particles that make up a rock.

thin section A slice of rock mounted on a glass slide and ground to a thickness of about 0.03 mm.

thrust fault A low-angle fault (45 degrees or less) in which the hanging wall has moved upward in relation to the footwall. Thrust faults are characterized by horizontal compression rather than by vertical displacement.

tidal flat A large, nearly horizontal area of land covered with water at high tide and exposed to the air at low tide. Tidal flats consist of fine-grained sediment (mostly mud, silt, and sand).

till Unsorted and unstratified glacial deposit.

tillite A rock formed by lithification of glacial till (unsorted, unstratified glacial sediment).

tombolo A beach or bar connecting an island to the mainland.

topography The shape and form of the Earth's surface.

transform fault A special type of strike-slip fault forming the boundary between two moving lithospheric plates, usually along an offset segment of the oceanic ridge. See also *passive plate margin.*

transpiration The process by which water vapor is released into the atmosphere by plants.

transverse dune An asymmetrical dune ridge that forms at right angles to the direction of prevailing winds.

travertine terrace A terrace formed from calcium carbonate deposited by water on a cave floor.

trellis drainage pattern A drainage pattern in which tributaries are arranged in a pattern similar to that of a garden trellis.

trench (marine geology) A narrow, elongate depression of the deep-ocean floor oriented parallel to the trend of a continent or an island arc.

tributary A stream flowing into or joining a larger stream.

tsunami A seismic sea wave; a long, low wave in the ocean caused by an earthquake, faulting, or a landslide on the sea floor. Its velocity can reach 800 km per hour. Tsunamis are commonly and incorrectly called tidal waves.

tuff A fine-grained rock composed of volcanic ash.

turbidity current A current in air, water, or any other fluid caused by differences in the amount of suspended matter (such as mud, silt, or volcanic dust). Marine turbidity currents, laden with suspended sediment, move rapidly down continental slopes and spread out over the abyssal floor.

turbulent flow A type of flow in which the path of motion is very irregular, with eddies and swirls.

Ultimate base level The lowest possible level to which a stream can erode the Earth's surface; sea level.

unconformity A discontinuity in the succession of rocks, containing a gap in the geologic record. A buried erosion surface. See also *angular unconformity, nonconformity.*

uniformitarianism The theory that geologic events are caused by natural processes, many of which are operating at the present time.

upwarp An arched or uplifted segment of the crust.

Valley glacier A glacier that is confined to a stream valley. Synonymous with *alpine glacier, mountain glacier.*

varve A pair of thin sedimentary layers, one relatively coarse-grained and light-colored, and the other relatively fine-grained and dark-colored, formed by deposition on a lake bottom during a period of one year. The coarse-grained layer is formed during spring runoff, and the fine-grained layer is formed during the winter when the surface of the lake is frozen.

ventifact A pebble or cobble shaped and polished by wind abrasion.

vesicle A small hole formed in a volcanic rock by a gas bubble that became trapped as the lava solidified.

viscosity The tendency within a body to resist flow. An increase in viscosity implies a decrease in fluidity, or ability to flow.

volatile 1 Capable of being readily vaporized. 2 A substance that can readily be vaporized, such as water or carbon dioxide.

volcanic ash Dust-size particles ejected from a volcano.

volcanic bomb A hard fragment of lava that was liquid or plastic at the time of ejection and acquired its form and surface markings during flight through the air. Volcanic bombs range from a few millimeters to more than a meter in diameter.

volcanic front The line in a volcanic arc system (parallel to a trench) along which volcanism abruptly begins.

volcanic neck The solidified magma that originally filled the vent or neck of an ancient volcano and has subsequently been exposed by erosion.

volcanism The processes by which magma and gases are transferred from the Earth's interior to the surface.

Wash A dry stream bed.

water table The upper surface of the zone of saturation.

wave base The lower limit of wave transportation and erosion, equal to half the wavelength.

wave-built terrace A terrace built up from wave-washed sediments. Wave-built terraces usually lie seaward of a wave-cut terrace.

wave crest The highest part of a wave.

wave-cut cliff A cliff formed along a coast by the undercutting action of waves and currents.

wave-cut platform A terrace cut across bedrock by wave erosion. Synonymous with *wave-cut terrace.*

wave-cut terrace See *wave-cut platform.*

wave height The vertical distance between a wave crest and the preceding trough.

wavelength The horizontal distance between similar points on two successive waves, measured perpendicular to the crest.

wave period The interval of time required for a wave crest to travel a distance equal to one wavelength; the interval of time required for two successive wave crests to pass a fixed point.

wave refraction The process by which a wave is bent or turned from its original direction. In sea waves, as a wave approaches a shore obliquely, part of it reaches the shallow water near the shore while the rest is still advancing in deeper water; the part of the wave in the shallower water moves more slowly than the part in the deeper water. In seismic waves, refraction results from the wave encountering material with a different density or composition.

wave trough The lowest part of a wave, between successive crests.

weathering The processes by which rocks are chemically altered or physically broken into fragments as a result of exposure to atmospheric agents and the pressures and temperatures at or near the Earth's surface, with little or no transportation of the loosened or altered materials.

welded tuff A rock formed from particles of volcanic ash that were hot enough to become fused together.

wind shadow The area behind an obstacle where air movement is not capable of moving material.

X-ray diffraction In mineralogy, the process of identifying mineral structures by exposing crystals to a beam of X-rays and studying the resulting diffraction patterns.

Yardang An elongate ridge carved by wind erosion.

yazoo stream A tributary stream that flows parallel to the main stream for a considerable distance before joining it. Such a tributary is forced to flow along the base of a natural levee formed by the main stream.

Zone of aeration The zone below the Earth's surface and above the water table, in which pore spaces are usually filled with air.

zone of saturation The zone in the subsurface in which all pore spaces are filled with water.

ANSWERS TO QUESTIONS

CHAPTER 1

Multiple Choice		True/False	Fill in
1. d	6. d	1. F	1. lithosphere
2. e	7. e	2. F	2. hydrosphere
3. c	8. e	3. T	3. biosphere
4. a	9. b	4. F	4. Venus
5. e	10. b	5. T	5. Mars

CHAPTER 2

Multiple Choice		True/False	Fill in
1. d	6. d	1. F	1. shield, stable platform, mountains
2. a	7. c	2. T	2. shields
3. d	8. b	3. F	3. stable platforms
4. e	9. b	4. F	4. lithosphere
5. c	10. e	5. F	5. trenches

CHAPTER 3

Multiple Choice		True/False	Fill in
1. d	6. a	1. T	1. sun
2. b	7. d	2. F	2. mantle
3. b	8. e	3. F	3. trench
4. d	9. e	4. T	4. oceanic ridge
5. b	10. c	5. T	5. gravity

CHAPTER 4

Multiple Choice		True/False	Fill in
1. b	6. c	1. T	1. X ray
2. b	7. b	2. T	2. quartz
3. d	8. c	3. T	3. silicate
4. b	9. c	4. T	4. calcite
5. b	10. c	5. T	5. calcite

CHAPTER 5

Multiple Choice		True/False	Fill in
1. a	6. d	1. F	1. basaltic
2. c	7. c	2. T	2. porphyritic
3. b	8. c	3. T	3. batholiths
4. d	9. c	4. F	4. basaltic
5. c	10. c	5. F	5. shield

CHAPTER 6

Multiple Choice		True/False	Fill in
1. b	6. b	1. T	1. stratification
2. d	7. d	2. T	2. shale
3. d	8. c	3. F	3. halite
4. d	9. c	4. T	4. graded
5. d	10. a	5. F	5. deltas

CHAPTER 7

Multiple Choice		True/False	Fill in
1. c	6. d	1. T	1. contact
2. c	7. a	2. F	2. slate
3. c	8. c	3. F	3. schist
4. d	9. d	4. T	4. slaty
5. c	10. d	5. F	5. gneiss

CHAPTER 8

Multiple Choice		True/False	Fill in
1. d	6. b	1. T	1. uniformitarianism
2. a	7. b	2. T	2. faunal succession
3. c	8. d	3. F	3. crosscutting relations
4. b	9. c	4. T	4. Precambrian
5. e	10. c	5. F	5. 4.5

CHAPTER 9

Multiple Choice		True/False	Fill in
1. b	6. e	1. T	1. mechanical
2. b	7. b	2. T	2. talus
3. b	8. b	3. F	3. spherical
4. d	9. b	4. F	4. ice wedging
5. b	10. d	5. F	5. chemical

CHAPTER 10

Multiple Choice		True/False	Fill in
1. a	6. c	1. F	1. erosion
2. b	7. c	2. F	2. braided
3. d	8. d	3. T	3. distributaries and splays
4. e	9. a	4. F	4. trellis
5. b	10. a	5. F	5. base level

CHAPTER 11

Multiple Choice		True/False	Fill in
1. a	6. e	1. T	1. aeration
2. e	7. b	2. F	2. water table
3. c	8. e	3. T	3. cone of depression
4. a	9. b	4. F	4. karst
5. d	10. a	5. F	5. stalactites

CHAPTER 12

Multiple Choice		True/False	Fill in
1. b	6. d	1. F	1. plucking
2. b	7. c	2. T	2. cirques
3. c	8. c	3. F	3. eskers
4. d	9. c	4. T	4. varves
5. c	10. c	5. T	5. northeast

CHAPTER 13

Multiple Choice		True/False	Fill in
1. a	6. a	1. T	1. wind
2. a	7. b	2. T	2. headlands
3. d	8. c	3. F	3. stack
4. c	9. a	4. T	4. rivers
5. b	10. b	5. F	5. atolls

CHAPTER 14

Multiple Choice		True/False	Fill in
1. c	6. d	1. F	1. barchan
2. c	7. a	2. T	2. longitudinal
3. d	8. d	3. F	3. ventifacts
4. e	9. d	4. T	4. lag
5. a	10. d	5. F	5. deflation basins

CHAPTER 15

Multiple Choice		True/False	Fill in
1. e	6. d	1. T	1. divergent
2. b	7. d	2. T	2. convergent
3. a	8. d	3. F	3. transform
4. b	9. c	4. F	4. subduction
5. c	10. e	5. F	5. transform fault

CHAPTER 16

Multiple Choice		True/False	Fill in
1. b	6. b	1. T	1. epicenter
2. b	7. c	2. T	2. magnitude
3. c	8. b	3. T	3. converging
4. a	9. e	4. T	4. shallow
5. c	10. c	5. T	5. S

CHAPTER 17

Multiple Choice		True/False	Fill in
1. a	6. e	1. T	1. divergent
2. d	7. a	2. F	2. Iceland
3. d	8. c	3. F	3. andesitic
4. d	9. c	4. T	4. andesitic
5. a	10. b	5. F	5. oceanic crust

CHAPTER 18

Multiple Choice		True/False	Fill in
1. d	6. d	1. F	1. midoceanic ridge
2. b	7. d	2. T	2. seamount
3. b	8. d	3. F	3. Atlantic
4. b	9. a	4. F	4. Pacific
5. b	10. d	5. T	5. rivers

CHAPTER 19

Multiple Choice		True/False	Fill in
1. a	6. a	1. F	1. mountain belt
2. b	7. b	2. T	2. eugeocline
3. b	8. e	3. T	3. thrusting
4. a	9. a	4. F	4. margins
5. b	10. d	5. T	5. accretion

ILLUSTRATION CREDITS

Chapter 1 (facing page photograph)   Courtesy of National Space Data Center, NASA.

Figures 1.1–1.4   Courtesy of National Space Data Center, NASA.

Figure 1.5   Courtesy Kit Peak Observatory, Arizona.

Figures 1.6–1.8   U.S. Geological Survey, Astrogeology Branch, Flagstaff, Arizona.

Chapter 2 (facing page photograph)   U.S. Geological Survey.

Figures 2.4–2.6   Courtesy National Space Data Center, NASA.

Chapter 3 (facing page photograph)   Courtesy National Space Data Center, NASA.

Figures 3.2–3.3   Courtesy National Space Data Center, NASA.

Figures 3.4, 3.5   U.S. Department of Agriculture, ASCS Western Aerial Photo Lab Salt Lake City, Utah.

Figure 3.6   Courtesy National Space Data Center, NASA.

Figure 3.7   Image Copyright by Earth Satellite Corp., Chevy Chase, Maryland, under GEOPIC trademark.

Figure 3.8   After P. J. Wyllie, 1976, *The Way the Earth Works* (New York: John Wiley and Sons).

Figure 3.11   Image Copyright by Earth Satellite Corp., Chevy Chase, Maryland, under GEOPIC trademark.

Figure 3.12   Courtesy of National Space Data Center, NASA.

Figure 3.14   Image Copyright by Earth Satellite Corp., Chevy Chase, Maryland, under GEOPIC trademark.

Figure 3.15   Courtesy of U.S. Geological Survey.

Chapter 4 (facing page photograph)   Image Copyright by Earth Satellite Corp., Chevy Chase, Maryland, under GEOPIC trademark.

Figure 4.1   Courtesy of M. Isaacson, Cornell University, and M. Ohtsuki, University of Chicago.

Chapter 5 (facing page photograph)   Image Copyright by Earth Satellite Corp., Chevy Chase, Maryland, under GEOPIC trademark.

Figure 5.1   Courtesy of U.S. Geological Survey.

Figure 5.5   Courtesy of U.S. Geological Survey.

Figure 5.10   U.S. Department of Agriculture, ASCS Western Aerial Photo Lab., Salt Lake City, Utah.

Figure 5.16   Courtesy of U.S. Geological Survey.

Chapter 6 (facing page photograph)   Image Copyright by Earth Satellite Corp., Chevy Chase, Maryland, under GEOPIC trademark.

Chapter 7 (facing page photograph)   Image Copyright by Earth Satellite Corp., Chevy Chase, Maryland, under GEOPIC trademark.

Figure 7.1   National Air Photo Library, Department of Energy Mines and Resources, Canada.

Chapter 8 (facing page photograph)   U.S. Geological Survey, Astrogeology Branch, Flagstaff, Arizona.

Chapter 9 (facing page photograph)   EROS Data Center, Sioux Falls, South Dakota.

Figure 9.11   After N. M. Strakov, 1967, *Principles of Lithogenesis,* vol. 1, trans. J. P. Fitzemms (Edinburgh: Oliver and Boyd).

Chapter 10 (facing page photograph)   Courtesy of General Electric.

Figure 10.20   Courtesy U.S. Geological Survey.

Figure 10.24   U.S. Department of Agriculture, ASCS Western Aerial Photo Lab., Salt Lake City, Utah.

Chapter 11 (facing page photograph)   Image Copyright by Earth Satellite Corp., Chevy Chase, Maryland, under GEOPIC trademark.

Figure 11.10   Courtesy of John Shelton.

Figure 11.13   Courtesy of David Herron.

Chapter 12 (facing page photograph)   U.S. Geological Survey.

Figure 12.5   U.S. Department of Agriculture, ASCS Western Aerial Photo Lab., Salt Lake City, Utah.

Figure 12.7   U.S. Department of Agriculture, ASCS Western Aerial Photo Lab., Salt Lake City, Utah.

Figure 12.9   Courtesy of J. D. Ives.

Figure 12.12   National Air Photo Library, Department of Energy, Mines, and Resources, Canada.

Figure 12.17   National Air Photo Library, Department of Energy, Mines, and Resources, Canada.

Figure 12.18   After J. L. Hough, 1958, *Geology of the Great Lakes* (Urbana, Illinois: University of Illinois Press).

Figure 12.19   After Teller, Lake Agassiz, Canadian Geological Survey.

Chapter 13 (facing page photograph)   Image Copyright by Earth Satellite Corp., Chevy Chase, Maryland, under GEOPIC trademark.

Figure 13.5   Courtesy of National Atmospheric and Oceanic Administration.

Figure 13.6   U.S. Department of Agriculture, ASCS Western Aerial Photo Lab., Salt Lake City, Utah.

Figure 13.7   U.S. Department of Agriculture, ASCS Western Aerial Photo Lab., Salt Lake City, Utah.

Figure 13.8   U.S. Geological Survey.

Figures 13.14–13.16   U.S. Department of Agriculture, ASCS Western Aerial Photo Lab., Salt Lake City, Utah.

Figure 13.19   Courtesy of John Shelton.

Figure 13.20   Bruce Colman Inc.

Chapter 14 (facing page photograph)   Image Copyright by Earth Satellite Corp., Chevy Chase, Maryland, under GEOPIC trademark.

Figure 14.3   U.S. Department of Agriculture, ASCS Western Aerial Photo Lab., Salt Lake City, Utah.

Figure 14.8   EROS Data Center, Sioux Falls, South Dakota.

Figure 14.10   Bruce Colman Inc.

Figure 14.15   Image Copyright by Earth Satellite Corp., Chevy Chase, Maryland, under GEOPIC trademark.

Chapter 15 (facing page photograph)   Image Copyright by Earth Satellite Corp., Chevy Chase, Maryland, under GEOPIC trademark.

Figure 15.1   Modified from L. Motz, 1979, *The Rediscovery of the Earth* (New York: Van Nostrand, Reinhold).

Figure 15.2   Modified after P. M. Hurley, 1969, The confirmation of continental drift, *Scientific American* 218(4) 52–64.

Figure 15.4   After American Association of Petroleum Geologists, 1928, *Theory of continental drift: A symposium* (Tulsa, Okla.: American Association of Petroleum Geologists), Figure 2.

Figure 15.5   After P. J. Wyllie, 1976, *The Way the Earth Works* (New York: John Wiley and Sons).

Figure 15.12   After B. Isacks, J. Oliver, and L. R. Sykes, 1968, Seismology and the new global tectonics, *Journal of Geophysical Research* 73(18) 5855–99.

Figure 15.19   U.S. Geological Survey.

Chapter 16 (facing page photograph)   Image Copyright by Earth Satellite Corp., Chevy Chase, Maryland, under GEOPIC trademark.

Figure 16.3   U.S. Geological Survey.

Figure 16.15   Courtesy of John H. Woodhouse.

Chapter 17 (facing page photograph)   U.S. Geological Survey, Astrogeology Branch, Flagstaff, Arizona.

Figures 17.3, 17.4   Courtesy of Woodshole, Oceanographic Institute.

Figure 17.5   U.S. Geological Survey.

Figure 17.9   U.S. Geological Survey.

Chapter 18 (facing page photograph)   Courtesy National Space Data Center, NASA.

Figure 18.2   Courtesy of Casey McDonald.

Figure 18.10   Courtesy of Woodshole Oceanographic Institute.

Figure 18.13   Courtesy of C. R. Scotese.

Chapter 19 (facing page photograph)   Courtesy of National Space Data Center, NASA.

Figure 19.1   National Air Photo Library, Department of Energy, Mines, and Resources, Canada.

Figure 19.3   Image Copyright, Earth Satellite Corp., Chevy Chase, Maryland, under GEOPIC trademark.

Figure 19.9   U.S. Geological Survey.

INDEX

Page numbers set in boldface type refer to illustrations.

aa flow, 192, 242
Abrasion, 192, 242
  ventifacts produced by, 242, **243**
  yardangs produced by, 242, **243**
Absolute time, 123
Abyssal floor, 25, **25**
Africa, **8**
  sand seas, 249
A horizon, **137**
Alluvial fan, 97, **97**
  characteristics, 166
  defined, 164
  formation, 164–166, **165**
Amphibole, 62, **63**
Amphibolite, 108–109
Andesite, 71
Angular unconformity, **117**
Antarctica, **8**
Anticline, 338
Appalachian Mountains, **23, 44**
  differential erosion, 159
Arctic Ocean, 324–325
Artesian water, 176, **176**
Asthenosphere, **19**
  convection, **39**
  defined, 18
Atlantic Ocean, 324
Atmosphere
  circulation, 2, **6**
  clouds, 2, **3–4, 6, 8**
  composition, 1, 2
  defined, 2, 14
  layers, 2
Atoll, **230**, 230–231
Atom
  composition, 50
  defined, 50
  helium, **51**
  hydrogen, **51**
  images, **51**
Atomic theory, 50

Backswamp, 161
Bar
  point, 160
  barrier, **266**
Barnes Ice Cap, Canada, **200**
Barrier islands, 97, **97, 226**
  defined, 226
  formation, 226, **226**
Barrier reef, **229**
Basalt, 71
Basement complex, **334**
Batholith, 79–80, **79–80**
Beach, sediment deposition on, 224–225
Becquerel, Henri, 123
Bedding plane, 89, **90**
B horizon, **137**

Biosphere, 1, **8**
  composition, 5
  defined, 2, 5, 14
Bonding, 51–52, **52**
Braided stream, 162, **162**
Butte, **158**

Calcite, 63
  cleavage, **56**
Caldera, 76
  Crater Lake, Oregon, **78**
Cambrian, 122
Canadian Shield
  composition, 21, **21**
  defined, 21
  eskers, **202**
  metamorphic rock characteristics, **104**
Cave mineral deposits, 181, **182**
Cementation, 96
Cenozoic, 122
Chemical weathering, 132
China
  earthquakes, 282–283
  loess deposits, 250, **251**
  tower karst topography, **179**, 180
C horizon, **137**
Clay minerals, **55**, 62
Cleavage
  calcite, **56**
  defined, 56
  halite, **56**
  mica, **56**
  slate, **108**
Climate, weathering affected by, 139–140, **140**
Cloud, 2, **3–4, 6, 8**
Coast. *See* Shorelines
Conglomerate, 91, **92**
Continental margin. *See also* Ocean, continental margins
  defined, 26
  location, 26
Continents
  drift
    glacial evidence, 257, **258**, 259
    paleoclimatic evidence, 259
    paleontologic evidence, 256, **257**
    structural evidence, 256–257, **258**
    theory development, 256
  elevation, 20, **20**
  environment types, 97, **97**
  evolution, 350–351
    accretion, **351**
  faults
    defined, 339
    normal, **339**
    strike-slip, **339**, 340, **341**
    thrust, **339**, 339–340

folded mountains, 22, **24**, 335, 337, 338–339, **338–339**
  Appalachian mountains, **23, 44**
  structure, **336–337**
shields, 21–22, **24, 334**, 334–335
  Canadian Shield, **21, 202**
stable platforms, 22, **24**, 335, **335**
  Nashville, Tennessee, **22**
Convection, 38
  asthenosphere, **39**
Coral reef, 228, **229**
Core, **19**
  composition, 18, 20
  defined, 18, 20
Crater, 7
Craton
  composition, 21
  defined, 21
Creep, **154**
  defined, 154
  types, 155, **155–156**
Crust
  composition, 18
  continental, 18, **19**
    characteristics, **335**
    structure, 24
  defined, 18
  oceanic, 18, **19**
    age, 23
    composition, 22, 324
    profile records to study, 23
    structure, 23, **24**, 324, **325**
Cuvier, Baron Georges, 116

Deflation basins, 240, 242, **242**
Delta, 97, **97–98**
  distributaries, 163, **164**
  formation, 162
  growth, 162–163
  Mississippi River, 32–33, **33**, 165
  splays, 162–163, **164**
Deposition
  glacial, 193
  groundwater, 181
  shoreline, 224
  stream, 160
  wind, 247
Desertification, 250
Differential erosion, 158
  Appalachian Mountains, **159**
  Bryce Canyon National Park, **159**
  Monument Valley, Utah, **159**
Dike, 80, **81**
Diorite, 72
Dissolution, 132
Distributary, 163, **164**
Divergent plate boundaries, 268
Dolomite, 63

Dolostone, 93
Drainage basins. *See* River systems
Dripstone, 181, **182**

Earth
   asthenosphere, 18, **19**, 39
   atmosphere, 1–2, **3–4, 6, 8,** 14
   biosphere, 1–2, 5, **8,** 14
   continents. *See* Continents
   core, 18, **19**, 20
   crust. *See* Crust, continental; Crust,
      oceanic
   formation, 12
   hydrosphere, 1–2, 5, **6,** 14
   internal structure, 288, **289–293,**
      291–292
   lithosphere, 2, 5, 14, **19, 20**
   mantle, 18, **19**
   Mars compared with, 12
   Mercury compared with, 7
   Moon compared with, 7
   oceans, 20, **20**
   Venus compared with, 7, 10–12
Earthquake, 42
   defined, 280
   elastic-rebound theory, 280
   epicenter, **281,** 281–282
   location, **42**
   magnitude, 282–283, **283**
      Richter scale, 282
   plate tectonic evidence, 265, **266**
   plate tectonic relationship with, 284,
      286
      convergent plate boundary seismicity,
         286–287, **287**
      divergent plate boundary seismicity,
         286, **286**
      global pattern seismicity, **285,**
         286
      intraplate seismicity, 287
      seismicity determining plate motion,
         287–288
   predicting, 283–284
      seismic gap, **285**
   primary wave, **280,** 280–281, 288,
      289–292, 291
   secondary wave, **280,** 280–281, 288,
      290–292, 291
   surface wave, **280,** 280–281
Electron, 50
Eolian systems, 238
Erosion
   coastal, 221
      process, 221–222
      sea arches, 222–223, **222–223**
      sea caves, 222–223, **222–223**
      sea stacks, 222–223, **222–223**
      wave-cut cliff, **221**
      wave-cut platform, **221,** 223
   differential, 158
   glacial, 192, 195, **197**
   groundwater, 178
   headward, 152
   stream, 150
   wind, 240
Esker, **201–202**
Everglades, 186, **187**
Exfoliation
   defined, 137–138
   Sierra Nevada, **139**

Fault
   defined, 339
   normal, **339**
   strike-slip, **339,** 340, **341**
   thrust, 339, **339–340**
Feldspar, 60, **61**
Fissure, **68,** 73
Floodplains, 160
   features, **160**
Fluvial environment, 97, **97–98**
Fold, **338**
Fossil, 89, **90**
Franciscan Formation, 38

Gabbro, 72
Geocline, 343
Geologic column, **122**
Geothermal energy, of groundwater, 178
Geyser, 176–178
   origin, **177**
Glacial system, 34
   Alaska, **34**
Glacier, 97, **97–98**
   causes, 211–212
   continental
      Antarctica, 200
      Barnes Ice Cap, Canada, **200**
      changes caused by, 199, **199**
      defined, 198–199
      drift evidence, 257, **258,** 259
      Greenland, 199–200, **200**
      land developed by, 200, **201–203,**
         203
   defined, 192
   deposition, **192,** 193
   erosion, 192, **192**
      roche moutonnée, **193**
      striations, **193**
   ice flow rates, 193–194
   isostatic adjustment and, 45
   Pleistocene. *See* Ice Age
   transporting rock, **192,** 193
   valley glaciers
      defined, 194
      land forms developed by, 195, **197,**
         198, **198**
      zone of ablation, 194–195, **195–196**
      zone of accumulation, 194–195,
         195–196
Gneiss, **105,** 107–108
Granite, 72
Gravity
   importance, 46
   isostatic adjustment, 45
      pack ice, **45**
Greenhouse effect
   Mars, 12
   Venus, 10–11
Greenland, 199–200, **200**
Groundwater, 170
   alteration, 184
      composition, **184,** 184–185
      saltwater encroachment, 185, **185**
      water table position, 185–186, **187**
   artesian water, 176, **176**
   artificial discharge, 173, **175**
   erosion by
      karst topography due to, 178–179,
         180, 181
      process, 178

geothermal energy, 178
geysers, 176–178
   origin, **177**
mineral deposition by, 181
   cave deposits, 181, **182**
   dripstone formation, 181, **182**
   Mammoth Hot Springs, Yellowstone
      National Park, **183**
   petrified wood, 181, **183**
movement, **172,** 173
   Snake River, 173, 175, **175**
natural discharge, 173, **174**
permeability, 170
porosity, 170, **171**
subsidence related to, 186
systems, 34
   Cape Canaveral, Florida, **35**
thermal springs, 176–178, **177**
water table, 170, **171**
   location, 171, **172,** 173, 185–186,
      187
Gypsum, 63, 93
   crystal formation, **56**
      hardness, **57**

Halite, 63
   cleavage, **56**
Helium
   atomic structure, **51**
Hess, H. H., 260
Himalaya Mountains, 349
Hornfel, 109
Hutton, James, 116
Hydraulic head, 173
Hydrogen
   atomic structure, **51**
Hydrologic system, **30,** 30–34, **32,**
   34–36, 36
Hydrolysis, 132
Hydrosphere, 1, **6**
   composition, 5
   defined, 2, 5, 14
   movement, 5

Ice Age, 31
   biological effects, 210
   drainage patterns, 205, **206**
   evidence for, 203–204
   ice position, 203–204, **204,** 205
   isostatic adjustment, 204
   lake formation, 205–207
      Great Lakes, **207**
      Lake Agassiz, 207, **208**
      North American shield area, **206**
      pluvial, 208, **209,** 210
   oceans affected by, 210
   sea level changes, 204–205
   wind effects, 210
Ice wedging, 130, **130,** 198
Igneous rock, 59, 60
   aphanitic texture, 69, **70**
      andesite, 71
      basalt, 71
      rhyolite, 72
   classification, 71, **71**
   glassy texture, 69, **70**
   magma
      basaltic, 68, 82–83, **82–83**
      composition, 68
      defined, 68

lava, **68**
silicic, 68, 82–83, **82–83**, 343–344, **344**
phaneritic texture, 69, **70**
diorite, 72
gabbro, 72
granite, 72
peridotite, 72, **72**
porphyritic-aphanitic texture, 69, **70**
porphyritic-phaneritic texture, 69, **70**
pyroclastic texture, 69, **70**
ash-flow tuff, 72
tuff, 72
Indian Ocean, 325
Intrusion, 79
batholiths, 79–80, **79–80**
dikes, 80, **81**
laccoliths, 79, 80, **81**
sills, **79**, 80, **81**
stocks, 80
Ion
defined, 50
electrical charges, 50–51
Ionic substitution, 54, **54**
Isostasy, 45, **45**, 46
Isotope, 50

Joint, 340, **341**

Karst topography
characteristics, 178
climatic conditions for, 178
evolution, 178–179, **180**
Kentucky, **179**, 181
tower karst, 178
China, **179**, 181

Laccolith, **79**, 80, **81**
Lagoon, 97, **97–98**
Lake Agassiz, evolution, 207, **208**
Lake Bonneville, **209**, 210
Landslides, 157–158
Lava, **68**, 74. *See also* Volcanoes
Limestone, 91, **92**, 93
Lithosphere, **19**
composition, 5, 20
defined, 2, 5, 14, 20
Loess
composition, 250, **251**
defined, 250
desert, 250
China, 250, **251**
glacial, 250

Magma. *See also* Volcanoes
basaltic, 68
generation of, **82–83**, 83
composition, 68
defined, 68
lava, 68
silicic, 68
generation of, **82–83**, 83, 343–344, **344**
Mammoth Hot Springs, Yellowstone National Park, **183**
Mantle, **19**
composition, 18
convection, 273–274
Hawaii, **274**
model, **273**

defined, 18
plumes, 274–276
volcanoes produced by, **275**
Marble, **105**, 108
Marine environment, 97, **97–98**
Mars
formation, 12
history, 12
size, 12, **14**
surface, 12, **13**
Mass movement, **154**, 155
Matter, states, 52
M-discontinuity, 18
Meander, 160
evolution, 161, **161**
Mechanical weathering, 130
Melange, 38
Melting, rock, 82–83, **82–83**
Meltwater, 194, **195, 201**
Mercury
formation, 12
history, 7
size, 12, **14**
surface, 7, **10**
Mesosphere, 19
composition, 20
defined, 20
Mesozoic, 122
Metaconglomerate, **105**, 109
Metamorphic rock, **59**, 60
alteration, 105, **105**
characteristics, **104**
chemically active fluids, 107
defined, 104
deformation, 104, **105**
distribution, 106
foliated, 107
gneiss, **105**, 107–108
schist, **105**, 107
slate, **105**, 107, **108**
grades, 109, **110**
nonfoliated, 108
amphibolite, 108–109
hornfels, 109
marble, **105**, 108
metaconglomerate, **105**, 109
quartz, **105**
quartzite, 108
origin, 109, **109, 111**
plate tectonics, 110–111, **111**
pressure changes, 106
temperature changes, 106
texture, 106
zonation, 109
Metamorphism, 104
Meteorite, 7
Mica, 60, **61**
cleavage, **56**
Milankovitch, M., 211
Mineral. *See also specific types*
breakdown, 57
clay, **55**, 62
composition, 54
ion substitution, 54, **54**
defined, 53
groundwater depositing, 181
cave deposits, 181, **182**
dripstone formation, 181, **182**
Mammoth Hot Springs, Yellowstone National Park, **183**

petrified wood, 181, **183**
growth, 57
inorganic solids, 53
physical properties, 54
cleavage, 56, **56**
color, 57
crystal form, 54, **55–56**, 56
hardness, 56–57, **57**
specific gravity, 57
streak, 57
rocks, 59
igneous, **59**
melting, 82–83, **82–83**
metamorphic, **59**
sedimentary, **59**
silicate, 58
silicon-oxygen tetrahedra, **58**
structure, 53
X-ray diffraction to determine, **53**
Mississippi Delta, **165**
Mississippi River
course, 32–33
delta, **33**, 165
Moho. *See* M-discontinuity
Mohs hardness scale, 57
Moon
formation, 12
history, 7
lava, **9**
size, 12, **14**
surface, 7, **9**
Moraine
Alberta, Canada, **202**
ground, **201**
lateral, **197**, 198, **198**
medial, **197**, 198
recessional, **201**
terminal, 198, **201**
Mountain. *See also* Continents, folded mountains
accretionary terrains, **349**, 349–350
oceanic plateaus, **350**
Appalachian, **23, 44**
belt evolution, 344, 346
erosion, **345**
isostatic adjustment, **345**
continental-oceanic plate convergence, 346–347, **347**
continental plate convergence, 347, **348**, 349
folded, 338–339, **338–339**
structure, 342, **342**
Himalaya, 349
igneous activity, 343–344, **344**
metamorphism, 343
oceanic plate convergence, 346, **346**
origin, 42–43
orogenesis defined, 337
rock sequences, 343
types, **343**
structural deformation, 338
Mount Pelee, 304
Mount Saint Helens, 304–307, **305–307**
Mount Vesuvius, **303**, 303–304

Naturalism, 116–117
Natural levee, 160–161, **161**
Neutron, 50

Niagara Falls
  erosion, **151**
  origin, **207**
Nile River, 149–150

Ocean. *See also* Plate tectonics
  abyssal floor, 25, **25**, 316–317, **317**
  Arctic, 324–325
  Atlantic, 324
  continental margins
    active, 318
    composition, 318
    defined, 26
    location, 26
    passive, 318
    slope, 318, **320**
  crust. *See* Crust, oceanic
  depth, 20, **20**
  floor mapping, 320–321, **322–323**
  floor topography, 39
  Indian, 325
  Pacific, 326
  plate movement history, 327, **328–329**, 329
  ridge, 25, **25**, 314, **315**
    Mid-Atlantic, **316**
    Murray Fracture Zone, **316**
  rifts, **24**
  seamounts, 25, 318
    Pacific Ocean, **319**
  seismic reflection profiles, 314, **315–317, 319**
  small basins, 326, **327**
  trenches, **24**, **25**, 25–26, 318
    Aleutian, **319**
Olivine, 62, **62**
Oolite, 91, **92**
Orogenesis. *See* Mountains
Oxidation, 132–133

Pacific Ocean, **6**, 6, 326
Pacific plate, 267
Paleomagnetism, 260–265
Paleozoic, **122**
Peridotite, 72
Permeablity, 170
Petrified wood, 181, **183**
Plateau, 158
Plate tectonics
  boundaries, 268
    convergent, 268–270, **270**
    divergent, 268, **269**
    oceanic ridges, 37–38, **38**, 314, **315–316**
    subduction zones, 37–38, **38**
    transform faults, 37–38, **38**, 270–271, **271**
  defined, 36
  earthquake evidence, 265, **266**
  fractures
    San Andreas Fault, 41, **41**, 280
  geography, 266–267, **267**
  importance, 36
  mantle convection, 273–274
    Hawaii, **274**
    model, **273**
  mantle plumes, 274–276
    volcanoes produced by, **275**
  metamorphic rocks, 110–111, **111**
  motion, **38**, 39, **272**, 272–273

ocean floor geology, 260
ocean floor sediment, 260, **261**
ocean floor topography, 39
  process, 36–38, **37**
  rifts, Red Sea, 39, **40**, 41
  rock magnetism, 260–262, **261**, **263–265**, 264–265
  theory, 36–37
  theory development, 260
Pleistocene, 122, 203, 210
Point bars, 160
Polarity, 262–265
Polymorphism, 53
Porosity, 170, **171**
Precambrian, 122
Proton, 50
Pyrite, **55**
Pyroxene, 62, **62**

Quartz, **55**, 60, **61–62, 105**
Quartzite, 108

Radiometric measurement
  isotopes used to determine, 124, **124**
  potassium-argon method, 124
  radioactive decay rate, **123**, 123–124
  radiocarbon method, 124
Radiometric time scale, 124–125
Recessional moraine, **201**
Red Sea, **8**
  rift, 39, **40**, 41
Reef
  atoll, **230**, 230–231
    origin, 231
  barrier, 229, **230**
  characteristics, 228
  ecology, 228–229
    Society Islands, French Polynesia, **229**
  fringing, 229, **230**
  platform, 231
Regolith, 136, **137**
  removal due to stream erosion, 150
  types, 137
Relative dating, 118
  crosscutting relations principle, 119, **120–121**
  faunal succession principle, 119
  inclusion principle, 119, **120**
  landscape development, 119
    example, 121, **121**
  standard geologic time scale, **122**, 122–123
  superposition principle, 118–119
Rhyolite, 72
Richter scale, 282
Ridge, oceanic, 25, **25**, 314, **315**
  Mid-Atlantic, **316**
  Murray Fracture Zone, **316**
Rift, Red Sea, 39, **40**, 41
River system, 31–33
  collecting system, 146, **147**
  defined, 146
  differential erosion, 158
    Appalachian Mountains, **159**
    Bryce Canyon National Park, **159**
    Monument Valley, Utah, **158**
  dispersing system, 146, **147**
  equilibrium, 148–149, **149**

Aswan Dam of the Nile River, 149–150
  urbanization affecting, 150
  importance, 146
  stream deposition, 160
    alluvial fans, 164–166, **165**
    backswamps, 161
    braided streams, 162, **162**
    deltas, 162–163, **164**
    floodplain deposits, 160, **160**
    meanders, 160, 161, **161**
    natural levees, 160–161, **161**
    point bars, 160
    stream terraces, 162, **163**
  stream erosion, 150
    downcutting, 150–152, **151**
    headward, 152, **152–157**, 155–158
    regolith removal, 150
  stream flow dynamics, 146–148, **147–148**
  stream valleys, 31, **32**
  transporting system, 146, **147**
Roche moutonnée, **193**
Rock salt, 93
Rocky Mountains, glaciers, 195
Rutherford, Lord, 123

Saltation, 148, **244**
San Andreas Fault, 41, **41**
Sand dune
  barchan, 247, **248**
  longitudinal, 247, **248**
  migration caused by winds, 245, **246**, 247
    Zion National Park, Utah, **246**
  parabolic, 247–248, **248**
  sand seas, 248
    Africa, **249**
  star, 247, **248**
  transverse, 247, **248**
Sandstone, 91, **92**
Saturation zone, 170–173, **171**
Schist, **105**, 107
Seamount, 25
Sedimentary rock, **59**, 60
  bedding planes, 89
  cementation, 96
  clastic, 91, **92**
  compaction, 96
  continental environments, 97, **97–98**
  cross bedding, 93–94, **95**
  cross section, **99**
  defined, 88
  deposition, 96
  graded bedding, 94–95, **95**
  importance, 99–100
  marine environments, 97, **97–98**
  mud cracks, 90, 95
  nonclastic, 91, **92**, 93
  physical features, 89, **90**
  ripple marks, 90, 95
  shoreline environments, 97, **97–98**
  stratification, 88–89, **90**, 93, 94
    Grand Canyon, **88–89**
  transport, 96, **97**
  weathering, 96
Shale, 91, **92**
Sheeting, 130
  Sierra Nevada, **131**
Shield, 21, **334–335**

Shoreline, 34, 97, **97**. *See also* Erosion, coastal; Reefs; Tides; Tsunamis; Waves
  classifications, 231, 232
    collision coasts, 231, **232**
    marginal sea coasts, 232, **232**
    passive-margin coasts, 231–232, **232**
    primary coasts, 232, **233**
    secondary coasts, 232, **234**
  evolution, **223**, 223
    equilibrium, **227**, 227–228
    erosion process, 227–228, **227–228**
  isostatic adjustment and, 45
  North Carolina, **35**
  sediment deposition
    barrier islands, 226, **226**
    beaches, 224–225
    process, 224, **224**
    spits, 225, **225**
    tombolos, 225, **225**
Silicate mineral, 58, **58**
Sill, **79**, 80, **81**
Siltstone, 91
Sinkhole, 178, **179–180**, 181
Slate, **105**, 107
  cleavage, **108**
Slope, **157**, 157–158
Slump, 154, **154**
Snake River, 173, 175, **175**
Society Islands, French Polynesia, reefs, **229**
Soil, 136, **137**
Solar heat, 2
Solution, 132, 178
South Polar ice cap, **8**
Specific gravity, 57
Spit, formation, 225, **225**
Splay, 162–163, **164**
Spring
  formation, **174**
  thermal, 176–178, **177**
Stable platform, 122, **336**
Stalactite, 181, **182**
Stock, 80
Stratosphere, 2
Stream deposition, 160
  alluvial fans
    characteristics, 166
    defined, 164
    formation, 164–166, **165**
  backswamps, 161
  braided streams, 162, **162**
  deltas
    distributaries, 163, **164**
    formation, 162
    growth, 162–163
    Mississippi River, **165**
    splays, 162–163, **164**
  floodplain deposits, 160, **160**
  meanders, 160, 161, **161**
  natural levees, 160–161, **161**
  point bars, 160
  stream terraces, 162
    evolution, **163**
Stream erosion, 150
  downcutting, 150–152, **151**
  headward, 152, **152–157**, 155–157
    examples, 157–158
  regolith removal, 150
Stream terrace, 162

evolution, **163**
Subduction zone, 36, 268–270
Submarine canyon, 320
  Monterey, **321**
Subsidence, 186
Sudan, Africa, dust storm, **245**

Talus cones, 131, **131**
Tectonic system. *See also* Plate tectonics
  boundaries
    oceanic ridges, 37–38, **38**
    subduction zones, 37–38, **38**
    transform faults, 37–38, **38**
  defined, 36
  earthquakes, 42
    location, **42**
  fractures, San Andreas Fault, 41, **41**
  importance, 36
  mountains
    Appalachian, **44**
    origin, 42–43
  ocean floor topography, 39
  plate motion, **38**, 39
  process, 36–38, **37**
  rifts, Red Sea, 39, **40**, 41
  theory, 36–37
  volcanoes, 52
    Japanese Island arc, **43**
    location, **42**
Tephra
  composite volcano produced by, 76, **77**
  composition, 75
  defined, 74
Tertiary, 122
Tidal flat, 97, **97**
Tide
  causes, 234, **235**
  coastal effects by, 233, 234
Time
  absolute, 118, 123
  discovery, 116
  geologic magnitude, 125, **125**
  relative, 118. *See also* Radiometric measurements; Radiometric time scale; Relative dating
  unconformities, 117–118
    angular, 118, **118**
    Siccar Point, Scotland, **117**
Tombolo, formation, 225, **225**
Transpiration, 31
Trench, ocean, **25**, 25–26
Troposphere, 2
Turbidity currents, 6–8
Tsunamis, 235
Tuff, 72
  ash-flow, 72
Turbidity current, 94-95, **95**

Unconformity, **117**, 118
Uniformitarianism, 116–117

Varves, 203, **203**
Ventifact, 242, **243**
Venus
  formation, 12
  history, 7
    water evaporation, 10
  size, 12, **14**
  surface, 7, 10–12, **11**

Volcanoes, 42
  basaltic eruptions
    aa, 73, **74**
    cinder cone, **75**
    Hawaii, **73**
    lava flow, 74, **74**, 76
    pahoehoe, 73–74, **74**
    process, 73
    shield, **75**, 76
  convergent plate margins, 302
    magma generation, **302**, 302–303
    Mount Pelee, 304
    Mount Saint Helens, 304–307, **305–307**
    Mount Vesuvius, **303**, 303–304
  distribution, **298**, 298–299
  divergent plate margins, 299
    basalt plateaus, 301, **301**
    Iceland, 300
    magma generation, **299**, 299–300
    seafloor observations, 300, **300**
  intraplate activity, 308, **308–309**
  Japanese Island arc, 42
  location, 42
  silicic eruptions
    ash flow, 76, **78**
    caldera, 76, **78**
    composite 76, **77**
    process, 76

Water. *See also* Groundwater; Hydrosphere; Meltwater; River systems
  chemical weathering affected by, 133, 139, **140**
  gas formation, 52
  liquid formation, 52
  solid formation, 52
Water table, 170, **171**
  location, 171, **172**, 173
    changes, 185–186, **187**
Wave. *See also* Tsunamis
  breakers, 217, **217**
  energy, 216
  longshore drift
    defined, 218
    process, 218–219, **219**
    Santa Barbara, California, 219–221, **220**
  motion process, 216, **216**
    base, **217**
  refraction, 217, **218**
Weathering
  bedding plane importance, 133, **135**
  chemical, 132
    dissolution, 132
    hydrolysis, 132
    oxidation, 132–133
  climate affecting, 139
    precipitation, 139, **140**
    temperature, 139–140, **140**
  differential, 136
    Bryce Canyon, Utah, **136**
  Egyptian pyramids, **141–142**
  granular disintegration, 134, **134–135**
  joint importance, 133, **134**
  mechanical, 130
    ice wedging, 130, **130**
    organic activity affecting, 130–131, **131**
    sheeting, 130, **131**

Weathering *(Continued)*
  rate, 140–141
    example, 141
  regolith produced from, 136, **137**
    types, 137
  spheroidal, 137–138, **138**
    exfoliation domes, **139**
Wegener, Alfred, 256
Wind, 36, **36**, 97, **97**
  causes, 240

desert, 240, **241**
erosion, 240
  abrasion, 242, **243**
  deflation, 240, 242, **242**
patterns, 240, **241**
sand dune migration caused by, 245,
    246, 247. *See also* Sand
    dunes
  Zion National Park, Utah, **246**
sediment transport by, 244, **244**

dust movement, 244–245,
    245
sand movement, 244, **244**

Yardang, 242, **243**

Zion National Park, Utah, cross bedding,
    246
Zone of aeration, 170, **171**
Zone of saturation, 170, **171**